THE HAND OF GOD

FROM OPPENHEIMER TO HYPERSONICS – A CRASH COURSE ON NUCLEAR WEAPONS AND HUMANKIND'S MOST DANGEROUS GAME

By Paul McCarthy

Copyright Information

Majic-12 Publishing
#1025
600 Greenbush Road STE 3
Rensselaer, NY 12144
United States
www.majic12publishing.com
Inquiries: info@majic12publishing.com

Printed in the United States of America

Publisher's Cataloging-in-Publication
(Provided by Cassidy Cataloguing Services, Inc.).

Names: McCarthy, Paul, 1988- author.
Title: The hand of God : from Oppenheimer to hypersonics -- a crash course on nuclear weapons and humankind's most dangerous game / by Paul McCarthy.
Description: First edition. | Rensselaer, NY : Majic-12 Publishing, [2023] | Includes bibliographical references.
Identifiers: ISBN: 979-8-9887603-0-6 (hardback) | 979-8-9887603-1-3 (paperback) | 979-8-9887603-2-0 (digital) | 979-8-9887603-3-7 (audio file) | LCCN: 2023913945
Subjects: LCSH: Nuclear weapons. | Tactical nuclear weapons. | Nuclear warfare. | Nuclear arms control. | Nuclear disarmament. | Radiation--Physiological effect. | Aerodynamics, Hypersonic. | Hypersonic planes, Military. | Stealth technology. | Military art and science. | Great powers--Foreign relations. | World politics. | BISAC: HISTORY / Military / Nuclear Warfare. | POLITICAL SCIENCE / World / General. | TECHNOLOGY & ENGINEERING / Military Science.
Classification: LCC: U264 .M33 2023 | DDC: 355.8/25119--dc23

Library of Congress Control Number: 2023913945

About the Author

Paul McCarthy is a passionate and experienced public policy expert and attorney working in the halls of New York State government. With a deep-rooted fascination for geopolitics, national defense, and law, McCarthy brings a wealth of knowledge to his captivating exploration of nuclear weapons.

Hailing from his home outside Albany, NY, alongside his fiancée Emma and their beloved dog, McCarthy finds solace in the pursuit of understanding. He obtained his BA in Political Science from the University at Albany and his JD from Chicago-Kent College of Law. At work, using his research, collaboration, negotiation, and critical analysis skills, he immerses himself in the complexities of state policy, helping to find solutions that work for all New Yorkers.

Beyond his professional pursuits, McCarthy enjoys life's simple pleasures. Engaging in strategy-based games, journeying into culinary exploration inspired by global cuisines, and vegging out in front of the latest and greatest TV shows and movies, he finds inspiration and joy in the ordinary moments.

With "The Hand of God," McCarthy invites readers on an insightful journey into the history and implications of nuclear weapons. Through this comprehensive perspective, he uncovers the profound dangers they pose and the delicate balance that holds our world in check.

Join Paul McCarthy as he unravels the enigmatic realm of nuclear weapons, illuminating the shadows and urging us all to contemplate the choices that shape our shared future.

Table of Contents

Foreword

Hello there! My name is Paul McCarthy. I am not a military strategist, scientist, general, diplomat, or weapons engineer. I am, however, a person with a passion for understanding the real-world forces that move and bend the tides of human history. Those lines of work and fields of understanding have a heavy bearing on those tides, so I strive to keep up with them as much as possible in my free time. While I am merely a casual observer and do not work directly in the fields that are the subject of this book, I am a public policy expert and an American attorney working in New York's government and political environment, with a personal history of study in political science, philosophy, and work on Capitol Hill, so my background experience is at least tangentially related to these subjects.

But enough about me; I am not nearly as fascinating as the topics we'll discuss in this work, so let's take a crack at laying out our approach here. As the title suggests, this is a book about nuclear weapons. It is neither a textbook nor a dramatized work of fiction but rather a casual exploration. I aim to discuss one of the most simultaneously fascinating, terrifying, and all-consuming drivers of the last century(ish) of scientific discovery, military weaponry, national strategy, and international politics: the nuclear weapon. I have attempted to do so in a way that is simultaneously easy to understand for every reading level yet captivating and informational enough for even those who work in a related field but seek more broad-based general knowledge of the topic. My objective here is to present a work of non-fiction that is both engaging enough for a casual reader who wants to expand their intellectual horizons and yet useful enough for somebody in a related field to pull open a specific section of the book for reference to understand a particular topic related to nuclear weaponry.

The book will be divided into relevant topical sections by part and chapter. Something that you may not be aware of if you don't know much about the field of nuclear weaponry or don't work in a related field is that there is a *lot* to cover here. To start, there is the history of the Bomb, stretching back into the early 20th Century (the 1900s) and late 19th Century (the 1800s), which we'll explore as we discuss the early concepts that underpin the technology. We'll explore the science, particularly the physics, that makes the technology possible. We'll learn the stories of the central figures involved in the authorization, funding, creation, proliferation (spread around the world), and research of the weapon and its associated technology: scientists, world leaders, diplomats, spies, activists, military leaders, and others. Understanding their thinking about the field is vital to grasping how we got to where we are today. A further critical discussion concerns the different types of weapons (strategic vs. tactical), the technologies that comprise these weapons (fission and fusion), and the associated derivatives like salted bombs and dirty bombs. If I'm using terminology you don't understand right now, don't worry; we'll get to all that later. Beyond these topics are political and international relations discussions about how nuclear weapons have shaped our world for the past several decades, their potential to shape our world in the future, and even more abstract debates about the morality of their use and existence.

My rule for this book is to keep the material interesting, fast-moving, and understandable for almost all audiences. If you can read, you should be able to read this book, and you should find it interesting from at least some perspective, even if you're not a nuclear physicist or an analyst monitoring the moves of a foreign power from the computer screens of the Pentagon.

A caveat, though– since this is neither a textbook nor the work of a genius nuclear physicist, it likely will not help you understand in extreme depth a specific issue area in the way that reading, for example, Leo Szilard's publications, would. Nor do I have the bandwidth here to be particularly analytical, so while I'll recite historical events and mention various topics, schools of thought, etc. I won't spend a great deal of time critiquing them or pulling them apart. Further, if you're a weapons expert, don't expect to dive into a chapter of this book on the topic of missiles and walk away with a broader scope of knowledge; that's unlikely to be the case, as I will be covering a wide swath of information on various topics and will not be able to do the deep dive that each subject deserves. For that, there are many scholarly works, biographies, historical pieces, interviews, and textbooks on every topic covered in this work, ranging from the Cuban Missile Crisis to game theory, should those be your interest areas. However, even if you share the same depth of knowledge about physics that Albert Einstein did during his career, you are likely to learn a thing or two about other aspects concerned with the nuclear age, such as secret research operations, international intrigue (espionage), weapons systems, or the defense activities of global powers. I aim to speak in predominantly general and broad terms that almost anybody can understand. While I am not an expert in any of these specific fields, I have spent a good deal of time on the research aspects of this book to ensure that the information is accurate and well-sourced and to give credit to the journalists, organizations, biographers, scientists, and academics who initially researched the information that stands as the foundation of this work. Though this is not a peer-reviewed work and does not pretend to be, I am confident that my readers will find the information accurate and verifiable through multiple primary sources.

Concerning this book's structure, things will occasionally be out of place and, in many cases, not chronological since we'll discuss many varied but associated topics. So information will sometimes be grouped thematically instead of chronologically. Still, I have done my best to explain the timeline of events and relate back where necessary to keep a coherent narrative for you, the reader. Also, at times, you may see a reference to something technical, like an "alpha particle," which is not initially described in an adequate way that a non-expert would understand. If you find that to be the case, I will likely describe the topic or idea in depth soon after when you have the foundational context to understand it more thoroughly. We'll build this context together as we work through the book. Fear not, though; it will all come together and make sense in the end.

As an additional note, there are aspects of this book that are not for the faint of heart. We will need to address existential considerations that could weigh on the most battle-hardened mind. I will discuss, from time to time, incidents, accidents, and even intentional acts that lead to tragic and sometimes grisly outcomes for those unfortunate to have been involved. Further, war and the loss of life are unavoidable topics when discussing this field, so we'll also explore those subjects.

On the book's title– I want to be clear that "The Hand of God" does not refer to any particular god of any specific religion. Nor would I want to suggest that I associate benevolence with the use of nuclear weapons. Instead, this terminology refers to the *power* of the technology– the physical effects generated by these weapons are rivaled only by cosmic forces and the divine abilities of the gods of myth and legend. Indeed, as a consequence of the gravity of this technology, the creators of the Bomb often referred to it using religious and divine terminology that simultaneously evokes curiosity, respect, and fear.

I should also note the *why*. What drove me to put this piece together? Nuclear weaponry and its associated concerns have existed for decades. What is the relevance of such a book at this point in history, and what can I add to the discussion? The answer is in the title; regardless of your religious preferences or lack thereof, I have a message for you: the Hand of God exists, and it is an ever-present force in our lives, whether we see it, are aware of it, or even want to begin to mentally accept the intensity and reach of its capabilities. The sheer gravity of its power affects society in a multitude of ways that our leaders, scientists, military officers, philosophers, and government functionaries need to understand and appreciate to a greater degree than almost any other grand force we grapple with in the world. My goal is to provide the average person a window into the importance of this subject in an easily understood and engaging fashion so that we can all intelligently and soberly discuss the topic and hold our leaders accountable for their decision-making regarding the phenomenal power surrounding this terrifying technology. As for our leaders and future leaders, I would consider this work to be essential reading; no mere mortal, as scholarly curious as Marie Curie, or as ambitious and civically minded as Franklin Delano Roosevelt, has any business waking the Kraken in its den without understanding how Rome may ultimately pay the irreversible price.

I want to acknowledge all the well-researched, knowledge-rich, and excellent sources, including books, journal articles, websites, and reports from private authors, academics, research institutions, foundations, and government agencies. This work would not have been possible without them. I've also occasionally included brief movie quotes in the text, as the topics covered in these classics are relevant to our exploration and provide a glimpse into the cultural effects of the Bomb; I would encourage the reader to check out these movies as they're not only classic works of cinema, they also in some cases provide a warning to all of us.

So many topics are covered here that it was difficult to delve into any of them comprehensively. For most of the issues and historical figures addressed in this literature, entire books, documentaries, and various research pieces have been created about them. I have an extensive works cited at the end of the book that is matched to the paragraphs of this book. If you're interested in a deeper understanding of any particular issue, I encourage you to check the works cited to dive down any particular rabbit hole.

  Finally, I have to note that due to the secrecy surrounding nuclear weapons and strategic planning, some of the information in this book, while well-sourced, is based on conjecture and public estimations of secret government activities. When it's the case that a fact or account is not firmly established, I will note that for the reader's benefit. Additionally, if you're reading this work long after I've written it, it's likely that additional declassifications and further revelations about this technology will contradict or clarify what I have written; I would encourage the reader to check contemporary sources to verify the accuracy of anything contained herein.

  Alright, let's get started, and thank you so much for lending me your attention.

Part I. Spooky Science: The Early History of Nuclear Research

Chapter 1. The Philosopher's Stone: Nuclear Physics

"The unleashed power of the atom has changed everything except our thinking. Thus, we are drifting toward catastrophe beyond conception. We shall require a substantially new manner of thinking if mankind is to survive." -Dr. Albert Einstein[1]

Regarding the history of nuclear weaponry, I will go back a little further than some conventional sources might. While many point to August of 1942 as the key birthdate of the technology, we should go back to the 1800s to establish a solid foundation. There, we'll take a peak at the formation of the field of nuclear physics, which stands as the foundation of the Bomb. Without nuclear physics, you don't have nuclear weaponry. This section of the book will be one of the most science-heavy. Still, I will try to keep the concepts and the wording as basic as possible so that everyone can understand the information necessary to process the foundational nature of what makes nuclear weaponry tick while, hopefully, not being overly dull.

Without getting too in the weeds, as this is not a science textbook and not intended necessarily to expand the horizons of nuclear physicists or historians of science, we can start with Dr. Einstein's famous 1905 equation, $E = mc^2$. In simple terms, this equation is about how much energy an atom has. The "E" stands for energy, the "m" stands for the mass (or weight) of the atom, and the "c²" (Latin *celeritas,* which means "speed") stands for the speed of light.[2]

The equation says that if you want to determine how much energy an atom has, you can do it by multiplying its mass by the speed of light squared. To put it another way, mass and energy are the same physical entity and can be changed into each other. The mc^2 part of the equation is the "rest energy" of an atom that is potentially available for conversion to other forms of energy. That sounds complicated, but all it really means is that the more mass an atom has, the more energy it has too.[3]

So, why is this equation so famous? When scientists figured out this equation, it helped them understand how the Sun and other stars produce energy. And as you'll learn later, the energy contained inside an atom is nearly unbelievable (a mass the size of a penny, if converted to energy, can power the New York City metropolitan area for a minimum of two years).[4] So in many ways, Einstein's equation set the foundation for utilizing the power contained within atoms.[5] As we now know, this mathematical foundation ultimately led to the creation of the most powerful weapon in the known universe.

But as I said, we'll step back a little further to the late 1800s and the work of a French engineer and physicist, Antoine Henri Becquerel. Becquerel was fascinated with and focused on the study of phosphorescence, which is the examination of materials that absorb and then emit light for an extended time period. Examples of phosphorescent material are glow-in-the-dark stickers or signs. As a kid, you may have had glow-in-the-dark stars and planets on your ceiling; those are typical examples of phosphorescent materials. Don't worry; they're not dangerously radioactive, as modern scientists and corporations know not to use such materials![6]

In late 1895, another scientist named Wilhelm Conrad Röntgen identified the X-Ray, a curious and little-understood phenomenon at the time that was named as such because "X" stands for something unknown in the field of mathematics. In short, as you probably know from your own doctors' and dentists' visits and the medical imaging you and your family may have undergone, X-Rays are a type of penetrating radiation that can pass through certain objects. They're useful for many things but are known most popularly for their medical applications. Although they had been produced in experiments as early as the late 1700s, they were not yet understood or specifically identified.[7]

Back to Becquerel, our French engineer. After learning that Röntgen had identified an invisible beam that could pass through paper, Becquerel wondered whether the phosphorescence (again, glowing) he had observed in certain materials was somehow associated with these mysterious "X-Rays." For example, he thought some uranium salts might emit X-Ray-like radiation when charged with bright sunlight. With regard to his hypothesis about X-rays, it turns out Becquerel was at least partly right; the uranium was emitting energy, but it had nothing to do with the Sun charging it. In fact, Becquerel discovered by May of 1896 that the energy produced by the uranium was independent of any outside (external) source of energy (like the sun). Interestingly, uranium's radioactive effects had also nearly been discovered in the mid-1850s through observational experiments on photographic materials (namely that the uranium would "expose" photographic material even in a dark room). Still, the science was not yet established, and the conclusions were not yet direct enough to identify radioactivity.[8]

We know now that uranium is a common radioactive element (elements being the basic building blocks of matter, such as hydrogen, carbon, helium, etc.). I won't get into the nature and description of uranium salts, as that's a technical discussion beyond our explorative scope here. For these purposes, though, if you're unfamiliar with radiation and radioactivity, radiation is the transmission of and emission of energy through particles or waves that travel through space or even materials. X-rays are a type of radiation everyone is familiar with. We'll talk more about radiation as we progress through the discovery of nuclear physics.[9]

Following in the footsteps of Röntgen and Becquerel was a daring and genius chemist and physicist by the name of Marie Curie, who discovered, using her husband's electrometer (a device used for detecting an electric charge), that energy coming from uranium could cause the air around a sample to conduct electricity. Dr. Curie observed that the energy emitted by the uranium was constant and that minerals with more uranium emitted stronger rays. She hypothesized that the rays coming from uranium could be unique to that uranium: that they were actually an atomic property of the element uranium itself, built into its atoms and part of its very nature. By 1898, Dr. Curie had discovered that thorium also emitted rays and that this emission was likely an atomic property of the element. That same year, Marie and Pierre Curie discovered the elements polonium and radium (Latin for "ray"). In addition to discovering various radioactive elements, they created the word "radioactivity" to describe the phenomenon they observed.[10,11]

Although not the subject of this book, what would eventually become an essential discovery to medical science, is the Curies' revelation that radium could destroy cancer cells more rapidly than healthy ones. Eventually, this would lead to advances in cancer treatment. I mention this aspect of the Curies' work because the effects of radiation on the human body and the environment are an important aspect of nuclear weaponry that we'll discuss later. By July 1934, Marie Curie would ultimately die from aplastic anemia, a severe blood disorder in which the body cannot produce adequate levels of blood cells, believed now to have been a product of her exposure to radioactive elements throughout her career. Dr. Curie's remains were so contaminated with radiation that she had to be sealed in lead lining; her papers from the 1890s are similarly packed in lead-lined boxes and require those consulting them to wear personal protective equipment (PPE).[12,13]

At the time, science did not have the foundation to comprehend the dangers of radiation exposure fully. While Dr. Becquerel's cause of death is unknown, he passed at age 55, and it was noted that he had burns on his skin, likely stemming from his handling of radioactive materials.[14]

Ernest Rutherford's studies moved the field rapidly forward around the same period as Dr. Curie's and Dr. Becquerel's experiments. His accomplishments would eventually lead him to be dubbed the father of nuclear physics. In 1899, he discovered new types of radiation: alpha and beta rays. Before Rutherford's work, atoms, the basic "building block of matter," were understood to be indivisible (that is, they were believed to be the smallest form of matter and could not be broken up). In 1903, he correctly observed through his radiation experiments that, in fact, "radioactivity' was a result of the disintegration of atoms into other matter. However, the nature of that matter was as yet not understood. By 1903, he had discovered a new type of radiation with far greater penetrating power, coined the "gamma ray." Gamma rays are similar to X-rays in that they are photons of pure energy. They have the same basic properties but come from different parts of the atom. While X-rays are emitted from processes outside the nucleus, gamma rays originate inside the nucleus. X-rays are also generally lower in energy and, therefore, less penetrating than gamma rays.[15,16,17,18]

Dr. Rutherford's most famous discovery came from the gold foil ("Geiger-Marsden Experiment") conducted in 1909 with Hans Geiger and Ernest Marsden. By firing alpha particles at gold foil, this experiment showed that atoms were built of a small, positively charged nucleus making up most of the atom's mass, surrounded by an expansive cloud of low-mass, negatively charged electrons. I'll explain alpha particles later, as they become important in the transition from hypothetical to practical nuclear physics in future decades. Rutherford also concluded that a hydrogen nucleus was a fundamental building block of all atomic nuclei (plural for the nucleus) and called it the "proton."[19,20]

These experiments led to the understanding that atomic nuclei also contained neutrons, which provide mass but have no charge. Thus, you have the proton, the neutron, and the electron, the basic building blocks of the atom. James Chadwick, who worked with Rutherford, is ultimately responsible for discovering the neutron. This discovery would be critically important to nuclear power and atomic weaponry, as neutrons are the proverbial "bullet" that sets off a nuclear reaction.[21]

Stepping way back into history for a moment, fundamental to understanding nuclear reactions is the concept of *transmutation,* which is the conversion of a basic chemical or element into one of an entirely different nature. While alchemy has a long history dating back as far as ancient Egypt, it's most commonly known for the search for the "Philosopher's Stone," first mentioned by Zosimus Alchemista of Panopolis, a Greco-Egyptian alchemist in the 4th Century AD. Historical analysis ties the roots of the concept of the Philosopher's Stone to ancient Greek philosophy. Regardless of its true origin, the Philosopher's Stone was believed, through a process called "chrysopoeia," to be capable of turning base metals such as mercury or lead into gold. We know now that lead and gold are basic elements, so no chemical process can turn one into another. That said, it was simply the case that the alchemists did not yet have the science available to them to understand how to transmute elements. The power of transmutation lies, instead, with nuclear technology. Enter Frederick Soddy.[22,23,24]

Soddy, who also worked with Rutherford, showed that the radioactive decay of radium would produce helium gas. This was demonstrated by taking a sample of radium and enclosing it in a glass envelope within an evacuated glass bulb (with no air or other material present). An analysis of the bulb later demonstrated helium to be present, showing that some of the radium had decayed into helium, which was not present in the original sample. This was an early demonstration of transmutation—one element turning into another element.[25,26]

In addition to his contributions to understanding transmutation, Soddy similarly discovered that certain elements can have different isotopes. Understanding isotopes is essential to understanding nuclear weaponry, and we'll discuss them further later, but the general idea is that isotopes of the same elements can have the same chemical properties while having different physical properties and masses. Most everyone knows about the element hydrogen. Hydrogen has an atomic weight of 1 (that is to say, it has 1 proton in its nucleus and no neutrons, so an atomic weight of 1) and multiple isotopes. One of these hydrogen isotopes is tritium (hydrogen-3), which has one proton (like hydrogen), but also two neutrons in its nucleus, giving it an atomic weight of 3. Tritium is often used to light up "EXIT" signs without the use of batteries or electricity.[27] Another isotope of hydrogen is deuterium (hydrogen-2), which has one proton and one neutron, giving it an atomic weight of 2. Many isotopes are not stable and fall apart in the process of radioactive decay. These radioactive isotopes (ones that decay, thus giving off energy) are known as "radioisotopes."[28,29]

As rapid understanding of radioactivity and the nature of the atom began to develop in the late 19th and early 20th centuries, scientists began to understand more and more about how atoms interacted with each other and the processes that were taking place inside atoms themselves. This understanding drove researchers to start to express concern and even fear about the potentially limitless power of the atom.

Chapter 2. The Era of the Atom: Limitless Power

"Before the last war began it was a matter of common knowledge that a man could carry about in a handbag an amount of latent energy sufficient to wreck half a city." - H.G. Wells, in "The World Set Free"[30]

One can seldom control something one does not understand. As it turned out, Ernest Rutherford, mentioned earlier as one of the chief architects of the study of nuclear physics, was skeptical that atomic reactions had the potential for the production of useful energy. He had this to say on the topic:

> We might in these processes obtain very much more energy than the proton supplied, but on the average we could not expect to obtain energy in this way. It was a very poor and inefficient way of producing energy, and anyone who looked for a source of power in the transformation of the atoms was talking moonshine.[31]

It turns out that, despite Rutherford's incredible advances in the field, not everyone agreed with Rutherford. One such skeptic of his conclusion was Leo Szilard, a Hungarian-born physicist with Jewish parents who fled Germany in the early 1930s as the Nazi Party dug its claws into the German government and began its reign of terror.[32] When Szilard encountered Rutherford's public claim that energy could not be obtained from nuclear reactions, he was already an accomplished physicist, having submitted patent applications for the linear accelerator, the electron microscope, and the cyclotron.[33]

Cyclotrons are machines that accelerate particles, such as protons and electrons, to incredibly high speeds using magnetic and electric fields in a circular path; they allow scientists to study the properties of particles at different energies. They have multiple uses in physics research, nuclear medicine, and particle therapy, and can also be used to create radioisotopes. They do so by bombarding a target material with high-energy particles (essentially smashing the particles together). Cyclotrons have been used to further science's understanding of the structure of atoms and their component particles.[34]

I mention Szilard's conception of the cyclotron because the idea of particle bombardment would be necessary for his later work. Szilard believed that Rutherford was fundamentally wrong about the energy generation potential of nuclear reactions, and shortly after Rutherford's claim was published publicly, an irritated Szilard set out to prove him wrong. Building on his ideas about particle bombardment, Szilard began to play with a concept that would change the world: What if one could shoot a neutron at an atom, causing the atom to split and therefore knocking two more neutrons out of the atom, which then each hit another atom, knocking more neutrons out of those atoms, and so on?[35]

In a case of reality sometimes being stranger than fiction, it's likely that the inspiration for Szilard's idea for harnessing nuclear energy came from a lesser-known H.G. Wells novel by the title, "The World Set Free," which Wells dedicated to Frederick Soddy after happening upon Soddy's research into nuclear isotopes. In "The World Set Free," Wells concocted an eerily prescient tale of nuclear reactors creating massive stores of energy and the construction of "atomic bombs" that are eventually dropped from airplanes in a devastating war that eradicates the world's cities and leads to the collapse of society. Szilard would eventually opine in his memoir that Wells' work had a great impression on him.[36]

At the time that Szilard was dreaming of the concept of a nuclear chain reaction, we already knew that radioactive decay, discussed earlier, emitted energy. That energy is the radioactivity observed by Becquerel, Curie, and others like them, though those scientists had a very basic view of the workings of atoms at the time. As it turns out, the energy stored in an atomic nucleus's neutron/proton bonds is massive. When those bonds are split, a colossal amount of energy is released.

To understand this concept of energy release in chain reactions, picture a rubber band and a tower of bricks. When stretched and held in its extended position, a rubber band stores potential energy, much like the bonds in an atom. When you unhook the rubber band suddenly, it will violently snap, expelling the stored energy. Similarly, when hit just hard enough to topple it over, a tower of bricks will break apart and release the potential energy stored when the tower was constructed. This concept is similar to the energy release when an atom splits apart. Imagine the bricks in the tower are atoms breaking apart when struck by the expelled neutron from another atom in the chain reaction. And so on. Chadwick's discovery of the neutron was crucial to this concept because, since neutrons have no electric charge, they are not repelled by the electron cloud of an atom and can penetrate an atom's nucleus, splitting apart even the heaviest elements.[37]

To understand the potential energy stored in atoms and what Szilard could only begin to fathom in his early experiments, we can compare it to common fuels we use in everyday life. The energy stored in a uranium pellet, about 10 grams or the size of a thimble, equals around 2,000 pounds of coal, 149 gallons of oil, or 17,000 cubic feet of natural gas.[38] Through this illustration, one can begin to grasp the raw power of nuclear reactions.

Throughout the mid to late 1930s, Szilard hypothesized and experimented with ways to create a self-sustaining nuclear reaction. He used his knowledge of chemical chain reactions and extrapolated that knowledge to atomic physics. By 1934, he had filed for a patent (a government instrument/document giving the creator a right to use, make, or sell an invention exclusively) regarding the liberation and use of nuclear energy for power production *and other purposes* through the process of nuclear transmutation. The following year, he amended the patent to identify uranium and bromine as specific elements that could be used in neutron reactions to release multiple neutrons in the transmutation process. Again, to recap, the idea here is that, for example, one can fire a neutron at a particular atom (let's use uranium) and that multiple neutrons will split from that uranium atom, collide with two more uranium atoms, and so on in an exponentially increasing chain, all the while releasing incredible amounts of energy. However, Szilard was aware of this technology's potential danger after considering the outcomes described in Wells' fictional work. Although fission had not yet been demonstrated in practice, the theoretical calculations and science appeared sound. Humankind's propensity to weaponize technology troubled Szilard enough that he implored the British Government to make the patents secret. While myopic leaders initially dismissed Szilard's concerns, eventually, government authorities recognized the threat and hid the patents.[39]

By 1934, Irène Joliot-Curie (the daughter of Marie and Pierre Curie) and Frédéric Joliot had successfully bombarded elements with alpha particles and caused them to become radioactive. Alpha particles are two protons and two neutrons bound together in a positively charged particle identical to a helium-4 nucleus. Working jointly, the scientists bombarded boron, aluminum, and magnesium with alpha particles, obtaining radioactive isotopes of elements that are not ordinarily radioactive– elements like phosphorous, aluminum, and nitrogen. These artificial isotopes would become critically important to the development of fission (which was, at the time, the theoretical chain reaction that could potentially unleash the true power of the atom).[40]

Standing on the foundation built by those before him, a prodigious Italian physicist, Enrico Fermi, worked with his colleagues to bombard over 60 elements with neutrons. They analyzed these elements to record new radioactive isotopes that were produced. Their tests on uranium showed several fascinating outcomes, and they had trouble interpreting them. Some scientists believed that the experiments showed that Fermi had produced "transuranium elements," which are elements with a higher atomic number than uranium. Uranium, with an atomic weight of 92 (remember, atomic weight is the total number of neutrons and protons in the nucleus), is the heaviest element found in nature. To produce other elements, one needs to increase the nucleus's weight. It was unclear at the time whether Fermi had actually produced transuranium.[41]

By 1938, a group of German scientists, Otto Hahn, Lise Meitner, and Fritz Strassmann, were on the verge of a breakthrough. They were working to interpret and verify Fermi's work on uranium. While their early results seemed to confirm the thoughts about Fermi creating transuranium, later experiments would prove much more fateful. Meitner fled Germany in 1938 to escape the persecution of Jews by the Nazis, but Hahn and Strassmann continued to experiment. Toward the end of 1938, they had demonstrated conclusive evidence that one of the products created by the uranium bombardment experiments was a much lighter element: barium. Thus, they had proven that the uranium had split into two lighter atoms. Meitner, now in Sweden, worked with her nephew Otto Frisch to confirm the phenomenon that had taken place, naming it "nuclear fission."[42] What Szilard had hypothesized and Fermi had worked toward was now an established reality. Nuclear fission, the splitting of atoms, had become possible in the lab.[43]

Around this time, Fermi had noted with concern the growth of fascism and antisemitism in Italy; since his wife was Jewish, he recognized the importance of leaving Italy while the opportunity still presented itself. He won the 1938 Nobel Prize for Physics, which provided a convenient excuse to travel to the United States, where he settled and began a career at Columbia University. The news out of Germany of the confirmation of fission came within weeks of Fermi's arrival.[44]

While scientists were aware for years that atomic nuclei (plural for nucleus) could eject small chunks, either through natural radioactivity or through the process of intentional bombardment via a projectile, science had, up until this point, not yet demonstrated that an atom could split apart. As a result of the earlier alarms sounded by scientists like Szilard, the consequences of this discovery and its associated capacity to release as yet incomprehensible energy levels, for better or worse, were widely known. As the world careened into World War II, Fermi knew there was work to be done.

Chapter 3. Fermi's Pile: The Chain Reaction That Changed the World

"There is no likelihood man can ever tap the power of the atom. The glib supposition of utilizing atomic energy when our coal supply has run out is a completely unscientific Utopian dream, a childish bug-a-boo." -Robert Millikan, 1928[45]

The scientific proof of fission, combined with the outbreak of World War II in September 1939, led to a fear spreading like wildfire in scientific circles: Nuclear power held the potential for as yet unknown, but likely limitless, civilian and military purposes. Many of the architects of nuclear physics and fission research had fled Germany, and some had fled Europe altogether as they witnessed the horrors that had begun to surface in Europe at the hands of the Nazi regime.

In August 1939, Leo Szilard worked with Albert Einstein to draft a letter to President Theodore Roosevelt, known as the famous "Einstein-Szilard Letter." Einstein, Szilard, Edward Teller, Eugene Wigner, and several other scientists feared that the Nazis would use their scientists to develop an atomic bomb using newly discovered fission technology. An uncontrolled and rapid chain reaction could theoretically result in a concentrated explosive force that the world had never seen outside the likes of enormous volcanic detonations. In the hands of the Nazis, these scientists reasoned, this power could lead to global fascist dominion. As such, the only possible response was to act quickly to develop this theoretical weapons technology and put it in the hands of the Allied forces.[46]

In his letter, Einstein deduced that the work of Joliet, Fermi, and Szilard had made clear that a nuclear chain reaction in a large mass of uranium could potentially generate "vast amounts of power..." that could lead to the construction of "extremely powerful bombs of a new type... A single bomb of this type, carried by boat and exploded in a port, might very well destroy the whole port together with some of the surrounding territory." Warning that the United States had poor uranium mines with limited quantities and that Canada, Czechoslovakia, and the Belgian Congo offered much better options to secure uranium for the war effort, Einstein made two recommendations, both that he suggested could be implemented through a partnership between the U.S. government and an unofficial agent acting on behalf of the government. The first of these recommendations was to make government departments aware of further development and put forward proposals for government action concerning securing uranium for the United States. The second recommendation was to speed up experimental work, which was only funded by University laboratories at the time, by providing funds in conjunction with private sources and potentially through a partnership with industrial laboratories.[47]

Einstein ominously concludes the letter by noting that the Nazis had halted the sale of uranium from the Czechoslovakian mines that the Germans had captured. He further notes that this decision by the Nazis may stem from the fact that the son of the German Under-Secretary of State, Carl Friedrich Freiherr von Weizsäcker, was attached to the Kaiser-Wilhelm-Institute in Berlin, where some of the American work on uranium was being repeated.[48]

Given the war unfolding in Europe at the time, Roosevelt did not initially have the bandwidth to respond to Einstein's letter. Two months later, on October 19, 1939, Roosevelt's Secretary, Edwin M. Watson, wrote back, informing Einstein that the President had established a committee whose purpose would be to thoroughly investigate the issues Einstein raised. Alexander Sachs, a friend of the

President who brought the letter to President Roosevelt, would cooperate with the committee.[49]

By late October 1939, the Advisory Committee on Uranium had begun meeting. Composed of military and civilian representation, its charge was to work with Sachs to investigate the state of research on uranium and recommend a course of action to the federal government. By early 1940, the Committee concluded that it should fund limited research on isotopes in addition to Szilard's and Fermi's work on fission chain reactions taking place at Columbia University.[50]

Two isotopes were identified as good candidates for fissile material. "Fissile material" is composed of elements that can be used to sustain a nuclear chain reaction, or "nuclear fission." These two isotopes were uranium-235 (U-235) and uranium-238 (U-238), the most abundant forms of uranium in the natural world. However, enriched uranium (a concentrated, pure form of uranium) had to be obtained to conduct further research and potentially create a working bomb. For that, U-235 and U-238 needed to be separated. U-235 and U-238 were chemically identical, so they couldn't be separated by chemical means. The Committee, therefore, looked into various means of separating them and landed on multiple possibilities, including high-speed centrifuges, gaseous diffusion methods, and liquid thermal diffusion methods.[51]

By June 1940, President Roosevelt, recognizing the importance of collaboration between scientists and the military, transferred the Uranium Committee into the new National Defense Research Committee (NDRC). This was a scientific body without military membership. Since it wasn't dependent on the military, it had more accessible access to funds and more influence to direct resources toward uranium research. On this streamlining of resources, Vannevar Bush, the Director of the Carnegie Institution who at the time had pressed for the creation of the Committee, later commented:

There were those who protested that the action of setting up NDRC was an end run, a grab by which a small company of scientists and engineers, acting outside established channels, got hold of the authority and money for the program of developing new weapons. That, in fact, is exactly what it was. Moreover, it was the only way in which a broad program could be launched rapidly and on an adequate scale. To operate through established channels would have involved delays-and the hazard that independence might have been lost, that independence which was the central feature of the organization's success.[52]

Foreign-born scientists were barred from Committee membership, and the further publication of uranium research articles was blocked for national security reasons.[53]

By early 1941, Glenn T. Seaborg and his colleagues, working at Berkeley as part of the U.S. government's secret research efforts, had discovered several new elements through the process of bombardment (essentially, smashing atoms together). One of these new elements, produced by deuteron bombardment of uranium-238 (smashing an isotope of hydrogen into uranium-238), would be critical to the United States' war efforts. The new artificial element was plutonium.[54]

Between 1939 and 1942, a team of scientists, including Enrico Fermi, Leo Szilard, and their colleagues, acting on the new knowledge about how fission could be achieved, worked to conduct experiments at Columbia University using nuclear "piles" to achieve fission from neutron emission. These piles were called as such because they were essentially rudimentary attempts at creating a nuclear reactor using actual stacks of graphite and uranium blocks. The graphite in these piles acts as a moderator to slow down the reaction so that it can be sustained. Self-sustaining nuclear fission had not yet been achieved at the time of these experiments, but a breakthrough was on the horizon.[55] At this time, research on experimental piles was taking place at multiple universities, including Columbia, Princeton, and Chicago, while the plutonium discoveries were taking place at Berkeley. Following the attack on Pearl Harbor at the end of 1941, U.S. government leadership opted to concentrate experimental pile work at the University of Chicago. The "Metallurgical Lab," or "Met Lab," for short, was born there. The lab's name was a codename to preserve the secrecy of the classified research.[56,57]

Initially housing a staff of only 153 individuals in 1942, the Met Lab would eventually utilize over 2,000 staff and over 200,000 square feet of university space for its efforts, with University of Chicago Vice President E. T. Filbey remarking to the director of the Metallurgical Laboratory, Arthur Holly Compton, "We will turn the University inside out if necessary to help win the war. Victory is much more important than the survival of the University."[58] The massive national effort taking place in Chicago would pay off on December 2, 1942.

Beneath the University of Chicago's football field, Fermi and his colleagues had constructed a 25-feet high by 20-feet wide pile of bricks, most of which were graphite moderators, with some containing small amounts of uranium inside drilled holes. Vertical wooden rods coated with cadmium (a material understood to slow the movement of neutrons) were interspersed into the pile of graphite and uranium. Ultimately, 771,000 pounds of graphite were used to build 57 layers. The pile used 80,590 pounds of uranium oxide and 12,400 pounds of uranium metal. The scientists couldn't be sure their moderators could control the reaction and stop it from spiraling. Nonetheless, Fermi assured the team that the probability of an accident was minimum. When asked what he would do if anything went wrong, he replied, "I will walk away– leisurely."[59,60]

On that cold day in December, Fermi and other scientists looked on from a balcony while a member of their team, George Weil, slowly pulled out cadmium rods as Fermi directed. The scientists monitored the reaction by utilizing Geiger counters (a device used to measure radiation levels). A high enough measurement from the counters would indicate to the scientists that a self-sustaining reaction had occurred.[61]

As the rods were removed from the pile, uranium nuclei began to split; the neutrons ejected from the original atoms were absorbed by other uranium atoms, safely releasing energy each time. As Fermi ordered the final cadmium rod to be pulled, he looked at Arthur H. Compton and remarked, "This is going to do it. Now it will become self-sustaining…"

Physicist Herbert Anderson described the events in the lab that day as follows:

> At first you could hear the sound of the neutron counter, clickety-clack, clickety-clack. Then the clicks came more and more rapidly, and after a while they began to merge into a roar; the counter couldn't follow anymore. That was the moment to switch to the chart recorder. But when the switch was made, everyone watched in the sudden silence the mounting deflection of the recorder's pen. It was an awesome silence. Everyone realized the significance of that switch; we were in the high intensity regime and the counters were unable to cope with the situation anymore. Again and again, the scale of the recorder had to be changed to accommodate the neutron intensity which was increasing more and more rapidly. Suddenly Fermi raised his hand. "The pile has gone critical," he announced. No one present had any doubt about it.[62]

Upon realizing they had achieved a self-sustaining reaction, one of the scientists, Eugene Wigner, offered his colleagues a bottle of Chianti. The scientists poured the Chianti into paper cups and held them up in a silent toast. Military leadership was quickly notified, but since no codewords were established to securely communicate that they had achieved fission, the communication relayed over the phone was, "The Italian navigator has landed in the New World." Though the scientists shared a sense of accomplishment and achievement over their breakthrough, Szilard's reaction had a different tone. He stood on the balcony watching as the others left, before turning to Fermi, shaking his hand, and exclaiming that he thought the day would be remembered as a "black day in the history of mankind."[63]

The Met Lab scientists had achieved something some had previously thought impossible: the release of energy by an artificially-created nuclear chain reaction. This is known as a reaction that has achieved "criticality" or "gone critical." Though the experiment had only produced enough energy to power a small lightbulb, the theory became a reality. Self-sustaining, energy-producing nuclear fission was now no longer hypothetical; it was a practical technology.[64, 65]

Chapter 4. The Manhattan Project: The War Machine Looks to Science

"People must understand that science is inherently neither a potential for good nor for evil. It is a potential to be harnessed by man to do his bidding." -Glenn T. Seaborg[66]

Let's step back to 1931, before the criticality breakthrough in Chicago. In Britain, as the power of nuclear science began to emerge, Winston Churchill, not yet prime minister, signaled to the world the potential of atomic energy in a speech entitled "Fifty Years Hence:"

> We know enough to be sure that the scientific achievements of the next fifty years will be far greater, more rapid and more surprising, than those we have already experienced... High authorities tell us that new sources of power, vastly more important than any we yet know, will surely be discovered. Nuclear energy is incomparably greater than the molecular energy which we use today. The coal a man can get in a day can easily do five hundred times as much work as the man himself. Nuclear energy is at least one million times more powerful still. If the hydrogen atoms in a pound of water could be prevailed upon to combine together and form helium, they would suffice to drive a thousand horsepower engine for a whole year. If the electrons, those tiny planets of the atomic systems, were induced to combine with the nuclei in the hydrogen the horsepower liberated would be 120 times greater still. There is no question among scientists that this gigantic source of energy exists. What is lacking is

the match to set the bonfire alight, or it may be the detonator to cause the dynamite to explode. The Scientists are looking for this.[67]

Britain had become acutely aware of the possibility that the Nazis could develop nuclear weaponry. Two physicists working together under Mark Oliphant at Birmingham University, Otto Frisch and Rudolf Peierls, released a memorandum in 1940 entitled "On the Construction of a 'Super-Bomb,'" which warned as follows:

> The energy liberated in the explosion of [an atomic] super-bomb is about the same as that produced by the explosion of 1,000 tons of dynamite. This energy is liberated in a small volume, in which it will, for an instant, produce a temperature comparable to that in the interior of the sun. The blast from such an explosion would destroy life in a wide area. The size of this area is difficult to estimate, but it will probably cover the center of a big city...

> It is a property of these super-bombs that there exists a "critical size" of about one pound. A quantity of the separated uranium isotope that exceeds the critical amount is explosive; yet a quantity less than the critical amount is absolutely safe. The bomb would therefore be manufactured in two (or more) parts, each being less than the critical size, and in transport all danger of a premature explosion would be avoided if these parts were kept at a distance of a few inches from each other. The bomb would be provided with a mechanism that brings the two parts together when the bomb is intended to go off. Once the parts are joined to form a block which exceeds the critical amount, the effect of the penetrating radiation always present in the atmosphere will initiate the explosion within a second or so...

We have no information that the same idea has also occurred to other scientists but since all the theoretical data bearing on this problem are published, it is quite conceivable that Germany is, in fact, developing this weapon. Whether this is the case is difficult to find out, since the plant for the separation of isotopes need not be of such a size as to attract attention. Information that could be helpful in this respect would be data about the exploitation of the uranium mines under German control (mainly in Czechoslovakia) and about any recent German purchases of uranium abroad.[68]

In May 1940, the same year the memorandum was published, Churchill became Prime Minister. In response to the memorandum, he created a uranium subcommittee to advise the British government on the potential of nuclear weaponry. This committee would become known as the MAUD Committee. Many assume MAUD is a cryptic atomic acronym, but in reality, it was derived from a telegram sent by Danish physicist Niels Bohr to his colleague Otto Frisch, who was mentioned earlier. Bohr had concluded the telegram with instructions to pass on his words to "Cockroft and Maud Ray Kent," who was the caretaker of his children.[69,70] The MAUD committee conducted its work in such secret conditions that it was not even allowed to work with many of the early pioneers of atomic science, as the government labeled them "illegal aliens."[71]

By this time, the British were collaborating with the Americans on preliminary research. In March 1941, the MAUD Committee released a report detailing the possibility of nation-states producing and using atomic weaponry. It reached a tripartite conclusion. First, producing a uranium bomb was possible, and such a bomb would lead to decisive results in the war. Second, work on uranium research should be continued on the highest priority and on a scale necessary to produce a weapon in the shortest possible time. Finally, collaboration with the Americans on experimental work should continue and be extended.[72]

The British government concurred with the MAUD Committee's conclusion and prioritized research into the Bomb. At this point, it was determined that a pilot plant for the Separation of U-235 from U-238 should be built in the United Kingdom, with a full-scale plant in Canada (where substantial uranium deposits existed).[73, 74] To this end, the British established the Tube Alloys Program, the codename for the world's first nuclear weapons program. The existence of Tube Alloys and the resources that the British dedicated to the program sparked interest within the American government.[75]

By the early months of 1942, before the Met Lab's fission discovery, following a series of military setbacks for the Allied Powers in the Pacific Theater at the hands of the Japanese Empire, United States officials had tentatively decided to proceed with the development of an atomic bomb. Two routes seemed possible: either a device based on a uranium platform or one based on the newly-discovered plutonium, which showed, in laboratory experimentation, great promise for fission. Either way, the hurdles to a usable bomb were significant. First, nuclear fission had to be demonstrated as not just theoretically possible on paper, but practically possible. As we know already, this hurdle was met in December 1942. Additionally, a sufficient amount of weapons-grade uranium or plutonium had to be cultivated and separated to place inside a bomb.

Also prior to the fission discovery in Chicago, Roosevelt had approved military coordination of the scientific experimentation, including, but not limited to, the activities at UC Berkeley and the Metallurgical Laboratory in Chicago. By mid-1942, the Army Corps of Engineers was tasked with coordinating with the Office of Scientific Research and Development (OSRD) to spearhead the project to build the Bomb, and 60% of the Corps' 1943 proposed budget ($54 million at the time, or $969 Million in 2023) was dedicated solely to this purpose. At this point, the research had become, unmistakably, a military operation.[76]

In August 1942, the Army Corps of Engineers established the Manhattan Engineer District (MED) to manage the project; this name was reflective both of the location of the project's original headquarters in Manhattan and of the desire to choose a name that would not attract the attention of the Axis Powers. The following month, Colonel Leslie R. Groves was promoted to brigadier general and acted swiftly to accelerate the project.[77] Groves had excellent engineering credentials, renowned administrative abilities, and was famously known for overseeing the Pentagon facility construction project during World War II. Shortly before Groves took over the project, it became known as "The Manhattan Project."

As part of 1942's project acceleration, Groves recognized that given the pace of the war, if the United States was to make any use of the hypothetical atomic bomb, either as an actual weapon or the threat of one, it had to be ready to produce one should the science pan out. Most importantly, the United States had to beat Adolf Hitler and the Nazi Regime in the race to build the Bomb, because U.S. leadership knew that if it was beat at this monumental task, it could be Game Over for the Allies. Though self-sustaining fission had not yet been demonstrated in a lab setting, Groves ordered the construction of the manufacturing facilities that could be used to build the Bomb. To this end, the Army Corps of Engineers set upon a remote area of East Tennessee. The site provided clean water, a rail line, adequate roads, affordable land, and perhaps most importantly: vast amounts of power provided by The Tennessee Valley Authority (TVA) and its hydroelectric dams. The sprawling and remote nature of the land also allowed for the fissile material processing facilities to be built far enough apart that if one were to be destroyed, the other facilities would not be victims of collateral damage.

By the Autumn of 1942, residents were informed that they would need to vacate the area. Most residents accepted the government payouts for their land, and even those who challenged the valuations were gone within a year. Given the intentionally misleading title "Clinton Engineering Works (CEW)," the site was secure both by nature and design. A highway initially running through the site, State Highway 61, was closed and redirected. On the Northern side, the Army Corps of Engineers erected a massive fence protected by armed patrols. The Clinch River surrounded the site on three sides, which provided a natural barrier. All personnel entering the facility had to have passes, and engineers were segregated, to the extent possible, from scientists so that work was siloed; this intentionally made it difficult for personnel working on one aspect of the project to discuss it with those working on other elements.

A town called "Oak Ridge" was constructed near the site to house personnel. Originally designed to accommodate 8,000 people, the town would balloon to a massive population of 75,000 by the war's end. The town hosted avenues named for the various states, arranged alphabetically from North to South. It was so well planned that side streets were adorned with letters beginning with the same letter as their primary avenue to help newcomers to the area find their way. To this day, Oak Ridge sports this feature. For example, if you travel along Highland Avenue, you'll find associated streets by the name of "Holston Lane," "Hamilton Circle," "Hillside Road," "Henderson Lane," "Henley Road," and so on.[78]

Three major facilities comprised the site: X-10, Y-12, and K-25. X-10 and Y-12 both began construction in February 1943, while construction started on K-25 in June 1943. X-10, also known as the Clinton Laboratories (later renamed the "Oak Ridge National Laboratory"), was built to develop and test an experimental Graphite Reactor. The Y-12 National Security Complex was designed to use calutrons (a version of the cyclotron intended for electromagnetic uranium separation) to separate the highly fissile weapons grade uranium-235 from the more ubiquitous uranium-238. The K-25 plant, encompassing 44 acres, housed the equipment necessary for another uranium separation process: gaseous diffusion.[79]

Ultimately, K-25 and Y-12 were utilized in tandem because the military and its scientists were unsure which method would yield enough enriched uranium to produce a weapon, if producing a weapon was even possible. Challenges at both facilities regarding scaling, maintenance, and equipment failures cast doubt on whether either facility could provide a viable path for production. By the end of Summer 1943, leadership had decided that they would not attempt to obtain enriched uranium through K-25 by itself; instead, K-25 would produce material that was around 50% uranium-235, which would then be fed into Y-12's equipment for final enrichment. Despite the facilities working in conjunction, the path to the Bomb was still unclear.[80]

Meanwhile, the processes at X-10 were an outgrowth from Fermi's Chicago Pile-1 (CP-1) uranium reactor, but were not intended to produce enriched uranium for an atomic bomb, even though CP-1 was a uranium reactor. Instead, X-10's mission was to build upon the fission success of CP-1 to use uranium to produce material for a hypothetical second type of bomb: a plutonium-fueled device. It had been established through analysis of CP-1 that the fission reactor in Chicago was producing potentially highly fissile plutonium from some of the irradiated uranium transmuting in the reactor.[81]

In June 1942, engineers and scientists associated with the Met Lab recognized that even if they were to achieve experimental fission at CP-1 (recall that they eventually did, in December of that year), they would need a scaled-up reactor to produce energy and fissile material. Such a reactor would need to produce vastly more fissile material than CP-1, so radiation containment shielding and cooling systems would be necessary to protect human life and the equipment and fissile materials involved in production. CP-1 had no radiation shielding, and while we won't get into the details of the collateral effects of that now, we'll discuss the hazards, side effects, and destructive capabilities of ionizing radiation later on. Scientists considered both liquids and gases for cooling purposes in a production-scale reactor– hydrogen and helium appeared to be good candidates for gaseous cooling, and water seemed to be a good candidate for liquid cooling. That said, water tended to corrode uranium, which posed challenges if it were to be used for cooling.

By Summer 1942, planning was underway for a helium-cooled pilot pile to be constructed by Stone & Webster in the Argonne Forest Preserve about a half hour outside Chicago. The plan was to construct a 460-ton cube of graphite interspersed with cartridges of uranium and graphite, cooled by helium circulated through the pile. Graphite surrounding the reactor would contain the radiation inside. Alternative liquid-cooled reactors were proposed, and Szilard even believed he could construct one using an electromagnetic pump (one he designed in conjunction with Albert Einstein) cooled by liquid metal. However, he had trouble obtaining the liquid metal required for such a reactor.

By October 1942, even before Fermi had achieved his self-sustaining reaction, General Groves had arrived in Chicago to resolve the issue of which pile design would be used. He gave the scientists in Chicago an ultimatum and directed them to pick a design, explaining that time was more valuable than money and that things had to move more quickly. He gave them one week to decide. Ultimately, leadership chose a multi-track solution. Fermi would continue experimenting on CP-1 while an intermediate pile with external cooling would be utilized at Argonne. The helium-cooled, plutonium-producing reactor would be built at Oak Ridge and operational by March 1944. Szilard would continue to research liquid metal cooling methods.[82]

Prior to X-10's full-scale operation, Groves went looking for a private sector partner, as he recognized the colossal challenges of rapidly transitioning from experimental nuclear fission to production-grade fission. The DuPont Corporation, by then over a century old and already involved as a partner to the United States in the war effort, showed promise as a materials and chemicals science company with a sound administrative structure and a massive bank of experienced scientists and engineers. It agreed to assist in the efforts on one condition: it would charge only a $1 fee to the United States for lending its expertise and staff. DuPont's posture may have been a result of the corporation having been accused of war profiteering during World War I through its gunpowder sales (which accounted for 40% of the gunpowder [1.5 billion pounds] used by the Allies during WWI).[83, 84]

DuPont ultimately expressed concerns about the risks of manufacturing plutonium on a large scale, and Groves similarly had concerns about placing a plutonium production facility so close to gaseous diffusion and electromagnetic separation plants. Oak Ridge was also relatively close to Knoxville, which could endanger many of Tennessee's citizens if a disaster occurred. Met Lab scientists and DuPont engineers believed the plutonium production facility required a site with at least 225 square miles. Hanford, Washington was ultimately chosen as it offered a remote location, a long construction season, and plenty of water to produce hydroelectric power for the facilities. To reduce the risks of radiation accidents, nothing was allowed within four miles of the separation complexes; rail lines, highways, laboratories, and municipalities would have to be even further away. With a plan for three or four plutonium production reactors and one or two chemical separation facilities, the "Hanford Engineer Works," codenamed Site W, was born.[85] It should be noted here that the choice of facilities in both Los Alamos, NM and Hanford, WA displaced multiple Native American populations from their ancestral lands, and the government's decision to do so remains a matter of discussion and controversy. At Oak Ridge, the only known Native American scientist who worked on the Manhattan Project, Herbert York, eventually served as the first director of the Lawrence Livermore National Laboratory.[86]

Meanwhile, Glenn T. Seaborg, who had moved from Berkeley to the Met Lab in Chicago, had been experimenting with plutonium extraction processes. DuPont had begun constructing plutonium isolation and experimentation facilities at Oak Ridge and Hanford even before Seaborg had successfully isolated it from uranium using his chemical processes. Concerns remained about providing adequate radiation safety, as plutonium was highly radioactive and therefore intrinsically more dangerous than uranium.[87]

The X-10 reactor, which began construction in February 1943 at Oak Ridge, would ultimately become DuPont's massive air-cooled experimental graphite reactor—an enormous graphite block with hundreds of channels filled with uranium slugs. As new slugs were forced into the reactor, slugs that had already been irradiated would drop out and fall into an underwater bucket, where they would undergo radioactive decay for weeks, after which point an underground channel would move them into a chemical separation facility. Due to its high level of radioactivity, the plutonium would need to be separated and removed with remote control equipment. On November 4, 1943, X-10 went critical for the first time.

Leadership opted for water cooling to scale up the experimental activity at X-10 to production at Hanford. While there was a debate between gaseous helium cooling and water cooling, ultimately, the water design won out for the Hanford facility. It was determined that the Hanford pile, containing 1,200 tons of graphite and 200 tons of uranium, would need 75,000 gallons of water per minute to stay cool while the process of plutonium production proceeded.[88]

Work at Hanford progressed deliberately. Between 1943 and 1944, plutonium production reactors were constructed at the site. These reactors were enormous; the B-Reactor, which stood at 120 feet high, was composed of over 17,000 cubic yards of concrete, almost 400 tons of structural steel, and 71,000 concrete bricks. The chemical separation buildings were similarly impressive, canyon-like structures that were 800 feet long, 80 feet high, and 65 feet wide, and contained forty processing pools. Seven feet of concrete shielding was required to protect the workers from the incredible amount of radiation given off by the piles, so these facilities had to be operated with remote control equipment; engineers relied on periscopes and TV monitors to conduct their work.

On September 13, 1944, Enrico Fermi activated the pile at the B-Reactor by placing the first uranium slug into the pile. Due to the scale of the reactor, the engineers and scientists had to pray that everything worked as planned since the incredible amount of radiation rendered maintenance on certain components inside the reactor impossible. Over the next couple of weeks, additional slugs were loaded until September 27, when the reactor achieved a power level higher than any reactor before it. Strangely though, the power began falling after only three hours. For some reason, the reaction was not self-sustaining, which puzzled and alarmed the scientists. Even more bizarrely, the reaction had restarted again by the next morning to the previous day's level, then dropped again.[89]

The scientists scrambled to address the problem, and had various theories about what it could be, from contaminants in the water used to cool the reactor to nitrogen in the air. It turned out that the bizarre case of the restarting reactor was caused by a previously undiscovered phenomenon called "xenon poisoning." Xenon-135 (Xe-135) is a fission by-product with a half-life of about 9.2 hours (a "half-life" is the time it takes an element to transmute through the process of radioactive decay), and it turned out that the Xe-135 had gradually built up in the reactor, absorbing neutrons that were necessary to fuel the chain reaction. When the xenon built up inside the reactor, it would shut down the reaction. It would then decay, and as it decayed, the neutrons in the chain reaction would no longer be absorbed. The reaction would then start again. This process continued in a cycle. Xe-135 will "burn away" in the reaction at a high enough power, but at low power, it builds up. Fortunately for the scientists at Hanford, DuPont had overengineered the reactor, adding 25% more tubes than thought necessary, which allowed Fermi to increase the reactor's power. This increased power burned away the Xe-135 and let the chain reaction become self-sustaining, eliminating the problem of xenon poisoning. As a side note, xenon poisoning was a central culprit in the Chernobyl power plant meltdown in the Soviet Union decades later.[90, 91]

At this point, once the xenon poisoning had been resolved, the reactor was in an operational state and achieving fission at the scale required to produce plutonium. The rods were sequestered for several weeks following irradiation so that they could undergo chemical treatment to make useful quantities of plutonium. While it was still unclear whether all of the effort, resources, and time spent to achieve fission and enrich uranium and plutonium would amount to any usable technology, the stage was set for initial attempts.

Chapter 5. Los Alamos: The Fat Man and the Thin Man

"If it could be tapped and controlled, what an agent it would be in shaping the world's destiny! The man who puts his hand on the lever by which a parsimonious nature regulates so jealously the output of this store of energy would possess a weapon by which he could destroy the Earth if he chose." - Frederick Soddy[92]

While research on fission and reactors was being conducted in Chicago and at Oak Ridge and production processes were being scaled up, the final puzzle piece remained the designs for the atomic bomb itself. This research was to occur at P.O. Box 1663, Sante Fe, New Mexico, the only location listed for all mail and official documents arriving at the secretive site, otherwise known as "Project Y."[93]

Project Y saw its genesis in the Spring of 1942. Selected by General Groves to spearhead this element of the Manhattan Project was none other than J. Robert Oppenheimer, a theoretical physicist and professor at the University of California at Berkeley. Oppenheimer recommended Los Alamos, New Mexico, as the site for Project Y, as it met critical conditions necessary for such a secret undertaking: a remote location that would provide protection from an enemy attack, it hosted a minimal human population which would be ideal in the case a nuclear catastrophe occurring during testing, and the isolated nature of the location would help protect the top secret information that would be handled there. Interestingly, while Los Alamos met all the criteria for the project, it was also a naturally picturesque location that Oppenheimer had spent time in during his youth. He believed the site's natural beauty would help the weapons scientists at the facilities better acclimate and cope with the difficult times they would face there.[94]

Oppenheimer was an interesting choice for this position for several reasons. First, he and Groves were not alike in their thinking; while Groves took a more militant approach to the project, Oppenheimer was insistent that Los Alamos maintain an academic environment that was conducive to the mental and emotional health of the scientists, engineers, and staff who would live and work there. Further, Oppenheimer had associations with a number of leftist groups that were later labeled communist sympathizing or front organizations.[95] It also didn't help that Oppenheimer's mistress, Jean Tatlock, and several people he was associated with were known to be communist sympathizers.[96] This would become a security concern as the Soviet Union rose to prominence and drove concerns among military leadership and the FBI that Oppenheimer was perhaps not the best choice to lead the activities at Los Alamos. Finally, Oppenheimer had not won a Nobel Prize, although many of the personnel he'd be overseeing had; this fact risked posing a credibility, respectability, and teambuilding problem for Oppenheimer. Nonetheless, Groves recognized Oppenheimer's abilities, charisma, and ambition and personally issued Oppenheimer's security clearance.[97]

Oppenheimer went quickly to work in early 1943, crossing the country to assemble a crack team of the nation's best scientists, engineers, and staff. Team members arrived from some of the best institutions nationwide, including Columbia, Iowa State, the Massachusetts Institute of Technology (MIT), Chicago, Princeton, Stanford, Purdue, and Columbia, while others joined from the Met Lab and the National Bureau of Standards. Los Alamos transformed virtually overnight, soon hosting some of the most advanced equipment and genius minds in the entire world. By that Spring, even British scientists were arriving to contribute to the effort.[98]

As a result of the fission experiments being conducted at the Met Lab, Oak Ridge, and other facilities, it was known that nuclear chain reactions and self-sustaining fission were capable of releasing significant amounts of energy. The challenge for the weapons designers at Los Alamos was to take the fission experimentation and determine the ideal shape and size of the mass required to induce nuclear ignition in a single, powerful, instantaneous atomic blast. The goal was to take a critical mass of fissile material and release the most energy possible before the explosion destroyed the mass. To achieve this, the scientists theorized, two subcritical masses of fissionable material would need to come together to create a supercritical mass that would explode. [99]

From a weapons design or energy production perspective, a subcritical mass is an amount of fissionable material insufficient in quantity, or of an inadequate geometrical configuration, to sustain a fission chain reaction.[100] By way of example, let's examine uranium-235 (U-235), which, as you'll recall, is the form of uranium isolated from U-238 that the Manhattan Project scientists determined held great potential for fission. A small amount of U-235, around a pound (.45 kg), is incapable of sustaining a nuclear chain reaction (fission); therefore, it's said to be "subcritical." The correct shape and configuration are also required to ensure that the uranium nuclei's ejected neutrons strike other nuclei. If more U-235 is grouped together and added to the mass, the chances that escaping neutrons from the core of the U-235 will strike other U-235 nuclei and cause a chain reaction is increased.[101] While fission experiments in the lab caused chain reactions and achieved criticality, Los Alamos sought a state some refer to as "supercriticality," a reaction that would release an as yet unfathomable amount of energy in a single, devastating explosion.

The scientists at Los Alamos surmised that supercriticality could be achieved by firing subcritical masses into each other using conventional explosives if the materials reached velocities of around 3,000 feet per second. The measurements and calculations had to be exact to make this happen, and the materials used would have to be extremely pure. Oppenheimer and his team imagined what is now known as a "gun-type" design for a uranium device. Another design was imagined for plutonium, the "implosion method." We'll discuss more on those designs later.[102]

During this period, Los Alamos was divided into four divisions: theoretical; experimental physics; chemistry and metallurgy; and ordnance design. The theorists worked out the general concepts that were required to lay the foundation for a device that would be practically implemented.

The Experimental Physics division used particle accelerators to determine which tamper materials would adequately push neutrons back into the core of the explosion to enhance its energy output (by creating more fission impacts before the bomb is finished exploding). In a nuclear weapon, a tamper material is a heavy, dense material that surrounds the fissionable material to hold the supercritical assembly together longer to increase the fission rate of the material and create a more powerful explosion. Uranium, beryllium, and tungsten are all good candidate materials for this purpose.[103]

The Chemistry Division set its sights on purifying the U-235 and plutonium, reducing them to metals, and processing the tamper material necessary to maximize the efficiency of the Bomb. The risk posed by impure uranium and plutonium was a phenomenon known as "predetonation" or "fizzle." For a fission weapon or "atomic bomb" to properly detonate at maximum efficiency, the core must explode in a supercritical state when the neutron collisions in the fission reaction are at their maximum.[104] Predetonation occurs if the explosion occurs when the neutron collisions are not at their maximum, which risks blowing apart the critical mass and "fizzling" out the weapon. Essentially, a fizzle results in a dud of a weapon that is barely more effective than a conventional (non-nuclear) explosive. There are exact calculations about the neutron activity required to avoid a predetonation, but for our purposes, it's only essential to understand that to prevent this failure in weapon ignition, very pure uranium or plutonium needed to be used, and the chemistry division's job was to ensure that. The metallurgical division's job was to turn the purified plutonium and U-235 into metal that would be used in the Bomb's core.

The Ordnance Engineering Division put its efforts into designing artillery pieces for the Bomb. You may recognize artillery weapons as large guns such as high-caliber howitzers used to shoot a large-bore projectile with great force in conventional warfare.[105] The artillery pieces at the Ordnance Engineering Division used standard specifications, but contained barrels designed to be very light, since they would only need to be fired once, unlike a regular artillery barrel which would facilitate multiple blasts in a conventional warfare setting. Because of its more radioactive nature, plutonium had a higher risk of predetonation, so more focus was put on that higher velocity design. The Division ordered multiple uranium guns and plutonium guns for testing.[106]

Of the two designs (gun-type vs. implosion-type), the Ordnance Engineering Division prioritized the gun-type design, because it seemed simpler and more practical than the implosion method. The gun-type method would fire a mass into another mass at great force. This is physically uncomplicated. Picture a mass of material held in place by a metal framework, with a small projectile aimed directly at its center from a gun nearby. That projectile is fired into the center, causing a violent collision. This was like shooting a target, which had been done many times before. The implosion device, however, would require that a symmetrical explosion take place surrounding all sides of a mass, causing each side to impact inward with the same exact force, at the same exact time, symmetrically. For this reason, early experiments on the implosion-type device were not as promising. That said, physicists at Los Alamos saw promise in the implosion method because it seemed workable in a uranium device, which wouldn't require the more complicated extreme purification of plutonium. Hence, Oppenheimer pushed ahead with work on the implosion experiments.[107]

These weapons designs were turned into models for what was hoped would ultimately become the atomic bomb. By the Spring of 1944, the Army Air Force began testing the bomb designs with the Boeing B-29 Superfortress. The B-29 was a long-range heavy bomber proposed in 1940 that entered military service in September 1942. Built around the United States at five production plants, it had a bomb capacity of 10 tons. Its use in the Pacific Theater was so widespread that, at times, it operated in flights of several hundred at a time.[108, 109] The two bomb designs being tested at this point were named after the Allied leadership: "Thin Man" was the plutonium gun design, named after U.S. President Roosevelt, and the "Fat Man" was the implosion prototype named after British Prime Minister Winston Churchill.

Despite early testing on prototypes and device designs, problems arose in the Summer of 1944. Glenn T. Seaborg had sounded the alarm that when plutonium-239 was irradiated, it would likely pick up an additional neutron, transmuting it into another isotope of plutonium, Pu-240. This isotope is particularly problematic because it is highly reactive and increases the likelihood of predetonation before the subcritical components in an atomic bomb experience fission. Measurements at the X-10 facility in Oak Ridge confirmed that plutonium-240 was present when the plutonium was irradiated. By July, it was ordered that work on the plutonium-gun "Thin Man" would be terminated, and efforts would be refocused on what initially appeared to be the more complicated uranium-based "Fat Man" design.

Because one avenue to the Bomb had now been eliminated, the estimated timeframe for delivery of the weapon to the military had to be adjusted. Projected delivery of a usable weapon was now slated for early 1945, if and only if the experiments indicated that the theories were sound and the devices would behave as expected. Though efforts had been refocused on a uranium-based implosion device, by late 1944, tests with uranium-235 indicated that even a gun-type uranium bomb might be feasible, so the plutonium problem would perhaps not be as concerning for the project's delivery goals of multiple weapons. At this point, there were serious doubts that plutonium could be used at all in the war effort because of the challenges that had yet to be overcome.[110]

By December 1944, as progress was made at Los Alamos on the weaponization aspect of the project, production finally began roaring forward at Oak Ridge, with the number of engineering and technical staff now eclipsing the number of construction staff. Three American private contractors were chosen to lead the charge on different uranium and plutonium methods to boost the United States' victory efforts in World War II's European and Pacific theaters. By this point, DuPont had several hundred engineers working at the Oak Ridge X-10 site and the Hanford site in Washington on the plutonium process. Union Carbide, through its subsidiary, the Carbide and Carbon Chemicals Corporation, was hard at work on the gaseous diffusion refinement method.[111]

But it was ultimately Tennessee Eastman, a subsidiary of the Eastman Kodak Company from Rochester, NY, that would come through for Los Alamos via its electromagnetic cyclotrons at the K-12 facility in Oak Ridge. Tennessee Eastman was the chemical division of Eastman Kodak, based in Kingsport, Tennessee. Technical and supervisory personnel were sent from both Rochester, NY and Kingsport. The K-12 facility, driven by Eastman's efforts, was the only operation at Oak Ridge that was able to successfully produce weapons-grade uranium usable by Los Alamos during the war.[112] As the xenon poisoning issue had been resolved in the plutonium reactors, plutonium production increased five-fold at the plutonium enrichment facilities. It was evident by the Spring of 1945 that Los Alamos would have access to sufficient quantities of plutonium and uranium to produce the atomic bomb.[113]

Chapter 6. Oppenheimer's Monster: Medusa's Gaze Turns a Desert to Glass

"Now I am become Death, the destroyer of worlds." -J. Robert Oppenheimer, quoting the Bhagavad Gita[114]

By early May 1945, the European Theater had been pacified as the Nazi regime was exsanguinated by the Allied war machine. The Germans surrendered unconditionally. As it was becoming clear that all of the work at Oak Ridge, Los Alamos, and other secret facilities were nearing a viable nuclear device, President Truman formed an Interim Committee. The committee was composed of high-ranking officials, including scientific advisors and military representatives, to determine how to properly use atomic weapons and to chart the course of nuclear policy following the war. Ernest Lawrence, who served as a scientific advisor to the committee, suggested that a demonstration of the Bomb may convince Japan to surrender to the United States. It was determined, however, that there were too many risks to this strategy– the Japanese might position American prisoners of war in the blast zone, shoot down the plane, or worse, the device could be a dud, which would only embolden America's adversary. The Committee ultimately recommended that the atomic bomb be kept secret until it was eventually dropped on a "dual-target"-- a Japanese military facility surrounded by workers' homes.

Some of the Manhattan Project scientists did not feel that their opinions had been adequately considered in the decision-making process surrounding the use of atomic weapons. After being briefed on the Interim Committee's conclusions, some Met Lab scientists established the "Committee on the Social and Political Implications of the Atomic Bomb," chaired by James Franck and including Glenn T. Seaborg and Leo Szilard. Franck's committee argued that postwar international control of atomic power was the only way to prevent an arms race if the United States bombed Japan without warning. The Scientific Panel of the Interim Committee disagreed, and it was determined that the use of atomic weaponry would move forward without notice. Four cities were identified as initial potential targets.[115]

By mid-July 1945, a successor to the "Thin Man" design, but using uranium fuel instead of plutonium, was ready for deployment. Because there was only enough enriched, weapons-grade U-235 available for one bomb, and because the designers had high confidence in the gun-type design, the weapon, known as "Little Boy," was sent westward to be used without testing of a fully-assembled, fission-ready prototype. With enough plutonium production now in the works for the construction of multiple plutonium weapons over the coming weeks, Los Alamos resolved to conduct a test of the implosion design on a plutonium core.[116]

Oppenheimer was now prepared to conduct his "Trinity" Test, whose name was, according to a letter addressed to Leslie Groves decades later, inspired by the poems of John Donne, specifically "Hymn to God, My God, in My Sickness" and "Batter my heart, three person'd God." The latter's title is an obvious reference to the Divine Trinity, a central belief in Christian doctrine that holds that God is composed of the Father, the Son (Jesus Christ), and the Holy Ghost. In "Batter my heart," the writer expresses the belief that by being tied to God, he is paradoxically set free. Richard Rhodes, the historian who wrote *"The Making of the Atomic Bomb,"* surmised that, for Oppenheimer, "the bomb... was a weapon of death that might also end war and redeem mankind." In this way, the atomic bomb also presented a sort of darkly poetic paradox.[117]

Prior to atomic detonation, Manhattan Project officials realized that calibration would be helpful to have an energy output index with which to compare the power of the Bomb. To this end, on May 7, officials placed 100 tons, or 220,000 pounds (99,790 kg) of trinitrotoluene (TNT), a military high explosive, on a 20-foot-tall tower, arranged in an enormous cube of around 3,400 boxes shipped in via rail cars. To simulate the radioactive crater they expected would be created by the blasts, scientists at the test site made a slurry of radioactive sludge produced in the reactor at the Hanford facility and ran it through plastic tubes interspersed with the stacked TNT boxes. This detonation would be used to calibrate the equipment that would be used to detect thermal radiation, energy release, and other factors during the Trinity Test. When it was finally set off, the conventional explosion of the TNT was so large that it was visible 60 miles away at a military airfield. Today, nuclear weapons' blast yield (explosive power) is measured in tons of TNT, which stems from the data collection devices at the Trinity site being calibrated to the "100-ton shot." [118,119]

Final preparations were conducted between May and July 1945, 210 miles south of Los Alamos, in a corner of the Alamogordo Bombing Range known as the "Jornada del Muerto," or "Journey of Death," where the 100-ton blast had taken place. Observation bunkers were installed 10,000 yards away from the planned epicenter of the explosion so that officials could observe and measure the detonation. Evacuation preparations were made in the event that the weather didn't cooperate and radiation from the blast traveled to nearby communities. On July 13, the final assembly of "The Gadget," the plutonium implosion device that was the subject of the Trinity Test, was complete. Some have referred to the Gadget as a Medusa-like contraption, as it was a sort of hideous, metallic spherical monstrosity laden with countless wires and electrical nodules sticking out of its shell.

The site on the morning of July 16, 1945, the planned detonation date, was one of nervous anticipation. The aim was to detonate at 4 A.M. that day, but it was raining outside before the scheduled detonation time, so the scientists and officials began making small talk to pass the time. Fermi offered wagers on whether the blast would ignite the atmosphere, and, if so, whether it would destroy only New Mexico or, rather, the entire world. While Fermi was half-joking, the possibility of atmospheric ignition was no laughing matter. The theoretical physicists at Los Alamos had worked on the problem, asking whether it was possible that a fission bomb could generate temperatures so hot that it could set off a nuclear chain reaction in the atmosphere that would light up the entire world and end all life on Earth. Ultimately, and fortunately for you and me, the physicists determined that, for a number of reasons, the energy generated by a fission bomb would be unlikely to set the world on fire. If it had, you would not be reading this right now.[120]

Oppenheimer, for his part, made a friendly wager with George Kistiakowsky, the head of the National Defense Research Committee, that the Gadget would not work at all. Kistiakowsky bet an entire month's pay that the detonation would succeed, while Oppenheimer bet ten dollars that it wouldn't. Edward Teller, a student of Oppenheimer's who worked on the project, spent his time slopping sunscreen all over his skin in the early morning darkness and offering everyone around him the opportunity to take the same precautionary measure.[121]

By 4 A.M., the rain had stopped, and the decision was made to detonate the bomb at 5:30 A.M. As scientists and military officials watched from afar, a flash of light lit up the desert sky and slammed observers standing five miles away to the ground. As of that date, it was the largest artificially created explosion in history. In an instant, the 100-foot tower upon which the Gadget had sat was turned to vapor. Below the blast, a crater formed in the desert floor, littered with a mysterious green glass that would later be called "Trinitite." It's theorized that some of the sand was sucked into the massive fireball created by the Gadget and rained down as a new kind of radioactive glass.

Scientists would later learn that Trinitite forms at all ground-level nuclear detonations, and the glass itself can be used to identify the atomic device used in the blast forensically.[122]

The blast was so large and powerful that witness accounts came in up to 200 miles away. One forest ranger 150 miles away gave an account of seeing a flash of fire, an explosion, and black smoke. Another 150 miles north of the blast said it "[lit] up the sky like the sun." A Navy pilot flying at 10,000 feet about 130 miles away said the explosion lit up his cockpit and was like the Sun rising in the south. Air Traffic Control, upon receiving notification, gave him a simple warning: "Don't fly South." The Alamogordo Air Base issued a press release afterward explaining away what witnesses had seen as a remotely located ammunition magazine that exploded, saying, "There was no loss of life or limb to anyone."[123]

Hans Bethe, the head of Los Alamos's Theoretical Division, was blinded for almost a minute because he stared directly into the fireball. Initial reactions among the observers were of jubilation, pride, and accomplishment. But as the reality of their achievements sat in a more somber tone came across them. Kenneth Bainbridge, the test director, called it a "Foul and awesome display," adding to Oppenheimer, "Now we are all sons of bitches." Oppenheimer would later write that the experience reminded him of the legend of Prometheus, whom Zeus punished for giving humankind the knowledge of fire. This moment in time gave birth to Oppenheimer's most famous quote, as he recalled a line from the Bhagavad-Gita, "Now I am become Death, the destroyer of worlds."[124] The Bhagavad-Gita is an episode recorded in the Mahabharata, a Sanskrit poem of ancient India.[125] Brigadier General Thomas F. Farrell recalled that "The whole country was lighted by a searing light with the intensity many times that of the midday sun. It was golden, purple, violet, gray and blue. It lighted every peak, crevasse and ridge of the nearby mountain range with a clarity and beauty that cannot be described but must be seen to be imagined. It was that beauty the great poets dream about but describe most poorly and inadequately."

The mushroom cloud created by the blast eventually rose to 70,000 feet above the desert.[126] Despite the fact that the atomic fireball created by Trinity measured about 2,000 feet in diameter and instantly etched a crater five feet deep by 30 feet wide into the sand, understanding the power exerted by the blast is nearly impossible for the human mind except by way of comparison.[127] While sources differ on the blast yield, compared to the 100-ton shot of TNT, it is generally accepted that the blast yield of the Trinity Test ranged from between 18 to 25 kilotons of TNT.[128, 129, 130] That is to say, the blast was equivalent to the amount of energy that would be released if between 18,000 to 25,000 *tons* of TNT were detonated all at once. This is equivalent to between 36,000,000 and 50,000,000 pounds (16,329,325-22679618 kg) of TNT.

With the successful detonation at Trinity, humankind had entered a new era: The nuclear age.

Chapter 7. The Bomb Unleashed: The 215,000 Souls of Hiroshima and Nagasaki

"Hundreds and hundreds of the dead were so badly burned in the terrific heat generated by the bomb that it was not even possible to tell whether they were men or women, old or young. Of thousands of others, nearer the centre of the explosion, there was no trace. They vanished. The theory in Hiroshima is that the atomic heat was so great that they burned instantly to ashes – except that there were no ashes."
-Wilfred Burchett[131]

Decades of scientific discovery, research, theory, and the dedication of exceptional minds culminated in the completion of the Manhattan Project and gave birth to the Bomb. All told, the project cost approximately $1.9 Billion. In 2023 dollars, accounting for inflation, it would have cost roughly $41.8 Billion, or over five times the cost of all United States heavy field artillery expenditures for the entirety of World War II.[132] The United States government now possessed and controlled the cosmic power of the atom. With the recommendations of the Interim Committee in hand, President Harry Truman had made his decision.

The final targets would meet three requirements. First, cities that had suffered no more than minor damage from conventional bombing would be considered so that it would be evident that an atomic bomb caused the damage. Secondly, the target would have to be a city devoted primarily to military production. Finally, the city should not be a city with traditional cultural significance to Japan, such as Kyoto.

The primary objective, at least on paper, was to use the shock and awe of atomic weaponry to break the resolve of the Japanese Empire and guarantee its unconditional surrender to the Allied forces.[133] It was the Interim Committee's determination, and the position of Truman's advisors, that this goal could not be accomplished without the use of the Bomb and that the alternative would be a long and bloody conventional ground invasion that could cost countless lives.

Between July 17 to August 2, 1942, the Big Three (United States President Harry Truman, Soviet Premier Joseph Stalin, and British Prime Minister Clement Atlee) met in Potsdam, Germany, for what was known as the "Potsdam Conference," to determine the course of the end of World War II and to determine the postwar European borders.[134] The Soviets had not declared war on Japan, so they did not join the rest of the Allied powers. Following discussions concerning how to deal with Germany and its former allies, the United States, Great Britain, and China released the "Potsdam Declaration," which concluded as follows:

> We call upon the government of Japan to proclaim now the unconditional surrender of all Japanese armed forces, and to provide proper and adequate assurances of their good faith in such action. The alternative for Japan is prompt and utter destruction.[135]

The Japanese Empire did not take heed and did not surrender. On August 6, 1945, at 2:45 A.M., three B-29 Superfortresses took to the sky from an airfield in the predawn darkness over the Pacific island of Tinian, approximately 1,500 miles off the coast of Japan. The lead bomber, known as the "Enola Gay," carried in its fuselage the 9,700 pound "Little Boy" uranium gun-type weapon that had been sent West for the war effort.[136] You'll recall from earlier reading that this weapon type had not yet been tested, as fissile material was scarce, but scientists had high confidence in the gun-type design. At 7:15 A.M, the crew of the Enola Gay armed the bomb and began an ascent to 31,000 feet. At 8:15 A.M, Little Boy was released from its restraints and took flight from the bomb bay. The B-29's pilot, Colonel Paul Tibbets, banked a sharp 155-degree turn and immediately headed in the opposite direction, as he had been informed he would have only 45 seconds before the device would detonate. The scientists at Los Alamos were not even entirely sure that the plane would have time to escape the shockwaves created by the blast.

Little Boy fell 6 miles in 43 seconds and detonated at an altitude of 2,000 feet. In a flash of light, between 70,000 and 80,000 souls left our realm.[137, 138] At first, the extent of the damage was unknown to Japanese officials. As it turned out, about an hour before the detonation, Japanese early warning radar had detected the approach of the American B-29s. However, the alert was lifted once it was determined that the number of incoming aircraft was minimal. It wasn't until after the blast that the Tokyo control operator of the Japanese Broadcasting Corporation noticed that the Hiroshima station had gone off the air. Soon after, the Tokyo railroad telegraph center became aware that the main line telegraph had gone offline North of Hiroshima. Reports began trickling in about a massive explosion in Hiroshima, and the Japanese military sent a plane to assess the situation. Upon reaching the city, the plane's operators circled in shock at the glimpse of what was once a city. A burning husk of Hiroshima's former greatness lay crumbling in devastation, shrouded in an enormous cloud of smoke.[139] The Enola Gay's co-pilot, Robert Lewis, wrote in his log, upon witnessing what had occurred, "My God, what have we done?"[140]

The devastation from the blast was immediately evident to those who witnessed and survived it. Birds burst to flames in mid-air, bodies were reduced to charcoal, and combustible materials up to 6,400 feet away from the blast's epicenter ("ground zero") were set on fire. 90% of all people present during detonation, half a mile or less from ground zero, were killed. Roughly half the city's population was dead or injured. Almost every structure within a mile of ground zero was utterly destroyed, those within three miles were damaged, and less than 10% of all buildings in the city survived without any damage. The fires consuming the city soon merged into one giant firestorm that engulfed nearly four and a half miles of the city, killing those trapped inside it. These were just the immediate effects.[141] The ongoing damage from the radiation emitted during the bombing is another issue, and one we'll explore later.

That same day, President Truman would address the nation, and the world, in a speech I will copy in its entirety due to its relevance to the history of nuclear weapons research and the applications of these weapons toward the entire enterprise of war, and, more controversially, that of peace:

> Sixteen hours ago an American airplane dropped one bomb on Hiroshima, an important Japanese Army base. That bomb had more power than 20,000 tons of T.N.T. It had more than two thousand times the blast power of the British 'Grand Slam' which is the largest bomb ever yet used in the history of warfare.

> The Japanese began the war from the air at Pearl Harbor. They have been repaid many fold. And the end is not yet. With this bomb we have now added a new and revolutionary increase in destruction to supplement the growing power of our armed forces. In their present form these bombs are now in production and even more powerful forms are in development.

> It is an atomic bomb. It is a harnessing of the power of the universe. The force from which the sun draws

its power has been loosed against those who brought war to the Far East.

Before 1939, it was the accepted belief of scientists that it was theoretically possible to release atomic energy. But no one knew any practical method of doing it. By 1942, however, we knew that the Germans were working feverishly to find a way to add atomic energy to the other engines of war with which they hoped to enslave the world. But they failed. We may be grateful to Providence that the Germans got the V-1's and V-2's late and in limited quantities and even more grateful that they did not get the atomic bomb at all.

The battle of the laboratories held fateful risks for us as well as the battles of the air, land, and sea, and we have now won the battle of the laboratories as we have won the other battles.

Beginning in 1940, before Pearl Harbor, scientific knowledge useful in war was pooled between the United States and Great Britain, and many priceless helps to our victories have come from that arrangement. Under that general policy the research on the atomic bomb was begun. With American and British scientists working together we entered the race of discovery against the Germans.

The United States had available the large number of scientists of distinction in the many needed areas of knowledge. It had the tremendous industrial and financial resources necessary for the project and they could be devoted to it without undue impairment of other vital war work. In the United States the laboratory work and the production plants, on which a substantial start had already been made, would be out of reach of enemy bombing, while at that time Britain was exposed to constant air attack and was still threatened with the possibility of invasion. For these reasons Prime Minister Churchill and President Roosevelt agreed that it was wise to carry on the project here. We now have two great plants and many lesser works devoted to the production of atomic power. Employment during peak construction numbered 125,000 and over 65,000 individuals are even now engaged in operating the plants. Many have worked there for two and a half years. Few know what they have been producing. They see great quantities of material going in and they see nothing coming out of these plants, for the physical size of the explosive charge is exceedingly small. We have spent two billion dollars on the greatest scientific gamble in history -- and won.

But the greatest marvel is not the size of the enterprise, its secrecy, nor its cost, but the achievement of scientific brains in putting together infinitely complex pieces of knowledge held by many men in different fields of science into a workable plan. And hardly less marvelous has been the capacity of industry to design and of labor to operate, the machines and methods to do things never done before so that the brainchild of many minds came forth in physical shape and performed as it was supposed to do. Both science and industry worked under the direction of the United States Army, which achieved a unique success in managing so diverse a problem in the advancement of knowledge in an amazingly short time. It is doubtful if such another combination could be got together in the world. What has been done is the greatest achievement of organized science in history. It was done under pressure and without failure.

We are now prepared to obliterate more rapidly and completely every productive enterprise the Japanese have above ground in any city. We shall destroy their docks, their factories, and their communications. Let there be no mistake; we shall completely destroy Japan's power to make war.

It was to spare the Japanese people from utter destruction that the ultimatum of July 26 was issued at Potsdam. Their leaders promptly rejected that ultimatum. If they do not now accept our terms they may expect a rain of ruin from the air, the like of which has never been seen on this earth. Behind this air attack will follow sea and land forces in such numbers and power as they have not yet seen and with the fighting skill of which they are already well aware.

The Secretary of War, who has kept in personal touch with all phases of the project, will immediately make public a statement giving further details.

His statement will give facts concerning the sites at Oak Ridge near Knoxville, Tennessee, and at Richland, near Pasco, Washington, and an installation near Santa Fe, New Mexico. Although the workers at the sites have been making materials to be used producing the greatest destructive force in history they have not themselves been in danger beyond that of many other occupations, for the utmost care has been taken of their safety.

The fact that we can release atomic energy ushers in a new era in man's understanding of nature's forces. Atomic energy may in the future supplement the power that now comes from coal, oil, and falling water, but at present it cannot be produced on a basis to compete with them commercially. Before that comes there must be a long period of intensive research. It has never been the habit of the scientists of this country or the policy of this government to withhold from the world scientific knowledge. Normally, therefore, everything about the work with atomic energy would be made public.

But under the present circumstances it is not intended to divulge the technical processes of production or all the military applications. Pending further examination of possible methods of protecting us and the rest of the world from the danger of sudden destruction.

I shall recommend that the Congress of the United States consider promptly the establishment of an appropriate commission to control the production and use of atomic power within the United States. I shall give further consideration and make further recommendations to the Congress as to how atomic power can become a powerful and forceful influence towards the maintenance of world peace.[142]

Although talks began soon afterward among Japanese leadership, the military, in particular, remained recalcitrant. Despite the annihilation of Hiroshima, no surrender came. But the die was cast, and orders had already been given earlier to deploy the next weapon as soon as it was available. By August 9th, the next device was ready, and American forces were already seeking to put pressure on the Japanese government through its civilians, dropping leaflets across the Empire that read:

We are in possession of the most destructive explosive ever devised by man. A single one of our newly developed atomic bombs is actually the equivalent in explosive power to what 2,000 of our giant B-29s can carry on a single mission. This awful fact is one for you to ponder and we solemnly assure you it is grimly accurate. We have just begun to use this weapon against your homeland. If you still have any doubt, make inquiry as to what happened to Hiroshima when just one atomic bomb fell on that city. Before using this bomb to destroy every resource of the military by which they are prolonging this useless war, we ask that you now petition the Emperor to end the war. Our president has outlined for you the thirteen consequences of an honorable surrender. We urge that you accept these consequences and begin the work of building a new, better and peace-loving Japan. You should take steps now to cease military resistance. Otherwise, we shall resolutely employ this bomb and all our other superior weapons to promptly and forcefully end the war.[143]

On August 9, 1945, only three days after the bombing at Hiroshima, A B-29 named *Bock's Car* took off carrying its 10,800-pound payload, a plutonium implosion-type weapon nicknamed the "Fat Man." The device was dropped over Nagasaki, Japan, at 11:01 A.M., exploding 1,650 feet above the city with a force of around 20-21 kilotons, or approximately 42 million pounds (over 19 million kg) of TNT. This was about 40% more powerful than the "Little Boy" bomb dropped on Hiroshima, which yielded around 15 kilotons, or approximately 30 million pounds (13.6 million kg) of TNT.[144]

Like Hiroshima, everything within about half a mile from ground zero was completely destroyed. Over one-third of the 50,000 buildings (or approximately 17,000) were seriously damaged by the blast or totally destroyed.[145] All told, 40,000 people died almost immediately due to the explosion, with 60,000 more injured.[146]

By the next day, Japanese Emperor Shōwa, more commonly known by his personal name "Hirohito" in English-speaking countries, convened a meeting of the Imperial Council. In the face of the Soviet Union declaring war on Japan the day prior, and the devastation of multiple Japanese cities by atomic weaponry, the Emperor broke a 3-3 tie of the Imperial Council. On August 12, the United States accepted the surrender, stipulating that the Emperor could remain only in a purely ceremonial capacity. However, resistance remained in the Japanese military, and by August 14, the Emperor had to step in once again. He called his cabinet together and declared that he could not "endure the thought of letting my people suffer any longer," continuing that if the war were not immediately concluded, "the whole nation would be reduced to ashes."[147] In a radio address on August 15, 1945, the Emperor declared to his people, and to the world, that the Japanese Empire must surrender in the face of "a new and most cruel bomb, the power of which to damage is indeed incalculable, taking the toll of many innocent lives. Should we continue to fight, it would not only result in an ultimate collapse and obliteration of the Japanese nation, but also it would lead to the total extinction of human civilization.[148]

For better or worse, Oppenheimer's creation had abruptly ended the bloodiest war in human history. Between 35 million to 60 million people lost their lives during the conflict, depending on how one measures the death toll.[149,150] 2,403 U.S. personnel, including soldiers, sailors, and civilians, were killed in the Japanese attack on Pearl Harbor that dragged the United States into World War II.[151] Between Hiroshima and Nagasaki, it is estimated that a range of between 130,000 to 215,000 people lost their lives from the explosive energy, fires, and structural damage caused by the blasts, and from the long-term side effects of radiation that would ravage the population in the years and decades following.[152]

The world and its peoples were entering a new era, not just because Germany, Japan, and their allies had been defeated, but because a new technology was about to reshape the global geopolitical landscape. Humankind had entered the age of atomic weaponry. But Oppenheimer's discoveries were not the end of the story for the development of this terrifying and intimidating new technology; instead, they were only just the beginning.

Part II. Cold War: The Eagle and the Bear Go MAD

Chapter 8. Operation Crossroads: The United States in Pursuit of Nuclear Hegemony

"If Congress joins with the military in creating a great international atomic armaments race, then the world is doomed to carnage and slaughter on a scale inconceivable to the average man, to a war that no nation can win... our only hope of survival is the immediate establishment of an international body with regulatory powers sufficiently strong to make production of atomic bombs by any nation impossible." -Richard Abrams[153]

Coming out of WWII, the United States knew it was only a matter of time until its international competitors would obtain the technology necessary to build the Bomb. In the meantime, it set its sights on learning more about the incomprehensible power it had discovered and simultaneously unleashed on the global stage. Emperor Hirohito surrendered on August 15, 1945. On August 16, a military advisor named Lewis Strauss, who would later become the chairman of the United States Atomic Energy Commission (AEC), pitched a series of tests to the Secretary of the Navy, James Forrestal.[154] By August 25, 1945, U.S. Senator Brian McMahon openly advocated testing atomic weaponry on captured Japanese ships.[155]

Since the existence of the Bomb was now internationally known, the necessity of secrecy surrounding these follow-up tests was significantly reduced. As a result, the Joint Chiefs of Staff decided to publicize the operation not just to the American press, but to the international media as well. More than 100 reporters attended, from nations as diverse as France, Egypt, Poland, the Soviet Union, and China.[156]

The tests were slated to take place in the summer of 1946, with the objective of better understanding the practical effects of nuclear weapons on military assets and potential radiological side effects resulting from the use of the weapons. Ultimately, Vice Admiral William H. P. Blandy, who was head of the operation, dubbed the series Operation "Crossroads," noting that "It was apparent… that warfare, perhaps civilization itself, had been brought to a turning point by this revolutionary weapon." Indeed, humankind had found itself at a crossroads, faced with profound choices about the arc of its future. The radiological concerns caused by the Trinity test, as well as scientists' newfound understanding of the dangers of radiation from lab experimentation, which we will discuss later on, drove the military to site the test at Bikini Atoll in the central Pacific Marshall Islands, as this atoll was very far from population centers. An indigenous (native) population of 162 people was displaced by the U.S. government in order to locate them a safe distance from the blasts and the resultant fallout.[157]

The ships assembled for the tests (clocking in at 242) and placed around Bikini Atoll would have accounted for, at the time, the sixth-largest navy in the entire world.[158] The target fleet comprised U.S. capital ships; surplus U.S. cruisers, destroyers, and submarines; auxiliary and amphibious vessels; and captured Japanese and German ships. One hundred fifty ships housed 42,000 men, mostly Navy personnel, assigned to Joint Task Force 1 (JTF 1), the unit that conducted the tests.[159]

Some objected to the cost of the test and the use of so many ships, despite their status as military surplus, because they had the potential for use operationally in other contexts. Nonetheless, United States leadership deemed it necessary to test its new weapon on naval assets to determine its potential efficacy in a maritime warfare scenario. Blandy explained to the United States Senate this objective: "The ultimate result of the tests, so far as the Navy is concerned, will be their translation into terms of United States sea power. Secondary purposes are to afford training for Army Air Forces personnel in the attack with the atomic bomb against ships and to determine the effect of the atomic bomb upon military installations and equipment."[160]

The planned tests did not come without detractors, as some were concerned about the costs, the risks, and the potential fallout of multiple large detonations. A group of scientists at Los Alamos, led by Henry Newsom, sent a memorandum to Oppenheimer's successor at the lab, warning that "the water near a recent surface explosion will be a witch's brew, and this will be true to a lesser extent for the other tests. There will probably be enough plutonium near the surface to poison the combined armed forces of the United States at their highest wartime strength. The fission products will be worse."[161]

Ultimately, three tests were scheduled as part of the operation: Able, Baker, and Charlie. Able was designed as an airdrop bombing, Baker was an underwater test at a relatively shallow depth, and Charlie was to be a deep water test far below the ocean's surface. On July 1, 1946, after staff had been evacuated to 10 nautical miles from Bikini Atoll, Able was dropped from a B-29 and exploded over the fleet. The explosion sunk five of the vessels. Able went as expected, except the bomb was accidentally dropped off the target by around 1,500-2,000 feet. Radioactivity was minimal, and the surviving target ships could be reboarded by the next day.[162] Unfortunately, because the Able shot (nicknamed "Gilda" after a movie released the same year) was off target, the bomb destroyed several ships carrying instrumentation, resulting in significant data loss via the destruction of cameras, spectrographs, and shockwave instruments.[163]

The Baker shot (nicknamed "Helen of Bikini"), like the Able shot, was a plutonium-based implosion-type device (similar to the "Fat Man" bomb dropped on Nagasaki, Japan), with a comparable yield to Able (23 kilotons, or 23,000 tons of TNT).[164] Its outcome, from a pure physics perspective, was far more destructive. On the morning of July 25, 1946, Baker was detonated while suspended 90 feet under the ocean surface. The underwater fireball, essentially a giant, hot gas bubble, reached the ocean floor and surface simultaneously. It left a crater on the ocean floor 30 feet deep and almost 2,000 feet wide. The water burst above the ocean surface, creating an enormous "spray dome" composed of approximately two million tons of water. The column of the burst was 6,000 feet tall and 2,000 feet wide with 300 feet thick walls. A tsunami generated by the blast created waves that were 94 feet high. Almost 4 miles away at Bikini Island beach, 15-foot waves violently ripped landing craft onto the beach, filling them with sand. Shortly after the detonation, falling water created a 900-foot base surge and blasted the target ships with irreparable radioactive contamination. Pigs and rats placed at the test site to gauge the bomb's effects on living fauna either died immediately in the blast or shortly afterward due to irradiation from the blast.

The Baker shot so contaminated the ships that the third test in the operation, Charlie, was canceled. Sailors initially attempted to clean up the vessels, but were stirring up radioactive material and, in the process, contaminating their clothing, skin, and likely their lungs. The Chief of the Medical Section during the Manhattan Project, Army Colonel Dr. Stafford Warren, determined that decontamination of the vessels for the Charlie test would be too dangerous and had little hope of effectiveness.[165] Warren concluded that even a reduction of "90% or more still leaves large and dangerous quantities of fission and alpha emitters scattered about… Contamination of personnel, clothing, hands, and even food can be demonstrated readily in every ship in the JTF-1 in increasing amounts day by day."[166]

The importance of Operation Crossroads is not that it was so highly publicized, that it constituted the first non-military uses of nuclear devices outside of the Trinity tests, nor even that the tests demonstrated efficacy against naval assets. What was so fundamental about Operation Crossroads were the conclusions that U.S. leadership drew about the future of warfare in an atomic world.

The most straightforward conclusions were that military equipment outside the direct blast zone could survive a blast and even come out unscathed by radioactive contamination in certain circumstances. The Able test demonstrated that a "clean" outcome could be achieved with certain air burst detonations. In others, though, such as the Baker water detonation, the long-term consequences via radiation would be far more impactful and long-term. Indeed, it was the Baker shot that led scientists and military strategists to the conclusion that the development of a tactical and strategic nuclear stockpile would bring the United States long-term security because the use of atomic weaponry would bring about extended and severe contamination that would far outlast the destruction and damage caused by an initial detonation… This would mean that foreign adversaries would face far greater consequences if they were subject to nuclear attack.[167] Indeed, Warren's own assessment upon leaving Bikini Atoll was, "I think the Navy now has an idea, a very little idea of what a source it can be and what a boomerang… Some very serious thinking is going on and many of the task force admit that war is unthinkable with atomic warfare."[168]

Ultimately, the military's Joint Chiefs of Staff Evaluation Board for Operation Crossroads concluded this of the tests: "National security dictates the adoption of a policy of instant readiness to defend ourselves vigorously of any threat of atomic weapon attack at any time… Therefore, so long as atomic bombs could conceivably be used against this country, the Board urges the continued production of atomic material and research development in all fields related to atomic warfare."

Over the following ten months, the Joint Chiefs' Evaluation Board studied volumes of technical findings and scientific data on the tests as it prepared its top-secret report, which was largely focused on the possibility of an atomic attack on the United States. In mid-1947, a final copy of the report was circulated to Rear Admiral Ralph A. Ofstie, a Navy board member, who shared it with a colleague who responded, "It scared the hell out of me." Ofstie wrote to the Board's chairman, saying, "If this should be the reaction of a highly experienced and keenly intelligent naval officer [it] suggests that somewhere or other we may have slipped a cog."[169]

In July 1947, the members of the Evaluation Board gave a top-secret briefing at the Pentagon in a secure room with almost all of the United States' top military officials. During the four-hour meeting, the Board informed the officials, "If used in numbers, atomic bombs not only can nullify any nation's military efforts, but can demolish its social and economic structure and prevent their reestablishment for long periods of time… With such weapons, especially if employed in conjunction with other weapons of mass destruction, as, for example, pathogenic bacteria, it is quite possible to depopulate vast areas of the earth's surface, leaving only vestigial remnants of man's material workers… [The Baker Shot] threw large masses of highly radioactive water onto the decks and into the hulls of vessels… These contaminated ships became radioactive stoves, and would have burned all living things aboard them with invisible and painless but deadly radiation." The Board concluded that a detonation like Baker on a large city ". . . would have not only an immediate lethal effect, but would establish a long term hazard through the contamination of structures by the deposition of radioactive particles." The

true weapon was not the blast itself, but the long-term
radiological devastation:

> We can form no adequate mental picture of the
> multiple disaster which would befall a modern city,
> blasted by one or more bombs and enveloped by
> radioactive mists. Of the survivors in contaminated
> areas, some would be doomed to die of radiation
> sickness in hours, some in days, and others in years.
> But, these areas, irregular in size and shape, as wind
> and topography might form them, would have no
> visible boundaries. No survivor would be certain he
> was not among the doomed and so, added to every
> terror of the moment, thousands would be stricken
> with a fear of death and the uncertainty of the time of
> its arrival. . . [The dead] would remain unburied and
> the wounded uncared for in the areas of heaviest
> contamination where certain death would lurk... In
> the face . . . of the bomb's demonstrated power to
> deliver death to tens of thousands, of primary military
> concern will be the bomb's potentiality to break the
> will of nations and of peoples by the stimulation of
> man's primordial fears, those of the unknown, the
> invisible, the mysterious. We may deduce from a
> wide variety of established facts that the effective
> exploitation of the bomb's psychological implications
> will take precedence of the application of its
> destructive and lethal effects in deciding the issue of
> war.[170]

The Board could have had no concept at the time just how
prescient and accurate its conclusions about the future of
nuclear arms development were. While Baker made evident
just how destructive a standard fission device could be, there
was potential for a far more powerful and dangerous weapon
on the horizon. As the United States headed toward Cold
War with its Russian adversaries, a concept dreamed of by

Edward Teller would come into the forefront; his obsession would give birth to something that would define the course of the entire twentieth century.

Chapter 9. Teller-Ulam: The Hand of God

"I begin to believe in only one civilising influence—the discovery one of these days of a destructive agent so terrible that War shall mean annihilation and men's fears will force them to keep the peace." -Wilkie Collins[171]

The casual references to Edward Teller in prior chapters of this book might lead one to believe he was of little consequence to the nuclear age, just another member of Oppenheimer's team, or just an interesting character to lighten up the story with his proclivity for the use of sunscreen during nuclear testing. Nothing could be further from the truth.

Edward Teller, born January 15, 1908, was a Hungarian-American physicist who worked on the Manhattan Project. Teller studied under Niels Bohr, a brilliant Danish-born physicist and early skeptic of the possibility of atomic weaponry, whose experiences during WWII drove him to dedicate himself to the Manhattan Project.[172] By 1935, Teller was a professor at George Washington University in Washington, D.C. Instrumental in the Einstein-Szilard letter to Roosevelt, warning of the possibility of Nazi atomics research, Teller would later join Enrico Fermi's team at the University of Chicago to assist in the Chicago Pile-1 research. It was at Berkeley that Teller first started working with Oppenheimer.[173]

One of the first scientists to arrive at Los Alamos in 1943, Teller was initially assigned to the Theoretical Division, working under Hans Bethe. Teller's initial assignment was to develop the implosion method, later used in the Fat Man bomb. However, this did not last long, as

Teller was reassigned the following year since he wasn't singularly focused on developing the implosion device. Rather, Teller had another concept on his mind— one that would profoundly alter nuclear science and the entire fabric of history. While the rest of Los Alamos was focused on developing fission science and the weaponry that would stem from it, Teller was obsessing over something potentially far more dangerous and powerful.

In 1920, Arthur Eddington, an English physicist, theorized that the fuel source of the stars could be a product of helium and hydrogen *fusing* together: "If, indeed, the sub-atomic energy in the stars is being freely used to maintain their great furnaces, it seems to bring a little nearer to fulfillment our dream of controlling this latent power for the well-being of the human race– or for its suicide."[174] Fission splits atoms. But fission, for Teller, was only the beginning. What if, he wondered, one could *fuse* atoms together instead of splitting them apart, thereby harnessing and controlling the power of the Sun and the stars themselves?

While the Manhattan Project's early research was ongoing, Teller set to work on his project, which he called the *Super*. The Super was a hypothetical thermonuclear bomb that would utilize nuclear fusion to create a potentially much stronger weapon than even those that were eventually utilized in Hiroshima and Nagasaki. Teller realized that an initial nuclear fission explosion could create adequate conditions to allow for the compression and, thereby, the *fusion* of hydrogen isotopes, which would release a massive amount of neutrons, creating an explosion vastly more powerful than a fission reaction alone.[175]

Teller did not have a great deal of support for his project to create the superweapon at Los Alamos. First, because the other scientists were focused on the immediate development of fission for the war, they needed all the personnel time and energy that they could get. This led to tensions as Teller was distracted by his research into the hypothetical weapon. Secondly, many of the other scientists expressed a deep sense of regret following the attacks on Japan, as the bombings took theory and science to their bloody, violent real-life conclusions.[176] For many of them, it was one thing to hypothesize and theorize about what such a technology could do (as one would with a sci-fi show or movie), while it was another thing entirely to watch the real-world devastation that such a power could visit upon their fellow humans. Some believe that Teller had no such reservations, but for his research, the end of the war led to a lack of support for further research on thermonuclear weaponry, and Teller, therefore, returned to the University of Chicago in 1946 to focus on academics.[177]

While the reservations of men like Oppenheimer (and the end of the war itself) had put a damper on research into fusion weaponry, a twist of fate would soon change things. Emil Klaus Julius Fuchs, also known as "Klaus Fuchs," was a German-born physicist working as part of Oppenheimer's team at the Los Alamos weapons site of the Manhattan Project.[178] According to MI5, Britain's domestic counterintelligence and security service, Fuchs was a German citizen and member of the German Communist Party in the 1930s. In 1933, when the Nazis ascended to head the German government, Fuchs fled Britain. He was later detained on the Isle of Man and sent to Canada as a citizen of an enemy nation. Fuchs was eventually released and returned to Edinburgh in 1941 to work on the Tube Alloys nuclear research program, which, as you'll recall, was Britain's predecessor to the Manhattan Project.[179]

Fuchs was committed to the communist cause, having been born of a Quaker minister with a strong socialist background. He joined the Communist Party of Germany at the University of Kiel, standing in opposition to Hitler. As the Nazis secured their power in Germany, conditions became hostile for members of various groups in Germany, including those who were members of left-wing parties.[180] Multiple members of Fuchs' family were arrested and incarcerated in Germany for speaking out against the Nazis, and his sister and brother-in-law helped Jews and opponents of the Nazi regime escape Germany and the associated Nazi persecution of the era. Harassment of Fuchs' mother and sister may have led them both to commit suicide. Fuchs' initial association with the Communist Party resulted from his belief that the communists were Germany's last bastion against Nazi supremacy. While there is debate about Fuchs' true motivations, some believe that he joined the Tube Alloys Program and eventually the Manhattan Project with a similar determination as many of the other brilliant physicists who escaped Europe at the time– to ensure that the Allies would get the Bomb first so that Hitler could not use it to dominate the world.[181]

Initially, Fuchs worked on the gaseous diffusion method of uranium enrichment for the British. Later working under Hans Bethe at the Theoretical Division of Los Alamos, Fuchs was assigned to the implosion project to determine a path forward for the plutonium bomb. Fuchs was one of the scientists present to witness the Trinity test. But research on the implosion method wasn't all Fuchs was up to at Los Alamos. No, there was something else afoot; Fuchs, later known as the "atom spy," was collecting intelligence on methods of nuclear weapons production for the Soviets.[182]

In 1943, the United States Army's Signal Intelligence Service began a secret program to decrypt Soviet diplomatic communications. The Signal Intelligence Service was the precursor to the National Security Agency (NSA), which is one of the United States' most secretive intelligence agencies, engaged in cybersecurity and signals intelligence (SIGINT), which collects and analyzes communications around the world to further the United States' national security interests.[183] The 1943 program was codenamed VENONA, and it took American cryptologists (those who decode encrypted communications) approximately two years to crack Soviet Russia's security service's encryption.[184] VENONA's decryptions forced the Americans and the British to face the harsh reality that the Soviets were in possession of the most dangerous information known to humankind: the designs for the atomic bomb. Fuchs, operating under the KGB codename "REST," had passed the Soviets everything they needed to replicate the technology.[185]

While the United States, Britain, and Russia had been allies in World War II, a fundamental ideological underpinning made Soviet possession of nuclear technology unacceptable to the Americans and the British. America and Britain were capitalist Western democracies, while a communist dictatorship led Soviet Russia. These basic views about how national and global economies should function bleed through to, and color politics and beliefs about, how nations should be run and how governments should interact with their people and with other countries. While Churchill initially wanted nothing to do with Stalin, deeply distrusting the Soviet leader, President Roosevelt was more pragmatic, urging Churchill to embrace a partnership, at least for the purpose of victory in WWII:

> We are all in agreement . . . as to the necessity of having the USSR as a fully accepted and equal member of an association of the great powers formed for the purpose of preventing international war… It should be possible to accomplish this by adjusting our differences through compromise by all the parties concerned and this ought to tide things over for a few years until the child learns to toddle.[186]

Soviet dedication to atomic weapons research was largely insignificant in WWII. Like the Germans, the Soviets had decided to dedicate their time, staff resources, and money toward developing technologies that would prove immediately useful in the battles of WWII. Nuclear weapons' expense and hypothetical nature drove all but the United States and Britain away. Russia had only about twenty physicists working on the program. This changed after 1945. At the Potsdam Conference (mentioned earlier as the meeting between the U.S., Russia, and Britain to determine the course of the end of WWII), Truman mentioned to Stalin that the United States "had a new weapon of unusual destructive force." Stalin played coy, remarking that he was "glad to hear it and hoped [the United States] would make good use of it against the Japanese." Stalin immediately went to his top advisors and informed them that, "We've got to work on Kurchatov and hurry things up."[187]

Igor Vacilyevich Kurchatov was the head of the Soviet nuclear weapons program. After fission was demonstrated by Hahn and Strassmann in 1938, Kurchatov urged the Soviets to engage in atomic research, submitting a plan focused on uranium in August 1940 to the Presidium of the Soviet Academy of Sciences. While Kurchatov was redirected to other projects in the early years of the war, he was shifted back to nuclear research beginning in early 1943 after Soviet intelligence discovered that the Americans and British were making headway in their own programs. Soviet research on the Bomb was greatly accelerated after the Soviets witnessed the devastating power of the atomic bomb in Hiroshima and Nagasaki.[188]

By the late 1940s, a top-secret site akin to Los Alamos was being built in the Soviet Union in a closed city called Arzamas-16, known today as *Sarov*. Throughout the years, the Soviets would build a number of these closed cities for clandestine nuclear activities. The Design Bureau No. 11 (KB-11 design bureau), established in the late 1940s, was charged with developing and manufacturing the first Soviet nuclear bomb. Arzamas-16 would eventually become home to two nuclear weapons facilities, the design institute, and the warhead assembly/disassembly facility. The city was so secretive that it was removed from all official Soviet maps and statistical documents. By 1948, the city was entirely isolated, and would not be recognized publicly as even existing until the year 1994.[189]

By 1948, Arzamas-16 was a flurry of clandestine activity. Like their American counterparts, scientists there were hard at work analyzing the potential uses of conventional weapons to build gun-type and implosion-type bombs. Eventually, the gun-type design was abandoned in favor of compression/implosion (similar to the Fat Man bomb). Dummy bombs of the hypothetical size and weight required to carry an atomic weapon were made, and a four-engine heavy bomber called the PE-8 was modified to carry the large bombs required for practice runs. Fissile materials uranium-235 and plutonium-239 had begun to arrive from mining and production facilities in the Urals, a mountainous region inside Russia that forms the conventional border between Asia and Europe. The Soviets also began to investigate a suitable test ground for a site away from Europe that would not trigger the suspicions of other world powers.[190]

After WWII ended, the United States Air Force organized the Office of Atomic Energy (AFOAT-1). Initially, the organization set up a seismographic network (to measure shock and vibrations), acoustic stations (to measure significant sounds that could come from a blast), and aircraft flights monitoring radiation in the atmosphere using filter papers. A private contractor, *Tracerlab*, was trained by Los Alamos to analyze radioactive elements collected, and a threshold for ambient radiation on the filter papers was set at 50 counts per minute (cpm)… Counts over 50 cpm would indicate radiological activity that natural phenomena could not generally explain and would effectively serve as an alarm that atmospheric radiation levels were likely the result of artificial fission.[191]

On September 3, 1949, a weather plane flying from Japan to Alaska discovered something very curious: a filter paper, exposed for 3 hours at an altitude of 18,000 feet, detected radioactivity of 85 cpm. This was 35 cpm past the established threshold of significance. A redundant filter paper on the same plane had over 100 counts above the same threshold. False alerts were not unusual but were typically able to be explained by other environmental phenomena. As it turns out, this was anything but environmental phenomena. While Tracerlab was analyzing the samples from the September 3 flights, a paper exposed on a weather flight east of Japan on September 5 displayed a count of over 1,000 cpm. This sent up alarm bells in the United States defense apparatus, and the paper was immediately sent to California for analysis. Fission products were present in the sample. Using information about the fission of plutonium and its half-lives, the analysts determined that a nuclear explosion had occurred in Soviet Russia between the dates of August 26-August 29, 1949. Once scientists knew when the blast had likely happened, they analyzed acoustic data from their monitoring stations to determine that it had occurred at Semipalatinsk in Kazakhstan.[192]

Due to the global significance of such a discovery, Truman was notified immediately after the tests were verified, and there was high confidence that the Soviets had achieved nuclear ignition. On September 23, 1949, President Truman addressed the American people:

> I believe the American people, to the fullest extent consistent with national security, are entitled to be informed of all developments in the field of atomic energy. That is my reason for making public the following information.
>
> We have evidence that within recent weeks an atomic explosion occurred in the U.S.S.R.
>
> Ever since atomic energy was first released by man, the eventual development of this new force by other nations was to be expected. This probability has always been taken into account by us.
>
> Nearly 4 years ago I pointed out that "scientific opinion appears to be practically unanimous that the essential theoretical knowledge upon which the

discovery is based is already widely known. There is also substantial agreement that foreign research can come abreast of our present theoretical knowledge in time." And, in the Three-Nation Declaration of the President of the United States and the Prime Ministers of the United Kingdom and of Canada, dated November 15, 1945, it was emphasized that no single nation could in fact have a monopoly of atomic weapons.

This recent development emphasizes once again, if indeed such emphasis were needed, the necessity for that truly effective enforceable international control of atomic energy which this Government and the large majority of the members of the United Nations support.[193]

The Russian response was swift, and two days later, on September 25, 1949, the Soviet News agency known as TASS sent a message to Pravda, the official paper of the Soviet Communist Party:

It is well known that the Soviet Union builds large scale works: hydroelectric plants, mines, canals, and highways that sometimes require new technology and big explosions. To the extent that these explosions often take place in various parts of the country, it is possible that they have attracted attention outside the borders of the Soviet Union.

Regarding the subject of atomic energy, TASS believes it necessary to remind everyone that on November 6, 1947, Foreign Minister Molotov asserted that there is no longer any secret regarding the atomic bomb. Molotov meant that the Soviet Union knew the secret of atomic weapons, and it already had them in its arsenal. American scientists assumed that Molotov was bluffing, that Russia could not obtain nuclear weapons earlier than 1952. However, they were mistaken, because the Soviet Union already had the secret in 1947. There is no basis for anxiety among foreign countries. The Soviet government, despite its possession of nuclear weapons, will stand by its unconditional resolve not to use them. Regarding the control of nuclear weapons, it is necessary to say that control will be needed in order to verify that nuclear weapons are not produced.[194]

On August 29, 1949, the Soviets detonated an implosion-type device at the top of a 123-foot tower at the Semipalatinsk test site. The test was named RDS-1, also known as "nuclear charge 501" (or, in the United States, as "JOE-1" after Joseph Stalin). The device, utilizing a 5 kg plutonium core surrounded by a uranium-238 moderator, detonated with a yield of 22 kilotons (22,000 tons of TNT).[195] For the Soviets, it was the dawn of their status as a nuclear power. For the United States, it was the loss of its monopoly on the Bomb.

The shock caused by the discovery of the Soviet nuclear detonation, combined with China's communist revolution, caused serious concern among Western leaders and the American people. Soviet access to nuclear weapons technology had unquestionably shifted the pendulum of global power, indicating that the United States, its allies, and their shared ideals of Western liberal democracy would not go unchallenged on the world stage. What should the response be? Perhaps intuitively: shift the pendulum of power back with a technology far more dangerous and powerful than the fission devices currently in existence.

For some, eclipsing the power of the fission bomb was the clear answer. For others, like Oppenheimer, development of thermonuclear weaponry was unconscionable because if the hypothetical calculations about its potential power were accurate, it could be a true world-ending weapon. The United States had a decision to make, and it set its top minds to doing so. The Atomic Energy Commission resolved to convene a special meeting of its General Advisory Committee (GAC) to investigate this decision. The GAC, chaired by Oppenheimer and composed of other prominent members such as Enrico Fermi, Glenn T. Seaborg, and James B. Conant, met on October 29-30, 1949. The Committee was tasked with considering whether the Atomic Energy Commission was doing enough for the United States in the areas of defense and security, and it was also asked to assess the hypothetical thermonuclear hydrogen bomb. By October 30, 1949, the Committee issued its report, which came in three parts and made the following conclusions:

1. **Recommendations Concerning Existing Fission Technologies:** Expand facilities that separate uranium isotopes and produce plutonium. Increase the supplies of uranium ore required to produce nuclear weaponry. Fission bomb production should be continued, especially those bombs that can be made for tactical purposes.

2. **Analysis of the Potential for the Thermonuclear Hydrogen Bomb:** It's probable that hydrogen bomb technology can be produced. However, like the original fission technology, technical challenges stand in the way, so much more research will be required before any hydrogen weapon can be produced. More likely than not, though, with significant resources, a thermonuclear bomb could likely be produced within five years. That said, ". . . it is clear that the use of this weapon would bring about the destruction of innumerable human lives; it is not a weapon which can be used exclusively for the destruction of material installations of military or semi-military purposes. Its

use therefore carries much further than the atomic bomb itself the policy of exterminating civilian populations."

3. **We *can* likely build the hydrogen bomb, but should we?** The GAC did not think so. "Although the members of the Advisory Committee are not unanimous in their proposals as to what should be done with regard to the super bomb, there are certain elements of unanimity among us. We all hope that by one means or another, the development of these weapons can be avoided. We are all reluctant to see the United States take the initiative in precipitating this development. We are all agreed that it would be wrong at the present moment to commit ourselves to an all out effort towards its development."[196]

The GAC's majority opinion, signed by Oppenheimer, concluded that the thermonuclear super bomb could be produced, but shouldn't. The Majority hoped that a public statement could be made about the hypothetical technology and that it could be described in general terms that the public would understand, expressing the United States' desire not to develop such a weapon… Some felt the commitment not to create the weapon should be absolute, while others believed that it should be conditioned on a Soviet commitment also not to do so.[197]

A minority opinion issued alongside Oppenheimer's, signed by Enrico Fermi and Isidor Isaac Rabi, another gifted physicist, was starker in its opposition:

[The Super's area of destruction] would run from 150 to approximately 1000 square miles or more… such a weapon goes far beyond any military objective and enters the range of very great natural catastrophes. By its very nature it cannot be confined to a military objective but becomes a weapon which in practical effect is almost one of genocide.

It is clear that the use of such a weapon cannot be justified on any ethical ground which gives a human being a certain individuality and dignity even if he happens to be a resident of an enemy country…

Any postwar situation resulting from such a weapon would leave unresolvable enmities for generations. A desirable peace cannot come from such an inhuman application of force. The postwar problems would dwarf the problems which confront us at present…

The fact that no limits exist to the destructiveness of this weapon makes its very existence and the knowledge of its construction a danger to humanity as a whole. It is necessarily an evil thing considered in any light.

For these reasons we believe it important for the President of the United States to tell the American public, and the world, that we think it wrong on fundamental ethical principles to initiate a program of development of such a weapon. At the same time it would be appropriate to invite the nations of the world to join us in a solemn pledge not to proceed in the development or construction of weapons of this category. If such a pledge were accepted even without control machinery, it appears highly probable that an advanced stage of development leading to a test by another power could be detected by available physical means. Furthermore, we have in our possession, in our stockpile of atomic bombs, the means for adequate "military" retaliation for the production or use of a "super."[198]

The Atomic Energy Commission ultimately endorsed the GAC report with a vote of 3-2. However, supporters of the development of fusion weapons lobbied hard for the project to go forward. Ultimately, in late 1949, Truman appointed a special committee of the National Security Council composed of Secretary of State Dean Acheson, Secretary of Defense Louis Johnson, and Chairman of the Atomic Energy Commission David Lilienthal. Lilienthal and Johnson favored a program to begin accelerated research in short order. Acheson was more hesitant in his approach, ultimately consulting Oppenheimer on the decision. In the end, though, Acheson realized that "the American people simply would not tolerate a policy of delaying nuclear research."[199]

On January 31, 1950, President Harry Truman issued the following statement:

> IT IS part of my responsibility as Commander in Chief of the Armed Forces to see to it that our country is able to defend itself against any possible aggressor. Accordingly, I have directed the Atomic Energy Commission to continue its work on all forms of atomic weapons, including the so-called hydrogen or superbomb. Like all other work in the field of atomic weapons, it is being and will be carried forward on a basis consistent with the overall objectives of our program for peace and security.
>
> This we shall continue to do until a satisfactory plan for international control of atomic energy is achieved. We shall also continue to examine all those factors that affect our program for peace and this country's security.[200]

With that, the President of the United States authorized research into producing what would become the most powerful weapon ever harnessed by our species. By March

10, 1950, Truman had issued the approvals necessary to begin construction of reactors to produce tritium, which, as you'll recall, is an isotope of hydrogen, that was believed to be essential to the production of thermonuclear fuel.[201]

Further fueling the United States' efforts to develop the hydrogen bomb was President Truman's receipt of a document referred to as the National Security Council Paper Number 68 (NSC-68) in April 1950. Created with input from the Defense Department (DoD), the State Department (DOS), the Central Intelligence Agency (CIA), and other defense agencies, NSC-68 would set the stage for the United States Cold War policy for decades. NSC-68's conclusions were the culmination of an analysis of America's Cold War diplomacy. The conclusions of the paper were not comforting for United States leadership. Ultimately, the agencies determined that in the wake of the collapse of the Axis Powers in World War II, the world was left with two great superpowers, the United States and Soviet Russia, which had amassed resources and defense technologies that would dwarf the rest of the world. Among these was nuclear power. NSC-68 noted that the Soviet Union was driven by "a new fanatic faith" in communist ideology, and that the Soviet Union, as a result, sought to impose its "absolute authority over the rest of the world." As a result, conflict between the superpowers in some form would be unavoidable. Consequently, the development of nuclear weapons would mean that every individual on the face of the planet "faces the ever-present possibility of annihilation…"[202]

NSC-68 further urged that the United States attempt to contain Soviet expansion, and that to do so, it would need to engage in a rapid military buildup of conventional forces and its nuclear arsenal, including but not limited to the development of the hypothetical hydrogen bomb. The United States would need to engage in covert means to achieve its goals. The price tag on these defense projects would be approximately $50 billion (or $630 billion in modern dollars). By contrast, at the time, the United States was only spending about $13 billion per year for defense purposes. Truman waited to act further on the document, aside from the continued research into the hydrogen bomb. However, the start of the Korean War in June 1950 led Truman to sign NSC-68 into policy, since the American public then had a sense of the potential cost and risks of Soviet expansion. The document was not declassified until 1975.[203] After adopting NSC-68, United States defense spending increased by over 350%.[204]

I can offer only so many practical details about the science and background of Edward Teller's research into the hydrogen bomb. Due to the intrinsic danger of the fusion bomb and the existential threat it poses to the world as the most destructive weapon ever created by humankind, the technical details of the design are, for good reason, considered "born secret." Data on this type of weaponry is embargoed pursuant to the Atomic Energy Act of 1954:

> The term "Restricted Data" means all data concerning (1) design, manufacture, or utilization of atomic weapons; (2) the production of special nuclear material; or (3) the use of special nuclear material in the production of energy, but shall not include data declassified or removed from the Restricted Data category pursuant to section 2162 of this title.[205]

Nuclear secrets enjoy a special legal designation as Restricted Data. The term "born secret" means that regardless of where the information originates– whether in a private laboratory or an enterprising inventor's garage– it is secret regardless, and the government will actively work to keep it that way.[206] Born Secret information is taken so seriously by the federal government that agents of the Atomic Energy Commission (AEC) once banned an issue of *Scientific American*, a magazine intended to provide scientific information and news in a way that regular laypeople can understand, from public consumption, and went so far as incinerating 3,000 copies of the publication. The issue's original publication had contained information about the hydrogen bomb, with a plea from Hans Bethe that the world find a way to "save humanity from this ultimate disaster" by reversing the President's decision to build thermonuclear weapons.[207]

Despite the details of the thermonuclear weapon, or the "hydrogen bomb," still holding top secret status, due to leaks, disclosures, and certain limited declassifications over the years, we have a basic sense of the design and its general workings.

The original design for the thermonuclear weapon was known as the "classical super." The original idea was to superheat deuterium to a temperature that would allow fusion ignition, which would then "burn" the rest of the material using a fusion reaction. Deuterium, as you may recall from earlier chapters, like tritium, is another isotope of hydrogen (one with one neutron and one proton, for an atomic mass of 2). Scientists would need to solve two problems to get this design to work. First, they would have to establish initial ignition conditions, and second, they would need to determine whether the combustion wave would be self-supporting once established.

Even given the nearly incomprehensible energy exerted by a fission bomb, there were problems with the classical super design. Approximately 80% of the energy in a fission bomb core is in the form of "soft X-rays." The problem is that the hydrogen isotopes (deuterium in this example) are virtually invisible to the radiation at normal densities. As such, it cannot heat the deuterium to an adequate temperature to engage fusion ignition. The second problem, whether a fusion combustion wave would be self-supporting (or "self-sustaining") as fission reactions were in classic atomic bombs, was a matter of various complicated physical processes such as energy production, means of energy transfer via neutrons and ions, and so on. It became apparent to scientists in early experimentation that heating deuterium via the classical super design was impossible without tritium as a starter fuel (imagine this as a type of nuclear accelerant that would heat the deuterium up to an adequate ignition temperature). As pure tritium was extremely rare and difficult to produce, it would be impractical for weapons deployment.[208]

It became apparent through early experimentation that the problems of energy production to energy loss and the spread of combustion conditions (dissipation of energy) would need to be solved in order to adequately contain the nuclear processes required to attain a fusion reaction. In short, early experiments showed that in the burning zone of these weapons, the energy emitted was lost to the fuel mass since it is extremely unlikely to be absorbed. To understand this problem, one must understand the "mean free path" (MFP) of particles in nuclear reactions.

The MFP is the average distance an object will move between collisions.[209] The MFP for the high-temperature photons released in the reaction necessary to fuse the deuterium was measured in miles/kilometers. This means the energy is effectively not concentrated enough to achieve fusion because the energy spreads over a wide area. The energy would need to be better contained. Further, most of the fusion energy was deposited in a large region outside the combustion zone, thereby causing problems sustaining the fusion reaction (by concentrating the energy).[210]

As calculations became more complex and better understood in the late 1940s, it became clear that adding more tritium to the deuterium was a potential solution. A 50/50 deuterium (hydrogen-2)/tritium (hydrogen-3) mix was proposed to solve this, but despite tritium's high cost, even it could not resolve the combustion and propagation issues, so fusion was still not cracked. Deuterium was not enough by itself to trigger fusion using the classical super design, nor was tritium feasible, even if it could hypothetically serve as the starter fuel for a deuterium fusion reaction.[211]

The design that eventually won the race to build the thermonuclear bomb is today known as the "Teller-Ulam" design, an approach Stanislaw Ulam and Edward Teller developed to ignite and sustain a nuclear fusion reaction. The general concept behind Teller-Ulam is as follows:

> Construct a two-stage weapon consisting of a primary fission weapon (such as a uranium implosion device like Fat Man) that acts as a blasting cap to set off an initial reaction. The fission device would be used to compress a secondary device, serving as the main fusion bomb. This "staged implosion" process would allow faster and longer-sustained compression of the fusion fuel in the bomb, which would hypothetically more reliably lead to thermonuclear ignition.[212]

Ulam developed the concept of the two-stage weapon to solve the compression problem. If fissile material could be adequately compressed, it could achieve greater temperatures in a smaller space, thereby allowing the fuel to burn (or, in this case, a nuclear chain reaction to continue) more efficiently before losing energy and dissipating. While the MFP without adequate pressure conditions was a matter of kilometers, calculations showed that the MFP for neutrons could be decreased by a factor of one thousand, and photons by a factor of one *million*, with adequate density and combustion conditions. Neutron heating would occur in a narrower zone, maximizing the fuel's burn region, and photon absorption would serve as a heating mechanism.[213]

Actually making the concept work was not so simple, though. Ulam initially imagined using the kinetic (physical movement) energy of the stage one fission explosion as the shockwave force to compress the second stage fusion fuel (deuterium). While physical containment of the kinetic shockwave in the second stage is possible in principle, it's exceedingly difficult in practice at the time scale required. A nuclear detonation happens on the order of picoseconds, or

0.000000000001 seconds.[214] Teller had a better idea, which was born of his general understanding of atomic physics after years of research at Los Alamos and in his academic work. It was not the kinetic energy that would serve as the catalyst to detonate the second stage, but instead, radiation. Thus, the concept of "radiation implosion" was born. After all, most of a fission bomb's energy is in the form of thermal radiation rather than kinetic energy.[215]

One can begin to understand how powerful a thermonuclear weapon is simply by understanding that a weapon on the scale of the Fat Man, which destroyed an entire *city*, serves as a mere blasting cap or precursor to the real detonation in a thermonuclear weapon. Here is a bit more on how the reaction in such a device operates:

Hydrogen atoms are heated up to incredibly high temperature inside a thermonuclear bomb. When compressed by the X-rays in such a weapon, the hydrogen will reach temperatures exceeding those in the center of the Sun itself. The temperature at the core of the Sun runs about 15 million degrees Kelvin, whereas in a nuclear detonation, temperatures can reach into the hundreds of millions of degrees. While it may be hard to comprehend how we can reach temperatures many times the temperature of the core of the Sun here on Earth, it's most easily understood through the realization that the Sun is very, very large, so the fusion reactions on the Sun are distributed over a large, dense space, and are constantly ongoing. Fusion reactions in a bomb take place in a much smaller volume, for a much shorter period of time, and therefore, the energy release *per unit volume* is greater than on the Sun or in its core.[216]

In a fusion bomb, the hydrogen is heated up and compressed to an extreme state. The phenomenal heat and pressure cause the hydrogen atoms to fuse together, creating helium. During the fusion reaction, like the fission reaction, energy is released. Only in this case, the energy release is far greater than in a fission bomb, hence the larger yield.[217]

One might be surprised to learn that styrofoam is a critical component of fusion weaponry. Key to the success of the two-stage Teller-Ulam design is what's known as the "tamper." In a two-stage weapon, styrofoam is used to encapsulate the fusion fuel in the secondary stage of the weapon. The tamper reflects the heat and radiation from the fission "primary," or first stage, back into the fusion fuel, thus increasing its temperature and pressure, allowing it to undergo fusion. Styrofoam is an ideal substance for this purpose because it's strong, lightweight, and a good insulator. Without styrofoam, the reaction of the thermonuclear stage would not be achievable at the temperatures and pressures required before the bomb blows itself completely apart, utilizing only fission.[218]

By early April 1951, Edward Teller submitted his report for a new design for the fusion bomb to the United States government.[219] Enter Operation Greenhouse, the third series of nuclear weapons tests in the Marshall Islands (April 1951 to June 1951). The series of weapons tests took place in the Enewetak Atoll, all situated on towers to simulate air bursts. The operation consisted of four tests: Dog, Easy, George, and Item.[220]

Operation Greenhouse, conducted by Joint Task Force 3 (JTF 3), was the first set of tests in the Pacific following the Soviet Union's nuclear detonation in 1949 (again, much sooner than the United States anticipated the Soviets could achieve atomic fission). Although all the bombs used in the Greenhouse tests were fission, an important aspect of the Operation Greenhouse tests was that it intended to serve as a proof of concept test for thermonuclear fusion while also increasing the destructive yield of nuclear weaponry and simultaneously reducing the weapons' weight. About 10,500 Department of Defense (DoD) personnel participated in the tests. A critical element of Operation Greenhouse was testing fallout in nearby areas and radiation exposure to personnel. But as far as the future of nuclear weaponry, it was Shot "George" that played the most pivotal role.[221]

While shots "Easy" and "Item" had yields of under 50 kilotons, the "Dog" shot had a yield of 81 kilotons, and "George" had a yield of 225 kilotons.[222] By contrast, the Trinity yield was estimated at a maximum of 25 kilotons, making George nine times more powerful than Trinity and approximately 15 times more powerful than the Little Boy bomb. But it wasn't George's remarkable yield that made it so crucial to the future of thermonuclear weapons research.

The "George" shot is important to the hydrogen bomb timeline because it was a "boosted fission" bomb that used a capsule of hydrogen fuel inside a heavy shell next to a fission device. The device was based on the "classical super" design. Though scientists did not expect it to result in a full thermonuclear detonation, its goal was to demonstrate to the public and U.S. leadership that the scientists were making progress on the development of thermonuclear weapons by showing that thermonuclear burn was possible. The design was similar to those designs earlier conceived by atomic spy Klaus Fuchs and prolific polymath John von Neumann in 1946. But since this was an early test design, an incredible amount of fissile material had to be used (hence the 200+ kiloton yield) on a relatively small amount of hydrogen fuel. A scientist working on the project likened the design to "using a blast furnace to light a match."[223]

By the time the George test came about, the conceptual Teller-Ulam design had already been developed. Because George was based on a more classical super design, its outcome could not indicate to the defense apparatus and its scientists whether an actual thermonuclear device would function. What it could tell them, though, was whether radiation implosion was a viable path forward, since some of the hydrogen would indeed be subject to the forces of ionizing radiation from the primary fission stage. And that it did. Teller was present at the detonation, noting, "We felt the heat of the blast on our faces… but we still did not know if the experiment had been a success. We did not know whether the heavy hydrogen had been ignited." Later that afternoon, as the scientists waited for feedback from the test devices, Teller bet one of his fellow scientists that the hydrogen had not undergone any fusion and that the experiment was a failure. Teller lost the bet; less than an ounce of deuterium and tritium (1/16 of one pound) had undergone nuclear fusion, producing a blast energy equivalent to 25 kilotons of TNT, significantly higher than the yield of the Hiroshima "Little Boy" bomb.[224]

      The significance of this cannot be overstated: About 142 pounds (64 kg) of highly enriched, weapons-grade uranium yielded a blast of about 15 kilotons in Hiroshima, Japan. In the George test, less than an *ounce* of fuel delivered around 10 kilotons more blast energy because it had exhibited fusion burn.[225] Less than an ounce of material, about the weight of a pencil, with the power to destroy a city.

      For a thermonuclear fusion bomb to work in practice, scientists needed to flip the equation. Instead of a large mass of fissile material igniting a small amount of fusion fuel, the Department of Defense would need a small mass of fissile material to ignite a large mass of fusion fuel. To execute the Teller-Ulam design, scientists used some of the earliest computers to help with the calculations necessary to get the bomb precisely right. However, we can't say much about that because most of the technical details are, again, classified. In June 1951, Teller, Oppenheimer, and a group of scientists met at the Institute for Advanced Study at Princeton. Several Atomic Energy Commission (AEC) members and the General Advisory Committee (GAC) were present for the meeting. The general conclusion was that Teller-Ulam was technically sound and likely provided the United States with the best chance at succeeding in igniting a self-sustaining fusion reaction. Two operations were planned, Ivy and Castle. Ivy would verify Teller-Ulam's viability in an experimental setting, while Operation Castle aimed to demonstrate that thermonuclear fusion would have suitable military uses.[226]

Several additional tests were conducted in 1951. A noteworthy detail is that in late 1951, the British conducted their first atomic detonation as part of Operation Hurricane. Though the British had cooperated with the United States as part of the Manhattan Project, the United States was driven to keep its nuclear hegemony and maintain a monopoly on atomic weaponry following the war; concerned about British intelligence leaks, the United States ended nuclear cooperation with the British. The British government, not hot to trot on ceding its influence in the world, and in the shadow of the Joe-1 detonation in 1949, pushed forward with nuclear research. In October 1951, Britain detonated a warhead using British and Canadian plutonium.[227] The device was estimated to have a yield of 25 kilotons. With that detonation, Britain joined the United States and the Soviet Union as a nuclear power.[228]

Speaking of these tests, as an interesting aside, Eastman Kodak's efforts in the Manhattan Project did not go unpunished, and showed how widespread nuclear contamination from testing can be. In 1945, Kodak scientists in Rochester, NY were forced to investigate mysterious fogging on X-ray film they had sold to medical clients. They traced the problem to the cardboard used to package the film, which was produced by a mill in Indiana. When they tested the packaging, they found it was contaminated with a new type of radioactive material that had not been previously encountered. The radioactive material was fallout from the Trinity test, the first atomic bomb test, which had been conducted in New Mexico just a few weeks earlier. This discovery led Kodak to develop new methods for detecting and measuring radioactive fallout.[229]

In 1951, Kodak scientists again played a key role in detecting radioactive fallout, this time from the first atomic bomb test at the Nevada Proving Ground. A Geiger counter at Kodak's headquarters in New York state measured radioactive readings that were 25 times higher than normal after a snowstorm. This discovery helped to confirm that the Nevada test had indeed been successful. Kodak was able to do this testing because it already had some of the most advanced radiation-detection equipment in the world, since it had devoted substantial efforts toward creating films to detect radiation exposure for workers developing the atomic bomb during WWII.[230,231]

But I digress; the United States was not content to share a level playing field with other nuclear players on the world stage. On November 1, 1952, in the Enewetak Atoll, through Operation Ivy, the United States planned to effectuate Edward Teller and Stanislaw Ulam's Teller-Ulam design and ultimately harness the powers of the cosmos.

The "Ivy Mike" device ("M" for Megaton) was housed in a colossal steel casing, 80 inches wide, 244 inches long, with walls around a foot thick, the largest single forging built at the time. The inside surface of the casing, nicknamed "the sausage," was a lead-lined surface along with sheets of polyethylene foam to form a radiation channel that would conduct heat from the primary device to the secondary device. It was the first fusion bomb in the world, and it weighed approximately 82 tons.[232]

At 7:15 AM on November 1, 1952, the world entered the age of fusion. Ivy Mike was detonated on Elugelab Island. A military report from the era noted, "The Shot, as witnessed aboard the various vessels at sea, is not easily described. Accompanied by a brilliant light, the heat wave was felt immediately at distances of thirty to thirty-five miles. The tremendous fireball, appearing on the horizon like the Sun when half-risen, quickly expanded after a momentary hover time." A cover story in TIME, after the test was confirmed to the public, described Mike as follows, "the force and horror of atomic weapons had entered a new dimension… the first full-dress H-Blast (Operation Ivy) had turned the mid-Pacific sandspit named Elugelab into a submarine crater."[233]

And that, it had. Elegulab is no more, vaporized in the blink of an eye by the gargantuan explosion unleashed on the world by Mike. The bomb left a crater 6,300 feet in diameter and 160 feet deep. To this day, Mike ranks 4th largest among all U.S. nuclear tests. The bomb's first stage (a TX-5 fission bomb) ignited a secondary stage consisting of deuterium fusion fuel stored in a cylindrical Dewar (thermos) flask to yield an explosive power of 10.4 Megatons.[234] While such a number is extremely difficult to comprehend, context may prove helpful. 10.4 Megatons is equivalent to 10.4 million *tons*, or 20 billion pounds (9 billion kg) of TNT. While fission bombs are measured in the thousands of tons of TNT, fusion bombs are measured in the *millions* of tons. Assuming Little Boy and Fat Man are in the ballpark of a 20-kiloton yield, Ivy Mike was on the order of *500* times more powerful than the devices that leveled entire cities in Japan, ending World War II.

The United States had built a bomb so powerful that it made the destructive capacity of the city destroyers it had wielded against Japan look like mere firecrackers. Now, the United States possessed The Super— thermonuclear technology so powerful that, if deployed in adequate numbers, could end life itself on our world. But the U.S. would not be alone for long.

Chapter 10. Intrigue: Conspiracy to Commit Espionage, Joe-4, and the Soviet Thermonuclear Program

"Leave them [the physicists] in peace. We can always shoot them later." -Soviet Premier Joseph Stalin, on allowing Soviet physicists to continue their work on the atomic bomb unimpeded by political witch hunts.[235]

Klaus Fuchs (the Atom Spy) was not alone in his activities. Upon his arrest, Fuchs confessed that he was working with American Harry Gold, who served as a courier for the Soviet agents receiving Fuchs' leaked info. Once captured, Gold implicated David Greenglass, a man who worked at Los Alamos. Greenglass then pointed the finger at his sister and brother-in-law, Ethel and Julius Rosenberg, indicating that they were the spies who controlled the operation.[236]

As it turned out, Julius Rosenberg and his spy network had included engineers, a military aviation scientist, a civil design engineer, a machinist, and others involved in the United States' research efforts. Greenglass, who blew the lid off the operation, served in the Army's Special Engineer Detachment (SED) and was a machinist at Los Alamos. Greenglass passed information to Julius, including information about components developed at Los Alamos for the uranium implosion bomb.[237]

Born in the early 1900s in New York City, the Rosenbergs were active members of the Communist Party of the United States of America (CPUSA). After earning a degree in electrical engineering, Julius Rosenberg obtained a job as a civilian engineer with the U.S. Army Signal Corps. Around this time, in the late 1930s, the Rosenbergs began working together to feed United States military secrets to the Soviet Union. Julius was ultimately discharged from the military in 1945 for lying about his association with the Communist Party.[238]

On July 17, 1950, the Federal Bureau of Investigation (FBI) arrested Julius; a month later, Ethel, too, was arrested. Their trial began in New York's Southern District (SDNY) federal court. The two were charged with conspiracy to commit espionage against the Republic by providing atomic secrets to the Union of Soviet Socialist Republics (USSR). An interesting note is that neither of the Rosenbergs could be charged with outright treason because the United States was not officially in a state of declared war against the USSR.[239]

A month later, on March 29, 1951, the Court convicted the Rosenbergs. On April 5, Judge Kaufman sentenced both Julius and Ethel Rosenberg to death. Their co-conspirator, a fellow engineer, Morton Sobell, was sentenced to 30 years imprisonment. Greenglass, for his part in the conspiracy, was sentenced to 15 years in prison. Judge Kaufman stated of his decision to impose the death penalty, "I consider your crimes worse than murder. I believe your conduct in putting into the hands of the Russians the A-bomb years before our best scientists predicted Russia would perfect the bomb has already caused, in my opinion, the Communist aggression in Korea, with the resultant casualties exceeding fifty thousand and who knows how many millions more of innocent people may pay the price of your treason." In early 1953, Greenglass wrote a letter to President Eisenhower, asking for leniency for the Rosenbergs, and that the President commute their sentences to a term of imprisonment. Eisenhower denied the request. Ultimately, some members of the public protested the decision to impose the death penalty, claiming that the Rosenbergs were being persecuted for their political beliefs and association with the Communist Party. Their legal team worked hard to secure leniency, appealing nine times to the United States Supreme Court (SCOTUS), but ultimately did not prevail.[240]

On June 19, 1953, President Eisenhower made the following statement, in part, to the American People:

> I am convinced that the only conclusion to be drawn from the history of this case is that the Rosenbergs have received the benefit of every safeguard which American justice can provide. There is no question in my mind that their original trial and the long series of appeals constitute the fullest measure of justice and due process of law. Throughout the innumerable complications and technicalities of this case, no judge has ever expressed any doubt that they committed most serious acts of espionage.
>
> Accordingly, only most extraordinary circumstances would warrant executive intervention in the case.
>
> I am not unmindful of the fact that this case has aroused grave concern both here and abroad in the minds of serious people, aside from the considerations of law. In this connection, I can only say that, by immeasurably increasing the chances of atomic war the Rosenbergs may have condemned to death tens of millions of innocent people all over the world. The execution of two human beings is a grave matter. But even graver is the thought of the millions of dead whose deaths may be directly attributable to what these spies have done.
>
> When democracy's enemies have been judged guilty of a crime as horrible as that of which the Rosenbergs were convicted;-when the legal processes of democracy have been marshalled to their maximum strength to protect the lives of convicted spies;-when in their most solemn judgment the tribunals of the United States have adjudged them guilty and the sentence just, I will not intervene in this matter.[241]

On June 19, 1953, the Rosenbergs were executed at Sing Sing Prison in Ossining, NY. Julius went first to the electric chair, with Ethel following soon after. Both President Eisenhower and SCOTUS had refused to intervene.[242] The Rosenbergs' trial and subsequent execution were notable for many reasons, not the least of which was that they took place during the Second Red Scare, a period of intense concern, even paranoia, about an internal communist threat to the United States, its government, and its secrets.[243] It was also the first espionage trial that ever resulted in a death sentence. Ethel was the first woman to be executed in America for a crime other than murder. At the same time, there is no direct evidence that Fuchs, the Rosenbergs, or any of their co-conspirators were directly responsible for transferring knowledge, blueprints, or other designs for fusion weaponry to the Soviets. However, there can be little doubt that the intelligence they gave the Soviets accelerated the Soviet fission program, and by extension, the development of the Soviet hydrogen bomb.

Following in the footsteps of the detonation of RDS-1/Joe-1, the Soviets' first fission bomb, they quickly moved to begin research into thermonuclear weaponry. Andrei Dmitrievich Sakharov, present for the detonation of RDS-1, and often called the "Father of the Soviet Hydrogen Bomb," focused his efforts on thermonuclear weapons research. He conceived a design called *"Sloika,"* or "Layer Cake." The model was similar to Teller's design, in which deuterium and uranium were used in alternating layers to ignite a fusion reaction. In 1950, Sakharov began work at the Arzamas-16 nuclear facility to move toward the creation of the hydrogen bomb.[244]

On August 8, 1953, Soviet Premier Georgy Malenkov announced to the Supreme Soviet of the USSR (the Soviet Union's Legislature) that the United States no longer possessed a monopoly on fusion technology:

We know that abroad the warmongers for a long time cherished illusions about the United States monopoly in the production of the atomic bomb. History has, however, shown that this was a profound delusion. The United States has long since ceased to have the monopoly in the matter of the production of atomic bombs. The transatlantic enemies of peace have recently found a fresh consolation. The United States, if you please, is in possession of a weapon still more powerful than the atom bomb and has the monopoly of the hydrogen bomb. This, evidently, could have been some sort of comfort for them had it been in keeping with reality. But this is not so. The government deems it necessary to report to the Supreme Soviet that the United States has no monopoly in the production of the hydrogen bomb either.[245]

Four days later, on August 12, 1953, Malenkov made good on his promise. At the Semipalatinsk test site, the Russians detonated a device known as "Joe-4," as it was the fourth Soviet nuclear detonation announced by the United States. Though the device only produced a yield of 400 kilotons (approximately 20 times larger than the Hiroshima/Nagasaki devices), and significantly smaller than Ivy Mike, it did indeed ignite nuclear fusion, which was responsible for 20% of the device's blast yield.[246] For his efforts in bringing the Soviet Union into the age of fusion, Sakharov became at age 32 the youngest person ever elected to the Soviet Academy of Sciences and received his first of three "Hero of Socialist Labor" titles, the highest civilian award that was issued in the Soviet Union.[247]

With the detonation of Joe-4, the world's two superpowers had moved past the theoretical research stage of thermonuclear weaponry. Sights now turned from proving that fusion could occur inside a bomb to making those bombs smaller, more efficient, and practical in a wartime scenario.

Compact and relatively lightweight systems would need to be engineered to scale down the bombs from the massive test devices used by the United States and the Soviet Union. After all, Ivy Mike's blast yield may have been impressive, but a 164,000-pound steel canister the size of a small building simply could not be carried by a bomber or effectively deployed in the field through any means.

Ivy Mike was a "wet" fusion bomb, meaning that the hydrogen in the bomb was in liquid form. Castle Bravo (detonated March 1, 1954), a test intended to scale down the bomb to a manageable size for a deliverable thermonuclear weapon, was a "dry" device, substantially reducing its weight and size. While still large, Bravo utilized a much smaller form factor than Mike, at 23,500 pounds (10,659 kg), or 12 tons. Scientists miscalculated the anticipated yield of Bravo, as they did not expect that the lithium deuteride, with a 40% content of lithium-6 isotope, would contribute so significantly to the yield of the detonation. Bravo, a much smaller, more efficient device than Mike, detonated with a yield of 15 million tons of TNT (15 Megatons), by low-end estimates of the Fat Man/Little Boy yields, approximately one *thousand* times more powerful than the fission bombs used in the attacks on Japan. The mushroom cloud grew to 4.5 miles wide and reached a height of 130,000 feet, or approximately 26 miles high. The crater left by the detonation was 6,510 feet and a depth of 250 feet. The test resulted in severe fallout, largely due to its unanticipated massive yield. While it was the largest device ever detonated by the United States of America, it is particularly notable in the history of nuclear development because it was the first fusion device capable of being delivered in war.[248]

On November 22, 1955, the Soviets caught up to America's achievement of producing a deployable thermonuclear bomb. The RDS-37, a radiation-implosion hydrogen bomb, was air-dropped from a Tupolev Tu-16A bomber at the Semipalatinsk test site. It was the Soviet Union's 24th nuclear weapons test and its first deployable thermonuclear bomb. Notably, the United States would not demonstrate a thermonuclear airdrop weapon until about six months later. The RDS-37 detonated at 1,550 meters (a little over 5,000 feet) above the ground after being slowed by a parachute to allow time for the Tupolev to escape. The shock wave of the detonation hit the plane 3 minutes and 44 seconds after the drop, causing the aircraft to accelerate 2.5Gs, although it was undamaged. The mushroom cloud reached approximately 8.7 miles (13-14 km) high, with a diameter of up to 18.6 miles (30 km). The yield was ultimately between 1.6 to 1.9 megatons; it had been designed for a yield of 3 megatons on delivery, but this was reduced for the test. Despite deliberately hobbling the bomb, the blast significantly damaged a town 47 miles away (75 kilometers). A child was killed when a building collapsed, and a soldier was killed when a trench caved in from the shock of the bomb. Windows exploded as far as 124 miles (200 km) from the detonation site. Nearly 50 people were injured in the town.[249]

In the latter half of the 1950s, the United States and the Soviet Union continued various test detonations intended to make their thermonuclear weapons more efficient and more practical.

Chapter 11. The Leviathan Spreads its Tentacles: Global Proliferation and the King of All Bombs

"All-out nuclear war would mean the destruction of contemporary civilization, hurl man back centuries, cause the deaths of hundreds of millions or billions of people, and, with a certain degree of probability, would cause man to be destroyed as a biological species and could even cause the annihilation of life on earth." -Andrei Sakharov, Father of the Soviet Hydrogen Bomb and Nobel Peace Prize Laureate Human Rights Advocate[250]

Other major powers around the world were watching as the United States and Soviet Russia began testing ever smaller, more sophisticated, and more destructive nuclear weapons and delivery systems. Those other nations, with the requisite natural and economic resources, technical knowledge, and political will, turned their sights toward developing atomic weaponry or accelerating their fission research into full-scale thermonuclear technology.

By the mid-1950s, the British resolved to conduct their own fusion weapons research with the Operation Grapple series. The series tested several weapons, both boosted fission and fusion in nature. As part of Operation Grapple X, the British Royal Air Force (RAF) dropped a device from a Valiant XD824 (a high-altitude British jet bomber) at close to 7,400 feet (2,250 meters). The weapon detonated successfully at 1.8 megatons, approximately 140 times the size of the Hiroshima Bomb. The British underestimated the anticipated blast yield by 80%, and shockwave damage destroyed buildings, equipment, and infrastructure.[251]

Perhaps the first major nuclear event of the 1960s, although certainly not the most talked about, was France's entrance into the club of global nuclear superpowers. Beginning in December 1954, the issue of atomic research was raised at the cabinet level in France, ultimately authorized by Prime Minister Pierre Mendès-France. The French government began procuring plutonium, assembling a device, and preparing a test site. In 1958, Charles de Gaulle, then Prime Minister of France, greenlit a detonation for early 1960. He regarded the attainment of the bomb as a symbol of independence for France and as a foundation for its geopolitical influence on the world stage.

On February 13, 1960, the French government detonated an atomic bomb atop a 344-foot (105-meter) tower in the Sahara, French Algeria. The plutonium bomb yielded between 60 to 70 kilotons. In the three decades from the mid-1960s to the mid-1990s, the French would conduct 194 tests in the Pacific, involving fission, boosted-fission, and two-stage Teller-Ulam style thermonuclear warheads.

Meanwhile, as the Americans focused on enhancing and miniaturizing their bombs to allow efficient delivery of nuclear payloads, the Soviets focused on building bigger, more destructive devices. Not to be outdone by other nations entering the thermonuclear sphere or otherwise beginning their own atomic research programs, and with America designing ever more advanced weapons systems, the Soviets decided it was time for some saber rattling. Born out of this effort was the RDS-220 (Codename "Vanya"), what would become known as the "Tsar Bomba," or "King of All Bombs." The Bomb's CIA designation was "JOE-111."[252]

On October 30, 1961, a TU-95V Soviet long-range bomber dropped the Tsar Bomba in a remote region on an island in the Arctic Ocean over Novaya Zemlya. As the Cold War became increasingly tense, the Soviets wanted to show that they could build a weapon of absolute, unparalleled power.[253] Tsar Bomba can be thought of as an enhanced fusion bomb. While a standard Teller-Ulam design uses two stages, Tsar Bomba may have used three, or alternatively, multiple second stages. It is unclear whether Tsar Bomba was a two-stage weapon that used several second stages to increase the fusion yield, or was actually a three-stage weapon. In any case, this is roughly how a three-stage weapon functions:

1. A primary fission reaction is initiated by a small atomic bomb that is detonated by conventional explosives, compressing a sphere of fissile material (uranium-235 or plutonium-239). This, in principle, is how a standard fission bomb works.
2. The fissile material undergoes a chain reaction, releasing neutrons that result in a fission chain reaction.
3. The energy released by the primary fission reaction compresses fusion fuel in the second stage (typically deuterium and tritium). These nuclei fuse together, forming helium nuclei. It's at this stage that we see the output of a standard thermonuclear fusion bomb.
4. The energy released by the fusion reaction in the second stage compresses fusion fuel in the third stage. This third stage is a mix of deuterium, tritium, and lithium-6. (Lithium-6 is an isotope of lithium, a metal most commonly known for its use in modern battery technology).
5. In the third stage, helium nuclei and tritium nuclei are formed from fission, so fusion and fission occur simultaneously at this stage. The fission reactions release more neutrons, which then cause more fusion reactions to occur. This fission/fusion cascade results in an enormous energy release.

It was determined while designing the bomb that it would need to be detonated in a weakened form. The design for Tsar Bomba could have theoretically yielded a 100-megaton blast (100 million tons of TNT). However, at 100 megatons, the detonation would have had two severe and significant drawbacks. First, the bomb would have produced approximately 25% of all fallout produced since the advent of the atomic bomb in 1945. Secondly, the plane dropping the bomb could not have survived with such a yield, and both the aircraft and crew would have been lost, even if the bomb had been dropped using a parachute to slow its detonation.[254]

Ultimately, to save the plane and its crew, Tsar Bomba was dropped from a parachute to slow its fall to detonation at 13,000 feet. This would allow the Tupolev Tu-95 bomber to get at least thirty miles away from the blast site. Even with these precautions, the blast yield was so incredible that the crew was warned that their chance of survival was around fifty percent. Fortunately for the crew, they made it just far enough outside the blast zone to survive by the skin of their teeth.[255]

In its final design, the bomb used a lead tamper instead of a uranium-238 fusion tamper. The lead tampers reduced fast fission in the weapon and brought its yield down. At full yield, the bomb would have utilized uranium-238 fusion tampers to increase the output by forcing more fission.[256] Even so, Tsar Bomba's air burst was destined to become, by far, the most colossal detonation of any weapon ever in human history. It remains so, to this day. At 11:32 AM on October 30, 1961, the device detonated with a yield of 50 to 57 *megatons* of TNT.[257] Similar to other superbomb detonations, fathoming such a number is impossible. Imagining even the blast power of 50 to 57 million *pounds* (even *one* million pounds) of TNT is pretty much impossible, much less to try to do so with that many tons. One way to process it is to illustrate that in that single blast, Tsar Bomba released ten times more energy than *every single ordnance* detonated throughout the course of World War II *combined*. Every bomb, every missile, every rocket, every bullet spent... eclipsed tenfold by a single device in an instant.[258] That, indeed, is the only way to grasp it– imagining all the chaos, the noise, the blasts, the abject destruction that occurred through the entire course of the war... and then concentrate that all into a single instant and multiply it by ten; *that* is the power of the thermonuclear weapon.

The device's blast flash lit up the sky for almost an entire minute. Once the blinding flash subsided, it was replaced by a fireball six miles wide. Within ten minutes, the fireball was replaced by a mushroom cloud that grew to a height of 43 miles, reaching into the stratosphere. The cloud had a diameter of approximately 60 miles wide at its top. Its 50-megaton yield was equivalent to the instantaneous detonation of 3,300 Fat Man city-destroyer atomic bombs, or 1/10 the *entire* yield of all nuclear weapons ever detonated by all nations combined.[259] Everything within a 30-mile radius was vaporized, and catastrophic damage was exhibited in up to a 150-mile radius.[260]

The shockwave from the bomb was felt in a settlement located 430 miles away from the blast. Windows exploded and shattered out of their frames 560 miles away from the explosion. Atmospheric focusing of the shockwave resulted in windows breaking in Finland and Norway. The shockwave shot around the entire Earth thrice and was *still* detectable after the third revolution on seismic sensors, registering a seismic magnitude between 5-5.25.[261]

The explosion leveled an uninhabited village called Severny, located 34 miles (55 km) from the blast. Estimates have concluded that the heat from the combustion alone would have caused third-degree burns up to 63 miles (100 km) away from ground zero.[262] Because of its size and weight, the device was impractical for mounting on a missile, and even with long-range bombers, delivery of the device on a target enemy would be severely hamstrung because its mass would limit the distance a bomber would be capable of traveling with it in its bay.[263]

Perhaps counterintuitively, the Tsar Bomba was one of the "cleanest" nuclear bombs ever detonated. This is because fusion, unlike fission, does not cause fallout in the way that fusion does. While Tsar Bomba still required fission to trigger its fusion reaction (so it still did produce some fallout), 97% of the weapon's yield came from fusion.[264] In this way, fusion weapons are much more physically destructive in the immediate sense but much cleaner in the long run, so they don't have as much long-term impact by way of cancer, etc., on animals and humans.

Even before the detonation of Tsar Bomba, China, too, became interested in joining the nuclear club by the mid-1950s. On January 15, 1955, Mao Zedong and his government resolved to obtain their own nuclear arsenal.[265] The Chinese bomb program would be known as "*Liang dan yi xing*" or "*Two Bombs, One Satellite.*" The program aimed to develop a nuclear bomb, a ballistic missile, and an artificial satellite. This phase of nuclear research lasted approximately 20 years, from the 1950s to the mid-1970s.[266] China entered into an agreement with the USSR to trade fissile material to the Soviets in return for their expertise and scientific knowledge. However, when Sino-Soviet relations (relations between China and Russia) cooled in 1960, and the Russians pulled their resources out of China, the Chinese continued with their own research. Their first detonation would take place on October 16, 1964.[267,268] Upon China's detonation of its first nuclear device, it issued the following statement:

> China exploded an atomic bomb at 15:00 hours on October 16, 1964, thereby successfully carrying out its first nuclear test. This is a major achievement of the Chinese people in their struggle to strengthen their national defence and oppose the U.S. imperialist policy of nuclear blackmail and nuclear threats.

To defend oneself is the inalienable right of every sovereign state. To safeguard world peace is the common task of all peace-loving countries. China cannot remain idle in the face of the ever increasing nuclear threats from the United States. China is conducting nuclear tests and developing nuclear weapons under compulsion.

The Chinese Government has consistently advocated the complete prohibition and thorough destruction of nuclear weapons. If this had been achieved, China need not have developed nuclear weapons. But our proposal has met with stubborn resistance from the U.S. imperialists. The Chinese Government pointed out long ago that the treaty on the partial halting of nuclear tests signed in Moscow in July 1963 by the United States, Britain and the Soviet Union was a big fraud to fool the people of the world, that it was an attempt to consolidate the nuclear monopoly of the three nuclear powers and tie the hands of all peace-loving countries, and that it had increased, and not decreased, the nuclear threat of U.S. imperialism against the people of China and of the whole world . . .

The atomic bomb is a paper tiger. This famous statement by Chairman Mao Tse-tung is known to all. This was our view in the past and this is still our view at present. China is developing nuclear weapons not because it believes in their omnipotence nor because it plans to use them. [To] the contrary, in developing nuclear weapons, China's aim is to break the nuclear monopoly of the nuclear powers and to eliminate nuclear weapons.

The Chinese Government is loyal to Marxism-Leninism and proletarian internationalism. We believe in the people. It is the people, and not any weapons, that decide the outcome of a war. The destiny of China is decided by the Chinese people, while the destiny of the world is decided by the people of the world, and not by nuclear weapons. China is developing nuclear weapons for defence and for protecting the Chinese people from U.S. threats to launch a nuclear war.

The Chinese Government hereby solemnly declares that China will never at any time or under any circumstances be the first to use nuclear weapons. . . .

The Chinese Government will, as always, exert every effort to promote, through international consultations, the realization of the lofty aim of complete prohibition and thorough destruction of nuclear weapons. Until that day comes, the Chinese Government and people will firmly and unswervingly follow their own path to strengthen their national defence, defend their motherland and safeguard world peace.

We are convinced that man, who creates nuclear weapons, will certainly be able to eliminate them.[269]

China had become the world's fifth nuclear power, behind the United States, Soviet Russia, Britain, and France. But it was not content to stop there, with the United States and Russia detonating their own hydrogen bombs. In just three years, China became the fastest country to move from fission weaponry to developing and detonating full-scale fusion weaponry. On June 17, 1967, a Chinese communiqué announced, in part:

China has got atom bombs and guided missiles, and she now has the hydrogen bomb. This greatly heightens the morale of the revolutionary people throughout the world and greatly deflates the arrogance of imperialism, modern revisionism and all reactionaries.

The success of China's hydrogen bomb test has further broken the nuclear monopoly of United States imperialism and Soviet revisionism and dealt a telling blow at their policy of nuclear blackmail...

Man is the factor that decides victory or defeat in war. The conducting of necessary and limited nuclear tests and the development of nuclear weapons by China are entirely for the purpose of defense, with the ultimate aim of abolishing nuclear weapons.

We solemnly declare once again that at no time and in no circumstances will China be the first to use nuclear weapons. We always mean what we say.

As in the past, the Chinese people and Government will continue to make common efforts and carry on an unswerving struggle together with all the other peace-loving people and countries of the world for the noble aim of completely prohibiting and thoroughly destroying nuclear weapons.[270]

The first hydrogen bomb tested by China was dropped from a Xian H-6, a copy of the Soviet Tupolev Tu-16. It was detonated at an altitude of 2,960 meters and had a blast yield of 3.3 megatons.[271]

France was up next to the thermonuclear bat. Between 1960 and 1996, France conducted 210 tests. Charles de Gaulle was not amused that China had beaten France to the hydrogen bomb. By 1968, France conducted its largest-ever test, detonating a two-stage thermonuclear device on August 24, 1968. Codenamed *Canopus*, the bomb had a yield of 2.6 megatons.[272] The device was detonated at Fangataufa Atoll in French Polynesia. It was suspended from a large hydrogen-filled balloon and detonated at 1,800 feet. With its detonation, France had become the fifth nation with hydrogen capabilities, after the United States, the Soviet Union, Britain, and China.[273]

Some years later, in 1974, Indian Prime Minister Indira Gandhi announced that he had directed his nation to conduct a "peaceful nuclear explosion." By then, only the United Nations Security Council's (UNSC) five permanent members (the United States, the USSR, Britain, France, and China) had tested and produced nuclear weaponry. On May 18, 1974, Pohkran-I, also known as "Smiling Buddha," was detonated at a remote Indian army base before 75 researchers and scientists. The bomb had a yield of 12-13 kilotons of TNT, slightly lower than the bombs used in Hiroshima and Nagasaki. It was detonated on the same day as Buddha Jayanti, which is the celebration of the birthday of Gautam Buddha (AKA Siddhartha Gautama, The Buddha), hence the Western name for the operation, "Smiling Buddha."[274,275]

Pakistan, India's neighbor, frequently in conflict with the nation, had been watching. In the 1960s, it took part in the United States' "Atoms for Peace" program, receiving a five-megawatt Pakistan Atomic Research Reactor (PARR-1) in 1962 for peaceful nuclear power research. But signs were emerging even in the mid-1960s that the Pakistani program would not be for strictly civilian purposes, with the Minister of Foreign Affairs Zulfikar Ali Bhutto declaring at the time that "If India builds the bomb, we will eat grass or leaves, even go hungry, but we will get one of our own."[276]

By 1971, war had broken out once again between India and Pakistan. India ultimately won, and the embarrassing defeat for Pakistan further drove its ambition toward developing the Bomb. Bhutto, soon to be elected Prime Minister, called together a meeting with top Pakistani scientists in 1972, directing them to build the Bomb. Munir Ahmad Khan, trained in the United States at the Illinois Institute of Technology (IIT), who had also worked at the Argonne National Laboratory and had served as head of reactor engineering for the International Atomic Energy Agency (IAEA), was named Chairman of the new Pakistani nuclear program.[277]

At this time, Canada provided a 137-megawatt heavy water reactor known as the Canada Deuterium Uranium (CANDU) reactor, installed at the Karachi Nuclear Power Plant (KANUPP) in Pakistan. The reactor was capable of producing weapons-grade plutonium. In response to the Indian nuclear tests, Pakistan created Project-706 in the mid-1970s to enable a full-scale weapons research program. On May 28, 1998, less than three weeks after Indian nuclear tests, as part of the test series Chagai-I, Pakistan detonated several fission devices totaling a yield of 9 kilotons, which, while seemingly small compared to other nations' tests, represented a massive step forward for the country into the realm of nuclear capabilities.[278]

Even earlier than Pakistan, the North Koreans had become interested in nuclear technology; their interest peaked upon seeing the entire Japanese Empire brought to its proverbial knees with the dropping of only two bombs. North Korea's leadership ordered research into nuclear weapons in the 1950s, believing that if it did not, it might fall behind South Korea's capabilities. The Soviets assisted the North Koreans, building and establishing a reactor for them. North Korea saw disarmament efforts (we'll discuss these later) in the 1970s as a boon to its nuclear research program. It became a member of the International Atomic Energy Agency (IAEA) in 1947; between 1957 and 1979, the regime stationed an atomic scientist at IAEA's head office in Geneva with a mission of procuring information on how to design a nuclear reactor that was capable of producing weapons-grade plutonium. Since then, over the course of decades, North Korea's reactor has produced only enough power for 23 days of electricity, leading experts to believe that North Korea had no other intention for the reactor except to build nuclear weapons.[279]

By the 1980s, the North Koreans had joined the Nuclear Nonproliferation Treaty (NPT) (also something we'll discuss later in the context of disarmament efforts), committing to halt the spread of nuclear weapons and technology, and to promote peaceful cooperation on atomic energy. By 1992, North and South Korea had agreed not to "test, manufacture, produce, receive, possess, store, deploy, or use nuclear weapons" and also reached an agreement on banning nuclear reprocessing and uranium enrichment. Previously, the United States had deployed approximately a hundred nuclear weapons in South Korea as a deterrent following the Korean War. By mid-1993, the North Koreans were already threatening to withdraw from the NPT, rejecting inspections of its nuclear facilities by the IAEA. After diplomatic talks with the United States, the first inspections occurred in March 1994. By late-1994, North Korea entered a joint agreement with the United States to freeze its weapons-grade plutonium enrichment; earlier that year, the Central Intelligence Agency (CIA) determined North Korea had produced at least one nuclear weapon.[280]

Diplomacy took place between the United States and North Korea until the early-2000's when United States President George Bush imposed new sanctions and declared North Korea a part of the "axis of evil." Shortly after that, Pyongyang admitted to running a secret uranium-enrichment program to fuel nuclear weapons, a violation of the NPT, and pulled out of the agreement. On October 9, 2006, after further Western attempts at diplomacy failed, North Korea conducted its first underground nuclear fission test, with an estimated yield of about two kilotons. The United Nations Security Council responded by issuing unanimous condemnations and trade sanctions.[281]

Having mentioned those nations with confirmed nuclear programs (the United States, Russia, Britain, France,

China, India, Pakistan, and North Korea), I'd be remiss if I didn't mention Israel's nuclear ambitions. In the final months of 1960, during the final hours of President Eisenhower's administration, the U.S. government learned that the Israelis had been building a covert nuclear reactor with the assistance of the French government. The reactor was located near Dimona in the Negev Desert, which Israeli officials had formerly characterized to United States officials as a "textile plant." According to declassified documents, the French supplied Israel with raw materials for the plant in exchange for receipt back to France of any plutonium produced by the plant. Upon learning of the facility, the United States considered applying economic pressure on Israel but ultimately decided against doing so because it wasn't entirely clear what U.S. resources, tax-deductible charities, or other aid had gone toward the project. Though U.S. officials were irritated with Israel's lack of candor on the issue, the Israeli government continued to maintain that the Dimona plant was for strictly peaceful purposes.[282]

By 1960, the Israelis had entered into an agreement with Norway to receive heavy water (deuterium) at the plant. The national security apparatus in the United States concluded at the time that "Israel is engaged in construction of a nuclear reactor complex in the Negev... plutonium production for weapons is at least one major purpose for this effort... Israel will produce some weapons grade plutonium in 1963-64 and possibly as early as 1962." Notably, much of the file noting this development *remains* classified today.[283]

By the late 1960s, United States Intelligence had concluded that Israel likely possessed nuclear weapons capabilities, noting that Israel had "all the components for a weapon... awaiting only final assembly and testing." The United States was put in an awkward position by this development because it was already working toward nuclear nonproliferation and disarmament. On September 26, 1969, President Richard Nixon met with Israeli Prime Minister Golda Meir at the White House. The State Department warned Nixon before the meeting that, "Israel might very well now have a nuclear bomb," but at base minimum, possessed "the technical ability and material resources to produce weapons grade uranium for a number of weapons." There is no official record yet publicly available about the nature and content of that meeting, but most believe that the two came to an understanding about Israel's nuclear program. As part of the agreement, Israel would first refrain from testing nuclear devices. Second, it would refrain from keeping a high profile on its weapons program, and finally, it would forego making announcements or declarations about the program.

In return, the United States would keep its plausible deniability and pay no heed to the alleged Israeli program. Nor would it pressure Israel to sign onto nuclear nonproliferation agreements, namely the Non-Proliferation Treaty (NPT).[284] It wasn't until approximately a decade later that the infamous *Vela Incident* occurred on September 22, 1979.

In the dawn hours of that morning, the US VELA 6911 satellite, built to detect nuclear detonations worldwide, detected from high orbit a double flash signal from the vast South Atlantic/Indian Ocean. By this point, the Limited Test Ban Treaty was in place, and this flash, if in fact a nuclear detonation, would violate the Treaty. Scientists and intelligence analysts concluded initially that a low-yield nuclear detonation most likely caused the signal. Though the United States' later conclusions would find that the flash was caused by a non-nuclear event, such as a striking meteoroid around the satellite, defense experts in the United States strongly contested this conclusion. Studies that inspired the debate remained classified into the 1980s; however, it was leaked in the years following the incident that the Central Intelligence Agency (CIA) determined with a probability north of 90% that a nuclear detonation caused the flash. Additionally, a highly classified Naval Research Laboratory (NRL) study indicated that data from the Navy's global hydro-acoustic sensor system and other classified systems appeared to corroborate and even pinpointed the location of the detonation. In 2010, former President Jimmy Carter published one of his diary notes from the time, which concluded, "We have a growing belief among our scientists that the Israelis did indeed conduct a nuclear test explosion in the ocean near the southern end of South Africa." Most of the VELA Incident documents *remain* classified as of the time of this writing.[285]

Finally, regarding the VELA Incident, it should be noted that U.S. officials had speculated that South Africa may have been involved in the test, alongside Israel. Analysts pondered whether South Africa had enough fissile material for a nuclear test, with some disagreement between them. While South Africa didn't have a complete testable device at that time, they potentially had sufficient fissile material for a test. Some believe an Israeli test with South African logistical support may be the most likely scenario, given doubts about South Africa's enrichment capabilities.[286]

In an interesting turn of events in 1985, Mordechai Vanunu, a former Israeli nuclear technician, provided unprecedented insights into the Israeli nuclear arsenal, including photographs he secretly took as a technician at the Dimona facility. Vanunu immigrated to Australia and shared materials with the London Sunday Times. He was subsequently captured by Israeli agents, tried, and imprisoned for the material he handed to the Sunday Times. The photographs from Vanunu revealed an allegedly sophisticated nuclear program featuring over 200 bombs, including boosted fission devices, neutron bombs, F-16 deliverable warheads, and advanced ballistic missiles. The boosted weapons depicted in the photographs indicated a level of sophistication that would have required nuclear testing to achieve. Vanunu's revelations also exposed an alleged underground plutonium separation facility where Israel was producing 40 kilograms of plutonium annually, surpassing previous estimates. The photographs showcased advanced designs, allowing Israel to build bombs with as little as 4 kilograms of plutonium. These revelations prompted increased estimates of the total Israeli nuclear stockpile. United States experts noted that the Israelis "can do anything we or the Soviets can do." Vanunu's disclosures exposed the technical details of the Israeli program and stockpile and led to a subtle acknowledgment of Israel's potent nuclear deterrent by the Israeli authorities.[287]

Entire books have been written about various nation-states' atomic weapons development programs. While we don't have the capacity to delve deeply into that history here, it's worth checking out if you're interested in any particular state's program. But on to other Cold War topics...

Chapter 12. Peace Through Fear: Thermonuclear Mutually Assured Destruction (MAD)

"A strange game. The only winning move is not to play." - War Operations Plan Response (WOPR) AKA "Joshua," War Games (1983)[288]

By the 1950s, as both the United States and the Soviet Union rapidly developed more advanced nuclear weapons, it became apparent to the world's superpowers that other nations would also soon attempt to achieve nuclear capabilities. As illustrated in the prior chapter, that fear rapidly became a reality. Countries around the world were developing and successfully achieving this limitless power once only possessed by mythical gods in religious texts and ancient fables– the ability to level cities and crush nations of *millions* in the blink of an eye; a power so fierce and intense that it could end all life itself.

Long before the first atomic bomb was conceived (approximately 70 years, in fact), Wilkie Collins, a 19th-century English author, had this to say of military conflict as he witnessed the events of the Franco-Prussian War:

> . . . but what is to be said of the progress of humanity? Here are the nations still ready to slaughter each other, at the command of one miserable wretch whose interest is to set them fighting! . . . I begin to believe in only one civilizing influence– the discovery one of these days, of a destructive agent so terrible that War shall mean annihilation, and men's fears shall force them to keep the peace.[289,290]

The creator of the Nobel Prize, Alfred Nobel, who also invented dynamite, similarly recognized the power of weaponry sufficient to exact instantaneous total war, saying of his dynamite factories in 1891: "Perhaps my factories will put an end to war sooner than your [peace] congresses: on the day that two army corps can mutually annihilate each other in a second, all civilised nations will surely recoil with horror and disband their troops."[291]

By the time the Eisenhower Administration had come to power in 1953, the White House was already reflecting on the expense of the Korean War and the conventional weapons expended to fuel the conflict. Eisenhower saw an opportunity to scale down the use of conventional weapons and utilize nuclear supremacy as a strong deterrent against war. The strategy developed in response to this need was known as "Massive Retaliation" or "Massive Retaliatory Power."[292] John Foster Dulles, United States Secretary of State from January 1953 until April 1959, and an avid anti-communist leader, gave a speech outlining the new concept of Massive Retaliation to the Council on Foreign Relations in New York on January 12, 1954. After asserting that the United States' involvement in Korea was an emergency action intended to drive off communist aggression on the peninsula, he noted, "Emergency measures are costly; they are superficial; they imply that the enemy has the initiative. They cannot be depended on to serve our long-time interests." He went on to say:

> The Soviet Communists are planning for what they call "an entire historical era," and we should do the same. They seek, through many types of maneuvers, gradually to divide and weaken the free nations by over-extending them in efforts which, as Lenin put it, are "beyond their strength, so that they come to practical bankruptcy." Then, said Lenin, "our victory

is assured." Then, said Stalin, will be "the moment for the decisive blow."

In the face of this strategy, measures cannot be judged adequate merely because they ward off an immediate danger. It is essential to do this, but it is also essential to do so without exhausting ourselves... It is not sound military strategy permanently to commit U.S. land forces to Asia to a degree that leaves us no strategic reserves... Also, it is not sound to become permanently committed to military expenditures so vast that they lead to "practical bankruptcy..."

We need allies and collective security. Our purpose is to make these relations more effective, less costly. This can be done by placing more reliance on deterrent power, and less dependence on local defensive power. This is accepted practice as far as local communities are concerned. We keep locks on our doors; but we do not have an armed guard in every home. We rely principally on a community security system so well equipped to punish any who break in and steal that, in fact, would-be-aggressors are generally deterred. That is the modern way of getting maximum protection at a bearable cost. What the Eisenhower Administration seeks is a similar international security system. We want, for ourselves and the other free nations, a maximum deterrent at a bearable cost.

Local defense will always be important. But there is no local defense which alone will contain the mighty land power of the Communist world. Local defenses must be reinforced by the further deterrent of massive retaliatory power. A potential aggressor must know that he cannot always prescribe battle conditions that

suit him. Otherwise, for example, a potential
aggressor, who is glutted with manpower, might be
tempted to attack in confidence that resistance would
be confined to manpower. He might be tempted to
attack in places where his superiority was decisive.
The way to deter aggression is for the free
community to be willing and able to respond
vigorously at places and with means of its own
choosing.[293]

Though Dulles chose his words carefully, wisely opting not
to say the quiet part out loud, the implication was clear:
conventional forces would no longer rule the day insofar as
United States policy toward conflict with communist powers
was concerned. Instead, the doctrine of Massive Retaliation
would cause conventional weapons to take a backseat to the
threat of thermonuclear war in the event that Soviet Russia or
China engaged in hostile actions across the world.

What Dulles was explaining in the early 1954 speech
was the shift to Eisenhower's "New Look" defense strategy,
outlined in a 1954 National Security Council paper, which
turned an eye toward deprioritizing conventional weapons
for nuclear weaponry, even to respond to communist attacks
which utilized only conventional weapons. The strategy
aimed to provide *deterrence* against an attack on the United
States and to meet the mutual protection obligations it shared
with its allies, without mortally harming the U.S. economy
through intensely expensive production and maintenance of
conventional weapons and by fighting multiple fronts in
different parts of the world. The New Look policy
strengthened investments in nuclear weaponry and the Air
Force to provide nuclear deterrence.[294]

By the mid-1950s, many Western military strategists
believed that the advent of the hydrogen bomb would mark
the end of conventional ground warfare. But while the United
States clearly had the lead in nuclear technology, the Soviets
were quickly catching up and developing their own arsenal.

Strategists believed the next war could potentially be fought between thermonuclear superpowers– the United States and Soviet Russia. With the power of the hydrogen bomb evident, the loss of life resulting from such a war would be incalculable. Only through deterrence and the development of massive nuclear arsenals, strategists declared, could a nuclear holocaust be prevented. Winston Churchill, upon seeing the gigantic arsenals being constructed by the United States and the U.S.S.R., famously commented, "If you go on with this nuclear arms race, all you are going to do is make the rubble bounce."[295]

It was during this period, as an outgrowth from Massive Retaliation and the Soviet ramp-up of its nuclear program, that the concept "Mutually Assured Destruction" (MAD) became prominent. The idea is simple: Humankind can avoid future war and conflict by deploying thermonuclear weaponry so powerful, and with outcomes so horrible, that its use would lead to the destruction of all humankind, or at the very least, the nuclear superpowers engaged in such a conflict. This is a sort of atomic tit-for-tat with apocalyptic outcomes. Ultimately, suppose one nation launches a first attack using nuclear weaponry. In that case, the other power will retaliate with its own nuclear weapons. In order to maintain both strategic and tactical advantage, the full force of the weapons will be used to attempt to wipe out the enemy power before one's own society is eradicated in nuclear fire. Any conflict could lead to the use of nuclear weaponry, so any conflict would carry the weight of the risk. This is the idea of *nuclear deterrence* (deterring/preventing one's adversaries from attacking by making them aware of the extreme and perhaps untenable consequences of doing so), and of MAD itself.

Realizing the extreme stakes involved in war planning and maintaining scientific human resources (and the associated technical/institutional knowledge) after witnessing the true power of science through the atomic detonations in World War II, thought leaders in the United States War

Department conceived of Project RAND (short for *Research and Development*), an organization formed to connect military planning with research and development decisions. Commanding General of the Army Air Force H.H. "Hap" Arnold said in a report to the Secretary of War, on the importance of the creation of RAND in 1945:

> "During this war, the Army, Army Air Forces, and the Navy have made unprecedented use of scientific and industrial resources. The conclusion is inescapable that we have not yet established the balance necessary to insure the continuance of teamwork among the military, other government agencies, industry, and the universities. Scientific planning must be years in advance of the actual research and development work."[296]

By the 1950s, the RAND Corporation became a resource for U.S. military planners on Cold War knowledge and game theory. Similar to board or video games, but with far higher stakes, this "game theory" considers potential outcomes in wartime scenarios and how military strategists might create the best setup for the best possible outcome in a wartime scenario (or perhaps, more ideally, to avoid war altogether). Scientists at RAND built a wealth of knowledge around game theoretic approaches to international politics, defense, acquisition, theories on the future of war, and perhaps most importantly, nuclear deterrence. Theorists at RAND discussed lofty scenarios in which potentially millions or billions could be killed by nuclear weapons or other technologies, seeking to broaden strategic knowledge about nuclear weapons, war, and the associated international politics that could prevent their use, or alternatively, lead directly or indirectly to their use. RAND created simulations called the "Cold War Games," wherein strategists would game out nuclear conflicts. The games had various outcomes; generally, mathematicians involved in the simulations would tend to launch nuclear weapons more quickly, while social scientists tended to exercise more restraint.[297]

Some saw the game scenarios as cold and calculating. Folk singer Malvina Reynolds, in characterizing the activities of RAND in 1961, even wrote a song about them called "The RAND Hymn," which spoke of the corporation using human lives as counters in the games.

Many thought that RAND's approach was unrealistic and perhaps even morally bankrupt. In Fred Kaplan's book, *The Wizards of Armageddon*, he described the RAND scientists as "rational analysts" attempting to "impose a rational order on something that many thought inherently irrational– nuclear war. They would invent a whole new language and vocabulary in their quest for rationality, and would thus condition an entire generation of political and military leaders to think about the bomb the way that the intellectual leaders of RAND thought about it."

The prevailing perception of RAND defense intellectuals can be categorized into two main groups. The first is a view of these figures as brilliant intellectuals who averted global catastrophe, and the alternative perspective, as evidenced by popular culture and post-Cold War writings on RAND, is one of malevolent actors lacking any moral compass.

Many have viewed the utilization of game theory by scientists at RAND as an exaggerated notion wherein an obscure subject was considered beyond the understanding of ordinary individuals, yet wielded to make influential decisions impacting everyone's lives. These views held that the game theorists at RAND were callous individuals who manipulated game theory to push the world toward annihilation.[298]

That said, hindsight is 20/20, and as of this writing, the world has not burned in the nuclear fires of a post-apocalyptic thermonuclear holocaust. So perhaps a second look is deserved for the social scientists and mathematicians at RAND who spent their days playing games that would calculate scenarios that would inform policymakers who would potentially extinguish millions of lives. Two major divisions at RAND were involved in the Cold War Games: the Social Sciences Division (SSD) and the Mathematical Analytics Division (MAD), not to be confused with "mutually assured destruction," although their work was certainly related to the concept. Ultimately, the Mathematical Analytics Division's model failed because it did not capture accurate operationalizations of human behavior and game theory. SSD created a game in response, and the social scientists won out. Eventually, universities and the State Department used the SSD model.[299]

The games were very complex and involved a lot of math and statistics. However, the basic idea was that the players would take on the roles of leaders of the United States and the Soviet Union and would make decisions about how to deploy nuclear weapons and other military forces. The games would then simulate the outcome of these decisions and show how the Cold War might play out. The games involved real-world factors like nuclear weapons and the balance of power between the United States and the Soviets. If leaders made the wrong decisions in the games, they could start a simulated war that would destroy the world.

As it turns out, the SSD social science approach to the games demonstrated an understanding of the dilemmas (strategic *and* ethical) that would affect the decision to go to, and follow through with, nuclear war. While moral discourse was effectively excluded from the rational discussions of the defense experts at RAND, the SSD version of the Cold War games evinces moral and ethical judgment in the way the games were played out. One RAND expert, Albert Wohlstetter, later confessed that one purpose of the games was to strike "The Delicate Balance of Terror" to prevent catastrophe. He admitted in a 1985 interview that some viewed the Soviets as "ostensibly dangerous... useful idiots, a comfortable enemy meant to fill the role of an indispensable adversary that a vacillating American society so sorely needed."[300] One can discern from this that RAND may have achieved a dual purpose from the games: first, to understand the mechanics of nuclear war and how best to maximize our outcomes in such a scenario; and secondly, to make evident to the public and policymakers the possibility of mutually assured destruction and even indicate to them how easily nuclear war could break out. The latter purpose may have been an information campaign to instill in everyone the stakes involved, thereby averting catastrophe through fear.

In 1964, Secretary of Defense Robert McNamara calculated that approximately 400 megatons of payload properly distributed, or about eight times the yield of Tsar Bomba, would be enough to achieve total destruction of (and thus, deterrence against) the Soviet Union. While we'll discuss missiles and delivery systems later on, this would have required approximately 300 missiles to achieve. By 1956, the United States had over 15,000 megatons of nuclear weaponry available to it, which under this estimate, would make abject pacification of the Soviet Union possible using approximately 2.7% of the total megatonnage in the United States' nuclear stockpile.[301]

The idea of Mutually Assured Destruction was that war would be prevented solely by the risk of total annihilation for both sides. But as the Cold War moved on, and RAND and others grappled with the possible causes and outcomes of conventional and nuclear war, policymakers shifted slightly in their approach. The early idea, even before nuclear weapons were conceived (as evidenced by statements by men like Albert Nobel), was that the existence of an all-powerful weapon would prevent all war by making it simply too dangerous. But as the Cold War progressed and policymakers witnessed proxy conflicts (wars and other limited conflicts where the main global superpowers don't directly fight each other), it became clear to them that, while thermonuclear weaponry may reduce or prevent *direct* confrontation between major powers, it would not prevent all conflict or war. We now know this to be true, given recent developments around the globe, particularly the Russo-Ukrainian War.

Recognizing this reality, by the 1960s, the Kennedy Administration had replaced Eisenhower's concept of "Massive Retaliation" with a new strategy of "Flexible Response." This latter strategy called for the United States to develop an approach and capacity that would enable it to defeat its adversaries in a limited nuclear war.[302] For example, imagine a Soviet invasion of Europe. Under Massive Retaliation, the United States would have three options:

1. Do nothing and allow Europe to fall;
2. Engage using only conventional weapons, hoping the Soviets will do the same without escalation to the use of nuclear arms; or
3. Annihilate the Soviet Union using the full capacity of the nuclear arsenal, in the hope that we would do enough damage before they could retaliate.

Flexible Response allows a fourth option, wherein multiple resources are available should a crisis occur, depending on the severity of the conflict. Flexibility would allow the United States to respond to communist incursions into the West in a way that wouldn't *necessitate* escalation to full nuclear war. Kennedy's goal was to deter wars, both general and limited, whether nuclear or conventional in nature. To be sure, stationing more conventional forces in Europe was costly, but Kennedy did not want to rely solely upon the threat of Mutually Assured Destruction since the risk that it would escalate the stakes and outcomes of war was too high, and because the Soviets might simply "call the bluff" of the

United States, reasoning that a conventional invasion would not escalate to nuclear war.[303]

Key to flexible response are considerations of "counter-value" and "counter-force," targeting; these opposing approaches contemplate which targets to hit with available nuclear weapons. Counterforce targeting focuses on military and industrial infrastructure, to neutralize an enemy's ability to wage war. Countervalue targeting, on the other hand, targets civilian populations. The key advantage of counterforce targeting is that it may dissuade an enemy from escalating to targeting cities and population centers. Countervalue targeting, though, has greater deterrent capacity, because in the event an adversary's cities are obliterated, their society itself may come to an end.[304]

A new concept of deterrence emerged in the early 1960s after the United States defense establishment realized through its studies and those of organizations like RAND that the most likely outcome of nuclear war would be the total destruction of both sides involved. This concept was known as *Secure Second Strike*, meaning that a nuclear power would maintain the ability to strike back using nuclear weapons against a first-strike aggressor; the hope was that this second strike would guarantee massive (hopefully mortal) damage to the enemy state. One possible advantage of a first strike is that a nation-state can eliminate all of the counterstrike forces available to its enemy, thus winning the war in one strike and achieving global dominance. Suppose the enemy maintains second strike capability, though. In that case, they have developed an arsenal and technologies capable of destroying the first-strike aggressor, even if their society and governmental institutions have been almost completely destroyed.[305] From here on out, world powers would ensure that their second strike capabilities were maintained to promise to their adversaries, "If you come for us, you will have no way of neutralizing our arsenals, and we'll bring you with us into that good night."

Chapter 13. The International Order: NATO, Warsaw, and the United Nations

"If we do not want to die together in war, we must learn to live together in peace." -President Harry S. Truman, addressing the United Nations General Assembly, April 25, 1945

No discussion of the Cold War or the 20th Century nuclear order would be complete without mention of the institutions that were set up following World War II in the interests of global cooperation, mutual defense, and peace. Understanding the Nuclear Arms Race and the buildup of nuclear weapons in the 20th Century would be impossible without understanding the alliances and institutions that caused proliferation and an increase in nuclear stockpiles, and subsequently attempted to manage them. Perhaps the most widely known of these organizations is the United Nations (UN). The UN's predecessor was the "League of Nations," formed after World War I. When World War II broke out, it was clear that the League of Nations had failed in its efforts to ensure global peace. As technology made the world smaller and the stakes of war higher, U.S. President Franklin D. Roosevelt and British Prime Minister Winston Churchill held a secret meeting in August 1941, to explore an international peace effort. As a result, they arrived at a declaration called the "Atlantic Charter," which outlined the ideal goals of war and laid the path for the creation of the UN.[306] The Charter had eight goals:

1. The United States and Britain would not seek to expand their territory following WWII;

2. Nations would not see changes to their territorial
   borders without the consent of their own peoples;
3. All peoples would enjoy the right to choose their own
   governments (a foundational democratic principle);
4. All nations would have access to the world's markets
   and resources. This goal would have permitted
   economic cooperation, which has been viewed as
   foundational to both prosperity and peace throughout
   history;
5. All nations would work together to promote economic
   growth and social welfare;
6. All people would be free from the threats of war and
   poverty;
7. The seas would be open to all countries and enjoy
   "freedom of navigation," regardless of their political
   alignment; and
8. Nations seeking to use force to expand their territories
   would be disarmed.[307]

The title "United Nations" was first used to describe the
member countries that allied against Germany, Japan, and
Italy. On January 1, 1942, the United States, the United
Kingdom, the Soviet Union, and many other nations signed
the Declaration of the United Nations, which initially
outlined the war goals of the Allied forces. On April 25,
1945, the United Nations Charter was established by leaders
at the United Nations Conference on International
Organization (UNCIO). It was ratified by 51 members on
October 24, 1945, following the end of WWII.[308] Article I of
the United Nations Charter lays out the purposes of the
United Nations:

1. **Maintain International Peace and Security.** These
   goals are primarily achieved through the Security

Council, the Peacekeeping Force, and the International Court of Justice.

2.**Develop Friendly Relations Among Nations.** This is primarily achieved through dialogue and resolutions in the General Assembly and the Economic and Social Council, which promotes economic and social development worldwide.

3.**Achieve International Cooperation in Solving International Problems.** The UN has acted through the years to attempt to reduce poverty, protect the environment, and improve health through its actions.

4.**Be a Center for Harmonizing the Actions of Nations in the Attainment of their Common Ends.** This goal is focused on enabling nations to cooperate on common challenges that they encounter and to prevent conflict and build trust through ongoing dialogue.[309]

The Security Council is the most influential body of the United Nations, originally consisting of five permanent members and 11 total members. The five permanent members are the United States, Russia, the United Kingdom, China, and France.

The five permanent members listed above have veto power on substantive matters, such as the deployment of United Nations Peacekeeping Forces to conflicts around the globe or to impose sanctions on a nation-state. The Security Council cannot take action on a substantive matter unless each of the five permanent member nations votes in the affirmative. This often resulted in stalemates in the Security Council because the United States and the Soviet Union could not agree on a measure. In one notable instance, the Soviets boycotted the Security Council in June 1950 over China's UN membership. As a result, the United States was able to push through resolutions at the Security Council, which allowed nearly one million global troops to fight in the Korean War under United Nations Command.[310]

On December 8, 1953, President Eisenhower, who had become growingly concerned about "the fearful atomic dilemma," gave a speech at the United Nations General Assembly titled "Atoms for Peace." In the speech, Eisenhower urged the creation of an International Atomic Energy Agency (IAEA) to promote the peaceful use of nuclear energy. He also proposed that the United States would share its nuclear technology with other nations if they'd use it only for peaceful power generation purposes, as opposed to building more nuclear weapons.[311]

Four years later, in 1957, the IAEA was established as a body of the United Nations. Eisenhower remarked upon U.S. ratification of the statute creating the IAEA that "The splitting of the atom may lead to the unifying of the entire divided world. We pray that it will. Let us hope that the atom will stand again for the true and all-pervasive meaning given it by the ancient Greeks– indivisible. When the world is such, then peace will be ours forever."[312]

The IAEA's objective, through the "Atoms for Peace" program, was to promote and control the atom, encourage its use to generate power, and expose and discourage its use for weapons production. Article II of the IAEA statute states, "The Agency shall seek to accelerate and enlarge the contribution of atomic energy to peace, health and prosperity throughout the world. It shall ensure, so far as it is able, that assistance provided by it or at its request or under its supervision or control is not used in such a way as to further any military purpose."[313]

The United Nations' organs, particularly the Security Council and the IAEA, would play significant roles during the Cold War and after. The United Nations generally allowed dialogue between nuclear-armed nations who otherwise would have only their own (sometimes disrupted or uncooperative) diplomatic avenues or defense apparatuses to rely upon. Simultaneously, the IAEA works to monitor the global nuclear situation and proliferation of nuclear weapons, and to establish international standards for the safe handling and production of nuclear materials.

Outside the UN, the Western nations, generally democratic and capitalist in their proclivities, and the Eastern communist nations, worked to form enormous military alliances to protect their interests and ensure collective defense. The two most powerful international alliances that have ever existed in the history of humankind are the North Atlantic Treaty Organization (NATO) and the Warsaw Treaty Organization (WTO), also commonly known as the "Warsaw Pact." Wasting little time after World War II, and seeking to create a bulwark against Soviet armies stationed in Eastern and Central Europe following the war's end, a number of nations joined together under the North Atlantic Treaty or "Washington Treaty," on April 4, 1949. The original member nations consisted of:

> The United States, the United Kingdom, France, Italy, Denmark, Canada, Portugal, Norway, the Netherlands, Iceland, Belgium, and Luxembourg.[314]

At the very core of NATO's purpose and existence is Article 5, which provides:

> "The Parties agree that an armed attack against one or more of them in Europe or North America shall be considered an attack against them all and consequently they agree that, if such an armed attack occurs, each of them, in exercise of the right of individual or collective self-defence recognized by Article 51 of the Charter of the United Nations, will assist the Party or Parties so attacked by taking forthwith, individually and in concert with the other Parties, such action as it deems necessary, including the use of armed force, to restore and maintain the security of the North Atlantic area.

Any such armed attack and all measures taken as a result thereof shall immediately be reported to the Security Council. Such measures shall be terminated when the Security Council has taken the measures necessary to restore and maintain international peace and security."[315]

Article 5 lays out what is known as the "mutual defense pact" of NATO. In short, an attack on one member nation of NATO is an attack on every member nation of NATO. During the Cold War, NATO had more than three million troops and 100 army divisions on the ground in Europe, with 1.7 million additional soldiers at a high state of readiness in the event of conflict with Warsaw forces. Of the 3 million stationed in Europe, 400,000 were U.S. personnel, as the United States constituted then, and still to this day, NATO's most powerful supporter and ally.[316] By 1967, at the height of the arms race, the United States, and by extension, NATO, controlled over 31,200 nuclear weapons[317]; in the 1960s, the NATO supply of fission and fusion bombs constituted close to 20,000 megatons of explosive power.[318] This is a power equivalent to 20 billion tons of TNT.

By 1952, the Soviets were growing concerned about the potential reunification and rearming of Germany. Stalin proposed that should Germany reunify, it should be made to remain a neutral nation, not allied with any particular defense organization. While scholars have disagreed over the decades about whether Stalin was sincere in the reunification attempt, the NATO states were already considering allowing West Germany to return to statehood and join the NATO alliance. They calculated that Stalin's offer was a ploy to maintain control over Germany as a whole, and rejected the offer.[319]

Readers might be surprised to learn that in 1954, the Soviets conveyed to Western leaders that the Soviet Union would be open to discussions about joining NATO. The proposal, sent to the United States, Britain, and France on

March 31, 1954, proposed that NATO "would cease to be a closed military alignment of states and would be open to other European countries which, together with the creation of an effective system of European collective security, would be of cardinal importance for the promotion of universal peace." By May, Western leaders rejected the proposal on the grounds that the USSR's membership in NATO would not be compatible with the Alliance's defensive and democratic aims.[320]

By 1955, the Soviets' fears had come to fruition. The Federal Republic of Germany (West Germany) entered NATO in early May 1955, and the Soviets would not stand by without an answer. In the years after WWII, the Soviets had already entered bilateral treaties with all of the Eastern European states, except for East Germany, a Soviet-occupied territory.[321] On May 14, 1955, the Eastern Bloc nations entered the Warsaw Pact, formally known as the "Warsaw Treaty of Friendship, Cooperation, and Mutual Assistance." The initial member states were as follows:

> The Union of Soviet Socialist Republics (USSR), Poland, Hungary, Albania, Bulgaria, East Germany, Romania, and Czechoslovakia.[322]

A primary driver of the West's (mostly the United States, as the biggest player) massive stockpiling of megatons upon megatons of fusion weaponry was its grave concern that Europe would be incapable of preventing a ground-based full conventional invasion of Europe by the Warsaw Pact nations. The West's thinking on this issue is perhaps best summarized by a backward-looking RAND analysis from 1989 (paraphrased, and bullets added for clarity):

The debate over the conventional balance during this 30-year period can be divided into three distinct phases:

1. **Phase 1 (1945-1961):** Marked by extremely pessimistic assessments of overwhelming Soviet conventional

(non-nuclear) strength. The NATO-Warsaw Pact balance sheet of conventional forces throughout this period was seen as continually grim for the West. Despite the alliance's best efforts to enhance its non-nuclear defenses, the gap never narrowed very much… In general, however, despite these pessimistic interpretations, the state of the conventional balance was not of particularly great concern to Western leaders at the time, given their preoccupation with nuclear issues and the fact that U.S. nuclear superiority (embodied in the doctrine of Massive Retaliation) counterbalanced any non-nuclear deficiencies.

2. **Phase 2 (1961-1969):** Findings showed a balance that was not nearly so lopsided as had previously been thought. Earlier estimates of Soviet conventional strength were felt to have been exaggerated, and it also appeared that the West had some distinct advantages of its own in regard to non-nuclear capabilities. Many came to believe during this time that NATO could, at a minimum, hold its own conventionally. This, in turn, buttressed Kennedy's push for NATO's adopting Flexible Response.

3. **Phase 3 (1969-1975):** … A tremendous Soviet quantitative buildup was detected, which by the mid-1970s resulted in an undeniable Eastern numerical superiority in conventional armaments. The West was still believed to hold a qualitative edge (more capable weapons), but some believed even this to be eroding.[323]

According to NATO's assessment of the balance of military forces in Central Europe as of 1975, "the Warsaw Pact had

considerable numerical superiority over the NATO forces
deployed in Central Europe." At the time, NATO was
fielding 27 military divisions in Central Europe against
Warsaw's 58; NATO had 6,100 battle tanks on the ground
against Warsaw's 19,000; NATO had 1,700 tactical aircraft
deployed, while Warsaw had 2,460.[324] Granted, the
importance of the *quality* of the forces available cannot be
overstated (this includes how modern the military hardware
is, how well troops are trained, whether ranks of the officer
leadership have had combat experience, where the military
assets are situated, etc.), but numbers so lopsided as a matter
of assets available were not comforting to NATO. It is
through this lens that one can understand why nuclear
weaponry was so crucial as a deterrent; NATO initially
believed that it could protect itself using the threat of all-out
nuclear war ("Massive Retaliation"), and later, perhaps by
using nuclear weapons in a limited capacity to level the
playing field ("Flexible Response.") Of course, Warsaw, for
its part, couldn't allow NATO to build up massive stockpiles
without an ability to counter them, and so it too grew its own
stockpiles, building bigger and more destructive bombs, such
as the Tsar Bomba.

    As it turns out, NATO's fears about Warsaw's
conventional and nuclear capabilities as a threat to Europe
were not unfounded. As a result of leaks from Poland's
government in the early 2000s, the world learned that
Warsaw had developed plans for full-scale invasions of
Europe, some using a combination of strategic nuclear
weapons and conventional forces. One such plan developed
by military strategists in Warsaw was "Seven Days to the
River Rhine." The plan was to use overwhelming
conventional forces while utilizing nuclear strikes against
Western European targets, including population centers in
Denmark, Belgium, the Netherlands, and Germany. The
capital of Belgium (Brussels), where NATO's headquarters
is located, was on the annihilation list as part of the *Seven
Days* plans. Such plans took into account that NATO would

likely respond to such a full-scale attack with tactical nuclear attacks on Warsaw forces heading toward the European front. Russia recognized that NATO might engage in a strategic nuclear assault against Eastern European cities such as Warsaw and Prague. Like NATO's own planning documents, *Seven Days* assumed that the adversary would engage the first strike. However, in planning documents like *Seven Days*, utilized by both NATO and Warsaw, the assumption that the other side would have made the first strike was perhaps only cover so as not to signal to the world in the event of a leak of the document that the alliance was planning a preemptive strike.[325] In other words, adding that assumption to the planning document signaled to the world, "Don't worry, this is only our plan to *react* if we're hit, not to preemptively invade." In reality, any number of small conflicts could quickly escalate into what would be considered an act of war by either alliance; in the real world, a full-scale invasion seldom comes without failed diplomacy, tragic and unexpected events, and subtle escalations that bubble over into more significant conflicts.

Notably, *Seven Days* did not include nuclear strikes on France or the United Kingdom (both, by this time, nuclear powers). Today, it is assumed that that was a deliberate choice of Warsaw not to "poke the bear" by utilizing an atomic attack on a nuclear power that would almost certainly result in full-scale nuclear retaliation, and possible annihilation of all of Europe and the Soviet Union. Ultimately, the *Seven Days'* goal was to quickly reach the Rhine in a very short time period, reducing NATO's chances of victory in Europe to almost nothing.[326]

The United States knew that an international system of nuclear arms control and defense could not function properly if it did not have its own house in order. Institutions built around controlling nuclear weaponry and its potential effects and outcomes weren't only international; they were domestic as well. Following WWII, Congress debated whether the military or the civilian world should have control over atomic energy. In 1946, the Atomic Energy Commission (AEC) was created to oversee the massive trove of information and scientific, military, hardware, and human assets that were born as a result of the incredible resource expenditures that went into the Manhattan Project (recall that, at its peak, the Manhattan Project employed 130,000 workers and had spent $2.2 Billion on research– close to $47 Billion today).[327]

During the dawn of the Cold War, the AEC focused its efforts on designing newer and more efficient nuclear weapons and developing reactors for naval propulsion. In 1954, the Atomic Energy Act provided for the creation of a commercial nuclear power industry, and AEC was charged with regulating that as well. During the 1950s, as the AEC sought out peaceful uses of atomic energy, it was able to repurpose isotopes from the X-10 Graphite Reactor at Oak Ridge for utilization in various fields– cancer therapy, radioactive tracers, imaging, etc. In the mid-1970s, the AEC was abolished and reformed into the Nuclear Regulatory Commission (NRC), and the Department of Energy (DOE) was created. While the NRC focuses mainly on regulatory activities regarding reactor oversight, materials safety oversight, materials licensing, and management of nuclear waste, the DOE is tasked with defense responsibilities, including the design, construction, and testing of nuclear weaponry, in addition to management of loosely knit federal energy-related programs.[328]

The Soviets also developed a nuclear regulatory agency of their own to manage their atomic affairs following WWII. In 1945, a decree was signed establishing a governing body responsible for managing uranium work under a somewhat clunkier name than the "Atomic Energy Commission": the Special Committee under the State Defence Committee of the USSR. The Ministry of Medium Machine-Building was established in 1953, and by 1957, was developing a program to seed the nuclear power industry for the USSR.[329] In the USSR, nuclear weapons capabilities were coordinated by the First Main Directorate at the Council of People's Commissars of the USSR and the Ministry of Medium Machine-Building. Upon the USSR's dissolution and the Russian Federation's formation after the Cold War, the Russian Federation Ministry of Atomic Energy (Minatom) was created, absorbing the responsibilities of the old Soviet agencies.[330] Minatom would eventually become the Federal Agency on Atomic Energy (Rosatom) in the early 2000s, absorbing the work and responsibilities of Minatom.[331]

Part III. Thanatos's Trident: The Nuclear Triad

Chapter 14. Tactical and Strategic Delivery Systems: Chariots of the Nuclear Warhead

"America's backstop to these nuclear threats remains the nuclear triad of land-based missiles, manned bombers and submarine-launched missiles. The deployed nuclear forces would reply to any threat or use of nuclear weapons against the United States or its allies. It's the reason we have a robust triad..." -United States Air Force General Paul J. Selva[332]

Picture, for a moment, a modern machine gun round like the 7.62x51mm NATO cartridge. It has a devastating ability to neutralize enemy targets and pacify adversaries. Now take one billion bullets for those rounds, send them back in time, and deliver them to Babylonian King Hammurabi with an offer, "Use this weapon, and you can change the tides of history." Will your offer be effective? Will the king use the bullets to conquer the globe and change the landscape of the world's development? Unlikely. Without the requisite machine guns and cartridges, and the modern engineering that goes into the metals used to construct them, the bullets you provided would be little more than curious trinkets to be admired by the Babylonian people. A weapon is only as good as its delivery system, and delivery systems must be reliable, accurate, and able to meet the needs of the particular battlefield they're deployed on.

So comes the distinction in delivery systems and types of weapons. The first distinction critical to understanding nuclear weapons systems is the delineation between *tactical* and *strategic* weapons. Tactical and strategic weapons can be compared in a number of ways; while on a grand scale, the use of tactical weapons is akin to a fistfight between world powers, the use of strategic weapons is more like a gunfight. While tactical weapons are like checkers, strategic weapons are like chess. Tactical nuclear weapons are nonstrategic in nature and are intended for battlefield use. They tend to be shorter range and have a significantly lower yield than strategic nuclear weapons. One can think of tactical weapons as serving the purpose of winning a single battle, while strategic weapons end the war, or prevent the war altogether. Tactical nuclear weapons can be particularly effective against tanks, troop brigades, armored vehicles, etc. Because their use is much shorter range, and generally a lower yield, their use is potentially more palatable and less politically problematic for a nation that opts to utilize them; the hope is that if tactical nuclear weapons ultimately must be used, they will not escalate a conflict to a full strategic exchange.[333]

There were various types of tactical nuclear weapons produced during the Cold War, which we'll discuss later, but what's important to know now is that *tens of thousands* of these weapons were produced. Because NATO was so concerned about Soviet conventional military might, tactical nuclear weapons were seen as a critical piece of the puzzle in deterring the Soviets, or even having a practical answer to a Soviet invasion without guaranteeing that use of nuclear armaments would result in full-blown nuclear war. The idea here is *proportionality*. While the use of tactical nuclear weapons during a conventional battle would absolutely be a significant *escalation* in the use of force, it might not necessarily be interpreted as a totally asymmetrical, disproportionate response that would take the world to a place of Mutually Assured Destruction through the use of strategic weapons, for example.[334]

The United States first developed tactical nuclear weapons in the 1950s, and they demonstrated an ability to eliminate enemy targets in a specific area without causing wide-ranging destruction or radioactive fallout. These systems could deliver warheads on targets as little as 2.5 miles away from the forces firing them. While many of these weapons had yields lower than the bombs dropped on Hiroshima and Nagasaki, others had yields significantly higher than Hiroshima and Nagasaki, but not close to the levels of the most potent, multi-megaton strategic fusion bombs. To date, tactical nuclear weapons have only ever been used in test scenarios, not on the battlefield, as the only two atomic bombs ever to be used in warfare were the Fat Man and the Little Boy.[335]

Strategic nuclear weapons are another story entirely. Like Fat Man and Little Boy, these are the large-scale weapons intended not for limited conflicts, but for total war. These are weapons capable of traveling thousands of miles, deployed across the world to level cities and demolish nation-states. As city-destroyers, Fat Man and Little Boy can perhaps be thought of as the first strategic nuclear weapons, although at the time of their use, the common parlance delineating tactical vs. strategic nuclear weapons was not quite developed.

Fundamentally, strategic weapons systems are designed to harm an enemy at the core of the enemy's political, economic, or military power. This means incapacitating the enemy's military bases, cities, transportation infrastructure, communications infrastructure, production facilities, or their capitals (seats of government). A strategic weapons system is a combination of both the nuclear explosive itself and the unique delivery system used to ensure that the nuclear warhead reaches its target reliably and without interference.

Strategic delivery systems are designed to cross continents, so that they can strike an enemy at home. An effective strategic weapons system must allow its user to strike its adversary (or adversaries) anywhere in the world in the event of a global conflict. For example, during the Cold War, NATO required its strategic thermonuclear weapons systems to be able to strike and destroy Warsaw Pact forces regardless of which nation-state's borders they were located within. During the Cold War, only five nations had weapons systems advanced enough to be considered adequate, real strategic weapons systems: the United States, Soviet Russia, the United Kingdom, China, and France. Of those five, only the United States and Russia had full-scale systems large enough and advanced enough to be capable of unleashing global thermonuclear war.[336]

The development of strategic delivery systems in the 20th Century was perhaps as compelling a technological, scientific, social, and military enterprise as the development of nuclear technology itself, and the development of these systems was woven through the fabric of the Cold War zeitgeist. The world's superpowers had harnessed and weaponized the incredible potential of the atom. Still, they needed a way to ensure that weapons could be stored, transported, and utilized in a safe, reliable, and efficient fashion.

The three legs of the global superpowers' strategic nuclear capabilities are known as the "Nuclear Triad." Here, we're talking only about the United States and Soviet Russia, because only those two nations had full-blown global scale strategic nuclear setups. It is known as the nuclear triad because it comprises a three-pronged nuclear capability that utilizes technologies by air, land and sea:[337]

1.**Air Leg** - Strategic, long-range bomber planes capable of dropping bombs and missiles equipped with nuclear warheads.

2.**Land Leg**- Nuclear missile silos or alternatively, land vehicles, equipped with intercontinental ballistic missiles (ICBMs).
3.**Sea Leg**- Nuclear submarines with a global range equipped with submarine-launched ballistic missiles (SLBMs) that can be deployed on submarines stationed in various parts of the world.

The theory behind the nuclear triad is that by spreading the assets that make up a country's extensive strategic nuclear arsenal across various weapons platforms and locations (some fixed, and some moveable), a nation can better guarantee that its weapons systems will survive an all-out nuclear "first strike" from its adversary and guarantee that the nation can secure a second strike against the adversary that launched the first strike. In short, the nuclear triad is an insurance policy to ensure that the weapons will continue to be useful even in the event of global thermonuclear war, and to deter a nation from engaging in a first strike by letting them know, "We'll hit you back, no matter how many bombs you hit us with." The United States and NATO primarily utilized the term during the Cold War, but the Soviet Union and Warsaw also adopted the same idea and similar weapons systems. The nuclear triad would guarantee Mutually Assured Destruction in the event that any aggressor nation was foolish or bold enough to attempt a nuclear war of aggression.[338]

The concept of the nuclear triad was developed in the 1960s as the arms race heated up between NATO and Warsaw. For the United States, the triad allowed each major branch of the U.S. military to serve a significant role in nuclear deterrence. Per a 2015 Congressional Research Service study on United States nuclear capabilities:

"ICBMs eventually had the accuracy and prompt responsiveness needed to attack hardened targets such as Soviet command posts and ICBM silos, [and] SLBMs had the survivability needed to complicate Soviet efforts to launch a disarming first strike and to retaliate if such an attack were attempted." The other component of the nuclear triad, strategic bombers, "could be dispersed quickly and launched to enhance their survivability, and they could be recalled to their bases if a crisis did not escalate into conflict."[339]

The United States, for its part, had a truly colossal number of strategic weapons at its disposal. By 1967, the United States possessed 2,268 strategic weapons systems, accounting for all of its intercontinental ballistic missiles (ICBMs), submarine-launched ballistic missiles (SLBMs), and heavy bombers. Through 1990, between 1,875 and 2,200 of these strategic weapons were maintained. Nuclear warheads intended for attachment to these delivery systems peaked at approximately 13,600 in 1987.[340]

Discussion of the nuclear triad necessitates mention of the Strategic Integrated Operational Plan (SIOP). Recently declassified documents reveal that the U.S. had an extensive nuclear weapons plan for several decades spanning the Cold War. The SIOP included so many nuclear weapons that top military commanders considered it a danger to both the U.S. and its enemies. The plan targeted numerous Soviet and Chinese cities and military installations, and its overkill capabilities alarmed President Eisenhower and other officials. The documents also show concerns about the potential fallout and hazards posed by the excessive use of nuclear weapons. Despite attempts to reform the plan later, its true current status remains hidden due to security classification and inconsistent release of information by the Defense Department. Some documents indicated that a full strategic strike under the SIOP would deliver 3,200 nuclear warheads to 1,060 target points in the Soviet Union, China, and their allies. Deaths in Soviet Russia alone would be north of 100 million, 44% of their population at the time.[341,342]

Chapter 15. The Triad's First Leg: Strategic Doomsday Bombers

"If only some of our people in England could see or imagine what Mr. Wright is now doing I am certain it would give them a terrible shock. A conquest of the air by any nation means more than the average man is willing to admit or even think about. That Wilbur Wright is in possession of a power which controls the fate of nations is beyond dispute." -Major B.F.S. Baden-Powell, October 6, 1908[343]

Aircraft are a relatively recent invention in the course of human history, but humankind has dreamed of taking to the skies for about as long as societal records exist. One of the species' most remarkable geniuses, Leonardo DaVinci, dreamed up complex flying machines in the 1400s, even going so far as to sketch them out in his notes. By the late 1700s, daring aeronauts were making uncontrolled flights in air balloons filled with hot air or hydrogen gas, which made them lighter than the air. But these air balloons couldn't be controlled and were subject to the whims of the winds. By the early 1800s, Sir George Cayley built the first real fixed-wing aircraft– a kite mounted on a stick with a moveable tail– essentially, what we would call today, an air glider.[344]

On December 17, 1903, near Kitty Hawk, North Carolina, the Wright Brothers made the first successful flight in aviation history of a self-propelled, heavier-than-air aircraft. The gas-powered, propeller-driven biplane stayed in flight for 12 seconds, spanning a flight of 120 feet.[345] It would take very little time until nations began using humankind's ability to take to the sky to wage war against each other. In December 1911, an Italian pilot on an observation mission during the Italo-Turkish War reached over the side of his airplane and dropped four grenades on Turkish targets. This is the first known example of aircraft being used for "bomber" capabilities.[346]

Before the 1903 flight, Count von Zeppelin, a retired German army officer, flew his first airship (or "dirigible") in the year 1900. These airships were lighter than air, filled with hydrogen, and had a "rigid frame" with a steel framework. When World War I started in 1914, the German armed forces had several zeppelins that could travel through the air at 85 miles per hour (137 km/h) and were capable of carrying two tons of bombs. As military deadlock set in on the Western Front, the Germans began using them against towns and cities in Britain.[347] At first, the British were defenseless against these zeppelin raids, which killed 557 people, injured many more, and instilled shock and horror in the British public.[348]

Though the British were initially incapable of countering the zeppelins, by 1916, they had developed a range of anti-airship defense capabilities, including guns, searchlights, and fighter aircraft. Zeppelins, ultimately, were incredibly vulnerable to explosive shells because they were filled with hydrogen, which, as we know from the infamous Hindenburg disaster, is violently flammable. Zeppelin raids were discontinued in 1917 after 77 of the 115 German Zeppelins had been shot down or completely disabled.[349]

The power of heavier-than-air fixed-wing aircraft, or "airplanes," soon eclipsed the Zeppelin airships. Using existing Zeppelin hangers, the German military began constructing enormous heavier-than-air, fixed-wing aircraft known as the Zeppelin-Staaken R-series. These bombers used closed cockpits, multiple engines, and a network of machine guns for defense against fighter aircraft. The bombers could deliver two tons of bombs to targets, which doesn't seem particularly impressive now, but was a remarkable achievement at the time as it allowed the heavier-than-air craft to match the bombing tonnage of the Zeppelin airships.[350] Other nations were quick to match the Germans' ingenuity, and developed their own bombers. The French Voison was one such smaller bomber, which could carry 130 pounds of small bombs that the observer would drop over the side onto enemy targets.[351] Even earlier than the Germans, though, the Russians built the first heavy bomber, the Sikorsky Ilya Mourometz V, capable of utilizing 3-7 machine guns and carrying 1,150 pounds of bombs.[352] As a side note, in a bit of irony, the French Voisin 5, the first true bomber used in WWI (albeit much smaller than the German and Russian bombers), was utilized for bombing German Zeppelin hangers on August 14, 1914.[353]

By the 1930s, after multiple nations had developed bombers for the WWI war efforts, technology began rapidly advancing. Countries switched to all-metal, monoplane construction, which allowed aircraft to assume a much more significant role in wartime operations.[354] While earlier "multiplanes" or "biplanes" had multiple sets of wings stacked on top of each other, the monoplanes had only one set of wings. This reduced airflow interference and significantly increased efficiency. Since WWII, the use of multiplanes has been almost entirely supplanted by the use of monoplanes.[355]

The first new type of bomber was the "dive bomber" utilized in World War II, a bomber that would engage a steep, almost vertical dive directly downward toward its target before releasing its bombs. During Germany's invasions of Poland and France at the dawn of WWII, the German Junkers JU 87 was used to obliterate French and Polish defenses and terrorize civilians.[356] The JU 87 used a dive bombing technique earlier developed by the U.S. Navy, wherein the bomber would release the bombs at low altitude during the dive bomb run before breaking away prior to aircraft impact, ensuring maximum accuracy on target.[357]

As WWII progressed forward, these early dive bombers were supplanted by bombers of increasingly higher payloads (the number of bombs they could carry), new bombsights, radio navigation, radar sighting, and incredible altitudes and ranges. By the war's end, Allied bombers could drop bombs on targets accurately at night at altitudes of over 20,000 feet (6,100 meters).[358] The most capable and iconic of these heavy bombers was the Boeing B-29 Superfortress, the long-range heavy bomber that made the fateful nuclear strikes on Japan. At the time, the B-29 was the world's heaviest production plane, designed to meet increasing range, bomb load, and defensive requirements. Boeing would ultimately build 2,766 B-29s at plants in Wichita, Kansas. Bell Aircraft would build 668 in Georgia, and the Glenn L. Martin Company would build 536 in Nebraska. Production ended in 1946.[359] The B-29 program was actually more expensive than the Manhattan Project, costing $3 billion ($66 billion accounting for inflation) vs. the Manhattan Project's $2 billion.[360]

While the *Enola Gay* is the most well-known of the B-29 superfortresses for its atomic deployment on Hiroshima, the B-29s were used with devastating conventional impact as well. In the Pacific Theater, as many as 1,000 Superfortresses bombed Tokyo at a time, leveling vast portions of the city. After WWII, the B-29 would be used for in-flight refueling, submarine patrol, rescue missions, and weather reconnaissance. It was even used later in the Korean War, but ultimately retired from service in September 1960.[361]

While Stalin and Soviet Russia possessed bombers, mostly medium-sized bombers, they did not have the capabilities or capacity of the B-29 Superfortress. As fate would have it, in 1944, three B-29s landed in Soviet territory during emergencies after completing their missions in Japan. The Soviets quickly took control of these aircraft and reverse-engineered the technology. The Soviets copied the B-29 and created the Tupolev Tu-4, a clone of the B-29. It took its first flight on May 19, 1947, after WWII had ended. The Tu-4 was such a carbon copy of the B-29 that, while it used Soviet weaponry, engines, and gauges, it had all the same rivets used in the B-29 and weapons systems located in the same places. Ultimately, 847 Tu-4s were produced.[362]

We should discuss the aircraft carrier. The U.S. Navy pioneered the use of the aircraft carrier on November 14, 1910, when civilian pilot Eugene Ely flew his Curtiss pusher airplane off a specially-built deck on the cruiser *Birmingham* in Hampton Roads, Virginia. On January 18, 1911, Ely landed on a special platform built on the battleship *Pennsylvania,* using wires attached to sandbags that worked as arresting gear to stop the aircraft. The U.S. Navy's first aircraft carrier was the Landley (CV-1), and it served as an unarmed test bed for deck and flight operations in the 1920s. The first U.S. Navy ship designed and built for purposes as an aircraft carrier was commissioned in 1934 and served in the Atlantic during WWII.[363]

But the U.S. Navy cannot claim the most devastating use of early aircraft carriers. In fact, Japan had learned during the 1930s war with China, that aircraft carriers permitted unrivaled power projection by allowing ships to carry long-range aircraft to far-off destinations where they could then launch from aircraft carriers to extend their range further.[364] On December 7, 1941, Japan launched one of the most infamous, massive attacks in military history, known today as "The Attack on Pearl Harbor," using all six of Japan's first-line aircraft carriers: the Akagi, Kaga, Soryu, Hiryu, Shokaku, and Zuikaku. These six carriers enabled the foreign deployment of over 420 embarked planes and constituted, at the time, the most powerful carrier task force ever assembled in history.[365] Though this attack was critical in devastating the United States Pacific Fleet, it had, unfortunately for the Japanese Empire, fulfilled the dark prophecy of Japanese Admiral Isoroku Yamamoto, who wrote in his diary at the time, "I fear all we have done is to awaken a sleeping giant and fill him with a terrible resolve."[366] Japan's ill-fated decision to attack Pearl Harbor ultimately set in motion the chain of events that led to the only use of atomic weaponry in wartime.

By 1947, the United States Air Force had become an independent service, split off from the Army and placed under the control of the National Military Establishment. The United States, facing a new and formidable adversary in the Soviet Union, determined that it needed an *intercontinental* strategic bomber. This would mean that the bomber could carry a payload between continents to bomb a foreign adversary in its homeland. The Convair B-36 Peacemaker was born of this effort; commissioned for operation in June 1948, the B-36 was powered by six Pratt & Whitney R-4360 engines driving propellers that allowed the plane to cruise at 230 miles per hour with burst speeds of 435 miles per hour (370 km/h). Unlike earlier bombers that could carry around two tons (4,000 pounds/1,814 kg) of payload, the B-36 was capable of carrying 86,000 pounds (39,000 kg) or 43 tons of bombs. With a range of 10,000 miles (16,093 km), a wingspan of 230 feet (70 m), and a top altitude of 45,700 feet (13,929 m), this aircraft would allow intercontinental bombing operations for strategic purposes, if necessary.[367] To put the range capabilities of this aircraft in perspective, the distance from Washington, D.C. to Moscow is about 4,860 miles (7,821 km).[368]

The Soviets' first foray into intercontinental bombers was the prototype Tupolev Tu-85. The Tu-85 saw its first flight on January 9, 1951. This prop-driven aircraft had a top speed of almost 400 miles per hour (644 km/h), a range of 7,460 miles (12,006 km), and a top altitude of 38,380 feet (11,698 m).[369] Ultimately, the TU-85 was replaced by the well-known Tupolev Tu-95 (NATO Codename "Bear"). Powered by turboprop engines, the Tu-95 went into service on November 12, 1952; the Tu-95A version of the aircraft could deliver nuclear payloads.[370] The Bear has a range of 6,524 miles (10,499 km), a maximum speed of 515 miles per hour (829 km/h), and is still in use as of the time of this writing.[371] Turboprops can be thought of as a hybrid between a traditional piston engine propeller and a jet engine. While turboprops still use propellers, the propellers are driven by a turbine, while jets use fan blades inside the engine housing.[372]

In the Western corner of the world, ultimately, it was the Soviets' technology that illustrated just how vulnerable piston-powered prop-driven bombers would be to jet-powered aircraft. On November 30, 1950, a U.S. Air Force B-29 Superfortress was attacking an air base in North Korea when it was damaged by a fighter so fast that the United States couldn't identify it, much less shoot it down using the machine guns onboard the Superfortress. Lockheed F-80 jets pursued the attacker, but it was too quick, and took off into the distance. As it turns out, it was a Mikoyan-Gurevich MiG-15, a Soviet jet fighter. With that attack, it became clear to the Americans that Soviet technology had spread to China and would become a problem in Korea. In one incident in 1951, known as "Black Tuesday," communist jet fighters were able to take out six of nine superfortresses flying over Korea.[373]

Jet-powered aircraft have several advantages over standard prop-driven aircraft, not the least of which is speed. Classic propeller-driven aircraft aside, turboprop-driven and jet-powered aircraft have unique advantages and disadvantages relative to each other, but we don't have the bandwidth to get into that here. To put the difference between standard propeller-driven aircraft and jet-powered aircraft into perspective, while the B-29 Superfortress had a top speed of around 280 miles per hour (451 km/h),[374] the MiG-15 fighter jet could hit 658 miles per hour (1,059 km/h).[375] While bomber aircraft would have escorts of fighters protecting them, and would almost always be slower than smaller, more nimble fighters, the asymmetrical defense capabilities of jet fighters vs prop-driven bombers became unsustainable in the long run. Bombers had to become faster and more capable of evading enemy countermeasures if they were to serve their long-term deployment and strategic purposes. While German jet bombers existed as early as 1944 (the first was the Arado Ar 234 B Blitz [Lightning]), this plane was not a strategic intercontinental bomber by any means, capable of only a 4,410 pound (2,000 kg) payload, with a maximum range of around 1,000 miles. Although it did have, for the time, an impressive top speed of 461 miles per hour (286 km/h).[376]

After political conflicts in the U.S. Congress and disputes over the capabilities of the Convair B-36, the United States ultimately replaced the B-36 with what would become the most long-lived and prolific bomber in the world: The B-52 Stratofortress. Designed by Boeing in 1948 and first flown in 1952, the Stratofortress was built to deliver nuclear payloads into Soviet territory. With a wingspan of 185 feet and eight jet engines, the B-52 could fly at a maximum speed of 595 miles per hour (958 km/h) at an altitude of 55,000 feet (16,764 m).[377] The B-52 can carry a payload of 70,000 pounds (31,751 kg) with a maximum range of 8,800 miles (14,162 km).[378] So capable, reliable, and adaptable, the United States Air Force currently expects to use the B-52, with modern upgrades, through the year 2050.[379]

The Soviets wouldn't go long without an intercontinental jet bomber of their own. By 1956, they fielded the Myasishchev M-4, a jet bomber powered by four turbojet engines, with a top speed of 550 miles per hour (885 km/h) at 40,000 feet (12,192 m). Its combat range reached up to 9,000 miles (14,484 km). The bomber's primary payload was two nuclear bombs.[380]

Back to aircraft carriers – while they were commonplace for major world powers by the 1950s, used to significant effect in World War II, the age of the jet aircraft necessitated the use of carriers that could operate for extended periods and sport the ability to carry many aircraft. The first of these was the Forrestal class. What distinguished the supercarriers from older aircraft carriers was that they were over 1,000 feet long and supported crew sizes of between 4,300 to 5,000. These supercarriers featured angled decks and steam-powered catapults, which would launch jets carrying heavier payloads from the decks of the ships.[381]

Commissioned on November 25, 1961, the USS *Enterprise* was the first nuclear reactor-powered aircraft supercarrier ever to be put on the seas and the only carrier of its class. So effective was the nuclear propulsion on the craft that it out-ran the destroyer escort assigned to it during its pre-acceptance trial on October 29, 1961.[382] By May 3, 1975, the USS *Nimitz*, the most intimidating supercarrier in the world, was put into service.[383] Nuclear-powered and carrying a crew of over 6,000 personnel, ten Nimitz-class carriers were built,[384] and all remain in operation as of this writing, with the original USS *Nimitz* currently scheduled for retirement in 2026.[385] The Nimitz carriers can carry up to 90 aircraft, costing between $4.5 to $6.2 billion each to construct.[386]

I mention aircraft carriers here not because they are, in and of themselves, a leg of the nuclear triad. Nor does their nuclear-powered propulsion system afford them particular significance for our present discussion, because this book is about nuclear weaponry, not nuclear propulsion or power generation. Instead, aircraft carriers and supercarriers are so important to the deployment discussion because they allow long-range force projection. And while the Navy *has* carried nuclear missiles aboard its aircraft carriers in the past,[387] carriers' importance to the nuclear triad has more to do with their ability to provide logistical and fighter support around the world to clear the way for and protect a nation's long-range strategic bombers. That said, bombers cannot land on aircraft carriers and heavy strategic bombers cannot take off from them, as they're too large.

By the early 1960s, bombers began to obtain supersonic capabilities (the ability to surpass Mach-1, the speed of sound). The first of these ultra-fast strategic bombers was the Convair B-58 Hustler. The B-58 was capable of carrying a single nuclear weapon in its fuselage. 30 were ultimately built for testing purposes, and 86 for operational service. The B-58 was flown between 1960 and 1970, and set 19 world speed and altitude records. It was capable of flying at 1,325 miles per hour (2,132 km/h), with a range of 4,400 miles (7,081 km), at a top altitude of 64,800 feet (19,751 m).[388] But the B-58 had drawbacks; while it was very fast, with an impressive climb rate, it also was difficult to maintain and load with weaponry. It also had a payload capacity of less than 20,000 pounds (9,071 kg).[389] The B-58 was eventually replaced by the Boeing B-1B Lancer in 1985, capable of carrying a payload of 75,000 pounds (34,019 kg) and traveling at over 900 miles per hour (559 km/h) (Mach 1.2). The B-1B Lancer is still in service today.[390]

The Soviet answer to the American B-58 was the Tupolev Tu-22, (NATO codename "Blinder"). Though the plane could travel at maximum speeds of around 1,243 miles per hour (2,000 km/h), it suffered from a lack of range, with a maximum of 2,300 miles (3,701 km). This limitation made it a medium-range bomber without true intercontinental capabilities.[391] Various versions of the Tu-22M (NATO codename "Backfire"), which took its first flight in 1969, would improve upon the original Tu-22 by adding a range of up to 4,350 miles (7,000 km).[392] Eventually, the Tupolev Tu-160 (NATO codename "Blackjack"), put in service in 1989, would eclipse the utility of these prior supersonic bombers, with a massive potential payload approaching 100,000 pounds (45,359 kg), a top speed of 1,243 miles per hour (2,000 km/h), a top altitude of over 60,000 feet (18,288 m), and a range of 8,699 miles (14,000 km), making it a true supersonic intercontinental bomber.[393]

In the late-20th Century, bombers underwent another evolution. They began as prop-driven, slow-moving goliaths. They evolved turboprop and jet-powered capabilities that gave them intercontinental range. Then, they moved into the realm of supersonic, faster than sound movement. Finally, the U.S. government took delivery of the Northrop Grumman B-2 Spirit on December 17, 1993, at Whiteman Air Force Base in Missouri. With a wingspan of 172 feet (52 m), the bomber is capable of carrying 40,000 pounds (18,144 kg) of payload, with a maximum altitude of 50,000 feet (15,240 m),[394] and a nautical range of 6,000 miles (9,656 km) (a nautical mile is slightly larger than a land-based mile).[395] While the B-2 Spirit cannot travel at supersonic speeds, it has a high subsonic speed of 628 miles per hour (1,011 km/h). The B-2, also known as the "Stealth Bomber," generates a tiny heat signature and advanced anti-radar stealth capabilities to make it invisible to all but the most advanced radar systems.[396] In an interesting piece of trivia that will prove reminiscent of our upcoming Triad Leg 2 discussion, the B-2 Spirit, one of the most advanced bombers ever created, is actually based on the flying wing design from a classified wooden Nazi jet-powered aircraft called the Horton HO 229 V3, which had a top speed in excess of 600 miles per hour (966 km/h), and could outrun a British Supermarine Spitfire (330 mph/531 km/h) and an American P-51 Mustang (437 mph/703 km/h).[397]

Now that we have a basic understanding of the bombers used to comprise this leg of the Triad, let's talk bombs. The United States Mark IV atomic bomb, with a yield of 1-31 kilotons, was based on a new implosion system built upon the early success of the "Fat Man" bomb. This weapon was the first fission bomb mass-produced using assembly lines.[398] But while there were other iterations of fission-only strategic bombs, such as the Mark V, which significantly improved upon the Mark IV by making it lighter; more efficient; and easier to maintain, test, and assemble, the United States soon moved on to deploying thermonuclear weapons, which were lighter, more efficient, and delivered far higher energy on detonation than any fission bomb.[399]

The first operational, mass-produced thermonuclear hydrogen bomb was the Mark 17 (MK-17), in service from 1954 to 1957. The loaded weight of the weapon was a massive 41,400 pounds. A 64-foot parachute stabilized the bomb and slowed its descent, giving bomber aircraft time to escape the detonation.[400] This enormous weapon had a design yield of 15 to 20 megatons, around 1,000 times more powerful than the fission bombs used in Japan.[401] Only the Convair B-36 was capable of carrying this massive weapon.[402]

For the Soviets' part, RDS-3 was their first mass-produced air drop bomb, composed of a combined core filling in the ratio of 1:3/plutonium:uranium. This 25% plutonium, 75% uranium mixture allowed the Soviets to save on scarce plutonium when manufacturing the bombs. It had a yield of approximately 42 kilotons, and the Soviets observed that when it was detonated in the air, the radioactive fallout from the weapon was significantly reduced vs ground-level detonations (approximately 109 times less fallout).[403] The Soviets, too, would soon equip their bombers with thermonuclear weaponry. One example was the Kh-20, an air-to-surface, jet-powered cruise missile with a 5,000-pound (2,267 kg) thermonuclear warhead. A primary weakness of the bomber leg of the Triad is that air-based bombers are susceptible to interception, countermeasures, and anti-air missiles. Cruise missiles like the Kh-20 enable a bomber to fly within a safe range of its target, outside enemy air defenses, and release the self-guided weapon toward its target. You can think of a cruise missile as a sort of independent plane loaded with high explosives that flies itself into a target from a safe distance, allowing the missile to travel the rest of the way with a far lower chance of interception or damage to the bomber.[404] First introduced in 1960, the Kh-20 had a maximum speed of 1,417 miles per hour (2,280 km/h), a maximum range of 400 miles (650 km), and an 800-kiloton yield, around 40 times as powerful as Fat Man and Little Boy.[405] This weapon illustrates the importance of the delivery system, as the yield wasn't as massive as the Mark 17, but the weapon exhibited far greater anti-interception capabilities than a "dumb bomb" without a guidance system or its own propulsion.[406]

As the arms race heated up and technology progressed, thermonuclear payloads increased relative to the size of the bombs carried by bombers. The Mk/B-41 nuclear bomb was the highest-yield nuclear weapon ever deployed by the U.S. Designed to be dropped from bombers, it also had the highest yield-to-weight ratio of any U.S. weapon design. With a yield of 25 megatons and a weight of only 10,670 pounds (4,840 kg), the weapon went into service around 1960. About 500 were produced, for a combined total yield of up to 12.5 gigatons (although many B-41s were of a lower yield, perhaps nine megatons, so the total arsenal yield was lower than 12.5 gigatons).[407] More modern bombs like the thermonuclear B83 have a lower yield but are far more manageable; the B83 has a yield of about 1.2 megatons (still around 60 times as powerful as the Hiroshima/Nagasaki bombs) but weighs only 2,408 pounds (1,092 kg), making it far more practical and allowing bombers to carry multiple B83s.[408]

On that point, a single bomber can carry multiple nuclear weapons. The B-52 Stratofortress, for example, carried up to 20 United States AGM-86B thermonuclear cruise missiles (these entered production in January 1977), tipped with 200-kiloton nuclear warheads. Like the Soviets' air-launched cruise missiles, these enabled U.S. strategic air assets to hit multiple targets from a safe launch area with a low risk of interception by jet fighters and anti-aircraft weaponry. At a yield of 200 kilotons and 20 on board, this would enable a B-52 to strike 20 different targets with a total yield of four megatons. Each of the 20 cruise missiles on board, independently, would be around 10 times stronger than the weapons used against Japan.[409] For its part, the Convair B-36 Peacemaker could carry up to two Mark 17 weapons, with a total yield of up to 40 megatons on board. This constituted 40 million tons of operational, war-ready TNT-equivalent worth of payload on a single plane, almost 15 times the combined yield of all bombs dropped by U.S. and British forces on Europe between 1940 and 1945.[410]

As with any weapons system, the air leg of the Triad has its advantages and its weaknesses. Advantages include:

1. Before a bomb is dropped, bombers can be recalled if the nation ordering a strike has changed its mind or made a mistake, thus potentially averting a nuclear holocaust;
2. Bombers can be airborne quickly, giving them time to survive a first strike before a first strike reaches its targets, making them resilient for purposes of a secure second strike;
3. Because of their recall capabilities, bombers can be used to threaten or show resolve against a potential target without actually dropping a weapon; and
4. Because they are mobile and numerous, they can ensure strategic penetration of an enemy target, while land-based assets might be neutralized or destroyed during a first strike scenario.[411]

Strategic bombers also have critical drawbacks that highlight the necessities of the other legs of the nuclear triad:

1. Distribution of air assets is a problem. Major air bases with long runways are required to get gigantic bombers airborne (for example, the B-2 Spirit exists at only one air base in the United States). In the event these airbases are compromised in a first strike, the strategic air capacity of a nation could be completely hamstrung;
2. Bombers are capable of carrying both conventional and nuclear payloads. While this is a flexible tactical advantage, it risks confusing an adversary who may think that an imminent conventional bombing is in fact a nuclear one, resulting in full-scale nuclear retaliation;

3. The other side to the "resolve" capability, mentioned in (3) above, is that, while bombers can be used to threaten an adversary and show resolve without actually dropping a nuclear weapon, this tactic seriously risks extreme escalation into full-scale nuclear war;

4. Strategic bomber response time is significantly slower than other legs of the nuclear triad; and

5. Most importantly, bombers are very susceptible to air-to-air, land-to-air, and sea-to-air interception capabilities fielded by most world powers (stealth bombers and supersonic bombers are designed to attempt to defeat these countermeasures, with varying levels of success).[412]

Chapter 16. The Triad's Second Leg: Operation Paperclip and the ICBM

"Thus, the first and most vital step in any American security system for the age of atomic bombs is to take measures to guarantee to ourselves in case of attack the possibility of retaliation in kind." -Bernard Brodie et al. in "The Absolute Weapon: Atomic Power and World Order"[413]

We'll take a step back to 1936, shortly prior to World War II, when the Nazis began developing a new offensive technology now known as the "long-range ballistic missile." Eventually, the Nazis' research efforts would culminate in the creation of the Vergeltungswaffe (in English, "Vengeance weapon") or "V-2 Rocket." The V-2 was a liquid alcohol and oxygen-fueled, long-range ballistic rocket equipped with a guidance system. It was also the first rocket to enter space, as evidenced by a Nazi test in 1944.[414] At 47 feet long and a weight of 28,000-29,000 pounds (12,701 kg-13,154 kg), the V-2 carried 1,600 pounds (725 kg) of high explosives and could travel 200 miles (320 km) at a peak altitude of 50 miles (80 km).[415]

The V-2 carried with it a bloody legacy. At least 10,000 Nazi concentration camp workers forced into slave labor died manufacturing the weapon.[416] Used by the Germans against the Allied forces, the V-2 was intended as a city-devastator weapon, and while more people were killed manufacturing the rocket than were by offensive attacks, it still inflicted a heavy toll. Numbers killed are estimated at around 5,500, with an additional 6,500 injured. The weapon destroyed 33,700 homes and other buildings.[417]

When the V-2 rocket first appeared, its impact was psychologically devastating.[418] As a weapon, it seemed as though to be something out of science fiction during World War II– a colossal bomb that rained from the sky to devastate land targets; for observers on the ground, it would seem to come from out of nowhere in space, due to its long range and the observer's inability to locate its launch source visually. After all, this technology could devastate and end lives from the sky during a war where plane technologies were in their relative infancy, and some battles were still being fought on horse-drawn cavalry.[419]

The significance of this weapon, a massive ballistic rocket that could deliver destructive payloads to foreign targets at great distances, did not go unnoticed by the United States and the Soviets. As World War II drew to a close, the United States deployed special operations intelligence teams behind its frontline soldiers as they cleaned up the smoldering mess left by Allied devastation on the German Third Reich. One such team was the Joint Intelligence Objectives Agency (JIOA), operating under the United States Joint Chiefs of Staff (JCS). Composed of the Army's director of intelligence, the chief of naval intelligence, the assistant chief of Air Staff-2, and a representative of the Department of State, the JIOA was responsible for operating the *foreign scientist program*, codenamed "Overcast," and subsequently renamed "Operation Paperclip."[420] The JIOA sprung into action to seize intelligence, such as information regarding Hitler's arsenal of nerve agents and the bubonic plague weapon his scientists were developing.[421]

One particularly beneficial discovery by the Allies was found in a toilet at Bonn University– a list of scientists and engineers working for the Nazi government. Around 1,600 German scientists and their families were brought to the United States to work on various technologies, including rockets, biological weapons, and chemical weapons. Operation Paperclip also sought to ensure that the intellectual assets of the Third Reich, known for its cutting-edge industrial and weapons technology, would not fall into the hands of the Soviets. Officially, President Truman had banned the JIOA from recruiting Nazi Party members and active Nazi supporters into the program. In reality, the JIOA and the Office of Strategic Services (OSS) bypassed the directive by destroying or obfuscating evidence of war crimes from the scientists' records.[422] Some believed that the strategic and technological necessity of acquiring the Nazis' knowledge and science outweighed the moral considerations surrounding the scientists' and engineers' past actions. In fact, "Operation Paperclip," also known as "Project Paperclip," got its name from the so-called paperclips that held together the dossiers and files about the scientists who possessed particularly problematic pasts from their time working for the Nazis; paperclips would mark the files of those ultimately chosen for participation in the program.[423,424]

Wernher von Braun was the most valuable human asset acquired through the efforts of Operation Paperclip. As technical director at the Peenemunde Army Research Center in Germany, von Braun was instrumental in the design, development, and production of the V-2 Rocket. Von Braun and his rocket scientist colleagues were initially brought to Fort Bliss, Texas, and White Sands Proving Grounds in New Mexico as "War Department Special Employees," charged with developing the U.S. Rocket Program.[425]

I'd be remiss if I didn't mention before going further that the United States' utilization of Dr. Wehner von Braun's abilities was not without extreme controversy, and for good reason. Von Braun was not only a Nazi but also an officer of the Shutzstaffel (Nazi SS). Though he was arrested by the Nazis in 1944 for remarks he made about the war effort and the rocket, his responsibility for the deaths surrounding V-2 production is a matter of debate, and some regard him as a war criminal, notwithstanding his contributions to the American Rocket Program.[426,427] Once the project first became public, notable objectors to Operation Paperclip were

Eleanor Roosevelt and Albert Einstein. Nonetheless, security and defense considerations won out over moral objections and concerns about the German scientists' loyalty.[428]

Long-range rocket testing was underway at Fort Bliss by 1944, and by August 1945, hundreds of freight cars with V-2 components were arriving at White Sands. Rocket testing on the reassembled V-2s proceeded quickly. The Army Ordnance Sub-Office, German scientists, the 1st Anti-Aircraft Guided Missile Battalion, and General Electric collaborated on the project. Early tests replaced the warheads with instruments for experimentation, such as upper atmosphere physics and chemistry testing equipment, spectroscopy, solar X-rays, cosmic radiation, biology, and earth photography. The program quickly began making advances in radio-controlled missiles, automatic piloting, gyroscope testing, and redundant safety systems.[429] By June 1949, the program had put the first mammal, a rhesus monkey named Albert II, into space.[430]

Now that we've talked about the V-2 and Operation Paperclip, I'll pause here to do a basic rundown of how an intercontinental ballistic missile (ICBM) works. These self-guided missiles travel up through the air or into outer space to enter a suborbital trajectory. A suborbital trajectory is a flight path that takes an object into space but doesn't reach orbit (generally at least 62 miles/100 km high). Eventually, objects in a suborbital trajectory will fall back to Earth. ICBMs must have a range of greater than 3,418 miles (5,500 km). Otherwise, they're classed into one of a number of categories of shorter range missiles, such as a short-range ballistic missile (SRBM) or a medium-range ballistic missile (MRBM). ICBMs can have multiple stages; generally, the final stage sees the re-entry vehicle, containing a nuclear warhead, separating from the other stages and falling back down toward Earth on a trajectory toward its intended target.[431,432]

By 1950, the Army had established the Guided Missile Center at the Redstone Arsenal in Huntsville,

Alabama. Von Braun was named Chief of the Redstone Arsenal Guided Missile Development Division. In Huntsville, von Braun and his team developed the Redstone and Jupiter missiles for the Army.[433] While his team is best known today for their activities as part of the National Aeronautics and Space Agency's (NASA) space program, the reality is that their first projects had defense applications. While the U.S. government's rockets were also used for civilian purposes, initial attention was on developing intercontinental means of delivering nuclear warheads. The first of these ballistic missiles was the *PGM-11 Redstone*, first launched from Cape Canaveral, Florida, on August 20, 1953. The rocket produced 77,200 pounds (35,017 kg) of thrust at liftoff through the use of a 780-horsepower turbopump and burned a combination of liquid oxygen and liquid fuel composed of 75% ethyl alcohol and 25% water. It had, for the time, state-of-the-art guidance systems, a range of 200 miles (322 km), and was capable of delivering a nuclear warhead to a faraway target. The V-2's descendent had improved significantly on its predecessor and was now nuclear capable.[434] The Redstone was equipped with the Mark 39 fusion bomb warhead, with a yield of 3.8 megatons.[435]

Redstone was the beginning of an era, and though it had both military and civilian applications, with a range of only 200 miles (322 km), it was not an intercontinental solution for the United States' nuclear ambitions. But there was trouble afoot for the Americans, and they knew they had to move fast. On October 4, 1957, the Soviets launched Sputnik, the first artificial satellite to orbit Earth. While many across the world admired the Russian space achievement, the United States government and its Central Intelligence Agency (CIA) knew better—Sputnik wasn't a mere test satellite—it was the beginning of the Soviets' ICBM program.[436] The American answer was the Atlas, already under development. With a range of 600 miles (957 km), it would take its first flight in June 1957. The Atlas D, operational from October 1959 to October 1964, would be the first true, thermonuclear intercontinental ballistic missile (ICBM), with a range of 5,500 miles (8,851 km) and carrying a warhead of 1.44 megatons. 30 Atlas Ds were deployed in the field, based out of Vandenburg Air Force Base in California, but were required to be housed above ground, creating logistical and security challenges.[437]

Eventually, these issues were solved with the Atlas-F, an 82-foot-tall ICBM deployed from 1962-1965, capable of traveling 7,000 miles (11,265 km),[438] tipped with a thermonuclear warhead with a yield of 3.75 megatons, and traveling at a speed of 18,030 miles per hour (29,016 km/h);[439] this is around 24 times the speed of sound.

The silos used to house the Atlas missiles were a massive achievement in themselves. Costing up to $18 million to build in the early 1960s (around $182 million in 2023 dollars), underground silo complexes were constructed to withstand attacks from foreign adversaries and effectively launch America's ICBMs. The underground Atlas complexes were truly enormous, with a 52-foot (15.95 m) inside diameter and a shaft 180 feet (55 m) deep. Three blast doors protected the underground Launch Control Center, and two hydraulically-controlled 90-ton doors sitting above the silo protected the ICBM. Initially, the U.S. government purchased 10-22 acres for each silo, with the inner 5 acres protected by an 8-foot (2.4 m) chain link fence topped with barbed wire. Each of the silo's seven floors consisted of approximately 2,000 square feet (186 sq m).[440]

Next up was the Titan ICBM. The Titan I began development in 1955 as a backup in case the Atlas failed, and it would eventually replace the Atlas. At 98 (30 m) feet tall, this ICBM was 16 feet (5 m) taller than the Atlas D, but weighed a staggering 40,000 pounds (18,144 kg) less. Titan I could carry interchangeable W38 or W49 warheads with 1.44 or 3.75 megaton yields. Using refined kerosene and liquid oxygen, three Titan missiles were located at each site, the first ready for launch after fueling within 15 minutes and the

2nd and 3rd within 7.5 minutes. The Titan II, the largest ICBM ever deployed by the U.S. Air Force, stood at 103 feet (32 m) tall, weighed 330,000 pounds (149,6856 kilograms), and could deliver a 9-megaton W53 warhead (3x the combined explosive power of all bombs used in WWII, including the atom bombs used in Japan) 9,300 miles (14,967 km) to its target. In addition to the increased range and more powerful payload, the Titan II ICBM could launch within one minute instead of fifteen, as it had a storable liquid propellant that would not have to be fueled before launch. Holding almost one-third of the Air Force's entire megatonnage, the Titan II served as a central component of the nuclear triad from 1963 to 1987.[441]

Beginning development in 1962, the Minuteman ICBM was the replacement for the Titan missile. It was the first solid-fueled ICBM deployed and provided reliability and maintenance advantages over the Titan missiles, which were liquid-fueled. Initially intended for a 10-year lifespan, the Minuteman missile continues to be in operation at the time of this writing, and there are 400 combat-capable Minuteman III missiles currently in the U.S. arsenal.[442] The Minuteman missiles carry thermonuclear warheads with yields over one megaton. One such warhead, the W56, is said to have one of the most efficient weight-to-yield ratios of any nuclear warhead ever manufactured– 4.9 kilotons per 1 kilogram of its weight.[443] Such a ratio allows one to really grasp the destructive efficiency gains of hydrogen bombs. Little Boy delivered around 15 kilotons, with a weight of 9,700 pounds (4,400 kg). The W56 could produce an equivalent yield (enough to destroy the city of Hiroshima), with *6.6 pounds* (3 kg) of total bomb material... Consider that a gallon of water weighs roughly 8.34 pounds (3.8 kg). W56 produced one kiloton of yield for under one-half pound (8 oz) of bomb material vs. one kiloton per 647 pounds of bomb material in Little Boy, a 1,470% efficiency increase).

The Minuteman III missile brought about another significant innovation in ICBM technology: The Multiple

Independently Targetable Reentry Vehicle (MIRV). The concept behind the MIRV is to create an ICBM with *multiple* warheads that can, upon atmospheric reentry, fly toward separate targets. There are a number of advantages to MIRVs:

1. They can enhance the first-strike capability of a nuclear force, allowing simultaneous strikes of multiple, diverse targets;[444]
2. If so desired, the nation utilizing a MIRV can reduce collateral damage by striking a target with a smaller nuclear warhead deployed from the MIRV, and thereby limit the overall damage to surrounding areas (although it's important to note that the Minuteman III MIRVs still yielded 170 kilotons each, over ten times the explosive capacity of "Little Boy");[445]
3. MIRVs are capable of defeating even advanced antiballistic missile systems, as these systems would likely be unable to destroy separate warheads flying to different locations simultaneously;[446] and
4. MIRVs allow penetration of a hardened target (like a silo or a bomb shelter). One set of calculations estimated the probability of destroying a silo with hardened defenses designed to withstand 300 pounds per square inch pressures using a single 50-kiloton warhead is about 62%. Three 50-kiloton warheads delivered via MIRVs increase the likelihood of silo neutralization to 95%.[447]

From a strategic perspective, the one countervailing disadvantage of MIRVs is that they risk putting all of one's eggs in one basket, as so eloquently put by the Carnegie Endowment for International Peace. During an imminent attack, a nation will be pressured to use their MIRVs as early as possible to prevent significant portions of their nuclear assets from being eliminated since multiple warheads will be located on each ICBM available for deployment. Alternatively, suppose warheads are placed on different ICBMs in different locations. In that case, leaders will have higher confidence that their enemy will be unable to destroy all their weapons as efficiently, thus providing more choices for responding and considering all strategic options (including, hopefully, de-escalation).[448]

In service from 1986 to 2005, the Peacekeeper missile, also referred to as the "MX ICBM," was the most advanced ICBM in the United States arsenal. The missile featured extreme accuracy, taking advantage of an inertial guidance system updated by navigation satellites and the ability to field between 10 to 12 MIRVs. The missile had a range of 7,000 miles (11,265 km) and each of its 10-12 warheads yielded 300 kilotons, for a total yield of 3.6 megatons. Multiple deployment systems were proposed for the Peacekeeper missiles in order to protect them from a first strike by Soviet ICBMs. Silos located over 1,000 feet (300 m) underground were proposed. Another option was launching the missiles from massive transport jets. And finally, moving the missiles between multiple shelters using trucks and railcars was also considered. Ultimately, the missiles were simply siloed because the other options under consideration were regarded as prohibitively expensive.[449]

The Soviets were no slouches in the ICBM category, fielding capable vehicles of their own. As early as May 13, 1946, the Soviets ordered resources to be committed to developing long-range ballistic and cruise missiles and began a program to copy the German V-2 rocket.[450] Pursuant to a Soviet government order signed by Stalin on February 13, 1953, Soviet engineers began creating an intercontinental ballistic missile platform with a target range of at least 4,971 miles (8,000 km).[451,452] By November 25, 1953, the Soviets proposed putting a nuclear warhead on a strategic rocket; such a warhead would be capable of striking the *"probable enemy"* –Europe, the Near and Middle-East, and Japan.[453] What eventually came of these efforts was the R-7 Semyorka, a ballistic platform that had substantially greater lifting capacity than the original U.S. ICBMs. The R-7 was used to launch Sputnik into space, and a variant of the R-7 put the first human being in space on October 4, 1957– Russian Cosmonaut Yuri Gagarin.[454] Humans and satellites weren't the R-7's only payload; it was also designed to carry a fusion warhead with a yield between 3-7 megatons.[455]

Throughout the years, the Soviets, like the Americans, continued to advance their ICBM systems, making them more efficient, using better fuel technologies and advanced guidance systems. Their MIRV systems were particularly intimidating, with NATO codenaming the SS-18 platform "SATAN." The SS-18 Mod 2 / R-36M included up to eight reentry vehicles, each with a warhead in the range of 500 kilotons to 1.5 megatons, for a total yield of up to 12 megatons on a single ICBM. With a range of over 5,500 miles (8,851 km), the SS-18 Mod 2 was placed into service in 1973.[456]

By the late 1970s, the Soviets were developing a land-based ICBM technology that the United States and NATO never ultimately put into service– road-mobile ICBMs on massive trucks. The RT-2PM Topol / SS-25 (NATO codename "Sickle"), had a range of 6,835 miles (11,000 km) and could carry a warhead with a yield of between 550-800 kilotons.[457] The SS-25 was built on the earlier RSD-10 Pioneer / SS-20 (NATO Codename "Saber"), but the SS-25 was an intermediate-range ballistic missile, not an intercontinental version, with a range of only 3,107 miles (5,000 km).[458]

Road-mobile ICBMs have their drawbacks, not the least of which are maintenance and upkeep costs. They're also not shielded and buried underground like siloed weapons, so they are more vulnerable if they can be effectively located. That said, the *if* is important; tracking road-mobile ICBMs is inherently more difficult for an adversary than finding, mapping, and remaining cognizant of the location of ground-based silos. Silos never move; once they're identified, they're a target. For years, this ability to field an ICBM that is difficult to locate and track was an advantage for the Warsaw Pact and the U.S.S.R. and provided them with enhanced deterrent capacity with their land-based ICBMs.

The Bulletin of the Atomic Scientists provides a rundown of the history of Russian ICBMs, which is a good summary of the general trajectory of the advancement of ICBM technology in the 20th Century:

1. **Stage One (1959-1965):** The nuclear superpowers' first foray into ICBMs based on the V-2 platform. These missiles generally had to be stored separately from their fuel and would then have to be fueled in the event of a launch, which would take considerable time;
2. **Stage Two (1965-1973):** Widespread deployment of ICBMs in underground silos, with the ability to store missiles fully fueled in their silos;
3. **Stage Three (1973-1985):** Marked by the introduction of the multiple independently targetable reentry vehicle (MIRV), super-hardened silos to withstand enemy attack, and enhanced missile survivability;
4. **Stage Four (1985-1991):** Enhanced survivability of ICBMS through the use of new mobile (although mobile ICBMs are not used by NATO) and silo-based ICBMs, in an age when missiles could hypothetically destroy even super-hardened silos; and
5. **Stage Five (1991-present):** Reduction of technologies as a result of modern treaties.[459]

From 1959 to the late 2000s, the United States produced approximately 3,160 ICBMs. Warhead yields for these missiles generally ranged from 170 kilotons to 9 megatons. The number of existing U.S. ICBM warheads peaked in 1990 at 2,440. Accuracy improved the arsenal; making a missile twice as accurate allows it to achieve the same probability of destroying a target using an eight times lower yield. Between 1960-1967, the U.S. built 1,180 underground missile silos and 57 aboveground launch sites; North Dakota had the most ICBMs (300), while Vermont had the fewest (2). For the Russians' part, since 1960, they have built at least 5,000 ICBMs, carrying warheads with yields ranging from 220 kilotons to 20 megatons. Soviet ICBMs were generally of higher yield than United States ones, with the cumulative megatonnage of Soviet ICBMs being 3-4 times higher than those of the United States. For perspective, by the 1970s, Soviet ICBM total megatonnage equaled approximately 450,000 Hiroshima-sized bombs.[460]

ICBMs, like each leg of the Triad, have unique advantages. These are as follows:

1. ICBMs are the most numerous kind of delivery vehicle available to the nuclear powers, and they are geographically spread in attack-resistant underground silos;
2. While bombers are bundled into relatively few storage points that may be vulnerable to attack, the distributed nature, number, and attack-resistance inherent in silos means that an adversary would have virtually no chance of destroying all ICBM forces even with a full-scale nuclear attack, meaning second strike capability would be maintained;
3. Because an adversary only has a chance of neutralizing ICBM forces with a full-scale nuclear attack, an assault on the ICBM forces would provide a clear

signal to a nation under attack that they are the subject of a first strike, and can therefore formulate a response with a complete understanding of the scenario; and

4. The cost per delivery vehicle for ICBMs is roughly one-fourth of that of other delivery systems in the Triad.[461]

No delivery system is perfect. The potential drawbacks of ICBM assets are:

1. ICBMs are not mobile unless they are road-based like the Soviets' Topol-M, so once an enemy identifies a silo, it becomes a target. While hardening silos to protect missiles is an option, advanced bunker-busting/penetrating missile technology can likely still destroy silos. This makes ICBMs vulnerable to first strike capabilities; and

2. Due to ICBM vulnerability to a first strike, if the perception exists (even a mistaken one) that a nation is the victim of a first strike from an adversary or there are other tense strategic scenarios at play, a leader may feel inclined to "use them or lose them" in a crisis, therefore making ICBMs prone to encouraging severe escalation, even in the event of a false warning of an impending attack from an adversary.[462]

Chapter 17. The Triad's Final Leg: Into the Depths - The Elusive Nuclear Submarine

"Once more, we play our dangerous game, a game of chess against our old adversary - The American Navy. For forty years, your fathers before you and your older brothers played this game and played it well. But today the game is different. We have the advantage." -Sean Connery as Captain Marko Ramius in "The Hunt for Red October" (1990)[463,464]

On to the most recent advent of the Triad, the nuclear submarine (used under the water's surface) vessel, or SSBN (Submersible Ship/Ballistic missiles/Nuclear powered). While usable submarines have existed since before airplanes, due to the technological challenges created by the harsh undersea environment, planes became prominent in advanced 20th Century warfare before subs. Like our yearning for flight, humankind has been enamored with the idea of undersea travel for about as long as society has been around, with the Athenians even using divers in clandestine military operations.[465] For many readers, it's a safe bet that their first introduction to submarines was in Jules Verne's "*20,000 Leagues Under the Sea*" (1869-1870). But even before Jules Verne imagined the complex, multi-staffed massive *Nautilus* that is in many ways reminiscent of modern subs, humankind was weaponizing them for war.

Before the Revolutionary War, an intrepid inventor from Connecticut named David Bushnell devised a crude submersible vessel called *Turtle*. The *Turtle* used a hand-crank to drive a screw-like oar to move the boat forward and backward underwater, featured air pipes to bring in breathable air from the surface, and ballast tanks that took on or dumped water to dive from, or ascend to, the water's surface. It even featured a primitive torpedo to attack enemy ships. George Washington and Thomas Jefferson encouraged the use of the *Turtle* against British ships. A couple of attempts were made, which ultimately failed, but came close enough that the attempts served as proof that submarines *could* be used as a weapon of war.[466]

By 1800, American Robert Fulton had built a submarine, *Nautilus*, that sported many modern features: adjustable diving planes for vertical maneuvering underwater, a dual propulsion system, and a compressed air system that permitted the submarine to travel for around four hours underwater. During the Civil War, the Confederate submarine *CSS H.L. Hunley*, powered by nine men driving a hand-cranked propeller, became the first submarine ever to sink an enemy ship when it destroyed the Union ship USS *Housatonic* using a spar torpedo. But submarines were still unsafe; the entire crew was lost in that attack for reasons that are still inconclusive. By 1888, the U.S. Navy recognized the long-term potential for submarines as a war weapon and held a design competition for an underwater vessel. This eventually resulted in John Holland creating the prototype for the USS *Holland (SS-1)* in 1900. The sub was powered by a small gasoline engine that turned a propeller while the boat cruised on the surface. The engine would run a generator to charge batteries which were then used for electric propulsion underneath the surface.[467]

This design had its drawbacks. Gasoline is very volatile and flammable, and as you can imagine, considering the existing challenges of battery technology even in the modern era, during the early 1900s, batteries were not viable for subs due to their bulk, inefficiency, and explosive nature. Enter German scientist Rudolph Diesel (sound familiar?), who created an alternative to the gas engine. The diesel engine used more stable fuel that could be stored more safely than gasoline, and the engine didn't require a spark to light the fuel. The diesel engines afforded sailors longer and safer surface cruising, but batteries were still needed for power. Diesel submarines would be used for almost 50 years after the advent of the diesel engine.[468]

In World War I, submarines would become real players in theaters of war. The German *Unterseeboot* ("Undersea Boat") or U-boat, diesel-powered undersea vessels, wreaked havoc on trans-Atlantic shipping while sinking 8,000 merchant vessels and warships, and taking tens of thousands of lives. U-boats could attack and sink ships 20 times their size using a combination of their deck guns and torpedoes. Though the last major power to design and field a submarine in 1906, the Germans' engineering prowess resulted in a vastly superior weapon to other subs, able to travel 5,000 miles (8,047 km) without refueling. On September 5, 1914, the U-boat fleet initiated its first strike on a British cruiser, sinking it and killing 250 sailors. That same month, the U-boat *U-9* would sink three British battle cruisers in the span of a single hour, killing almost 1,500 sailors. It was through the use of the U-boat that the world came to contend with the notion that submarines would become an incredibly dangerous and effective naval weapon. Despite Winston Churchill's assertion that "I do not believe this would ever be done by a civilized power," the Germans initiated unrestricted sub-attacks on merchant ships, even those from the neutral colonies, in the war zone around Great Britain.[469] It was ultimately the fateful sinking by a German U-Boat of the *RMS Lusitania* on May 7, 1915, with 128 Americans aboard and a total of 1,195 souls lost, in combination with the intercepted 1917 *Zimmerman Telegram*, that stated that Germany planned to return to unrestricted submarine warfare after a hiatus, that contributed significantly to the United States' entry into World War I.[470]

Submarine warfare in the context of the Cold War was born of the quiet nature of the Cold War itself. "Warfare" in the Cold War was generally marked not by the dropping of bombs, the firing of bullets, or the leveling of cities, but rather by stealth, espionage, and a global game of chess that relied upon deterring one's enemy, catching the adversary by surprise, and ensuring adequate means of response in the event of the Cold War turning hot. Two particular weaknesses with regard to the first two legs of the Triad are that they are visible and, as a result, ultimately vulnerable. Silos can only be hardened so much in the face of bunker busters, and planes are susceptible to a vast array of countermeasures that can render them useless for deterrence and strike capabilities. An advanced solution exists to this problem: the nuclear submarine.

U.S. Navy Captain Hyman G. Rickover, dubbed "the greatest engineer to ever live" by U.S. President Jimmy Carter,[471] joined the U.S. atomic program in 1946; by 1947, he was placed in charge of the Navy's nuclear-propulsion program and sought to design the first atomic submarine. On September 20, 1954, the first nuclear submarine, the USS *Nautilus* (noticing a theme with submarine naming yet?) AKA *SSN-571* entered the waters of the Thames River in Groton, Connecticut. Its nuclear reactor first powered the propulsion system on January 17, 1955. The reactor was powered by uranium, generating steam that would drive propulsion turbines, allowing the *Nautilus* to travel underwater for virtually unlimited periods with speeds above 20 knots (23 mph/37 km/h).[472]

While early diesel U-boats had a crew of 25 officers and seamen (increased to a 57-person crew in later iterations),[473] a length of 214 feet (65 m), 12 torpedoes, and a finite range, the *Nautilus* dwarfed it. With a length of 319 feet (97 m), a displacement of 3,180 tons, and a crew size of 104, nuclear submarines were already rendering diesel vessels of the recent past obsolete.[474,475] After proof of nuclear submarine viability through *Nautilus*, the U.S. Navy commissioned new subs, including the *Thresher* class, which would, unfortunately, experience one of the most severe Naval accidents in U.S. history when the submarine was lost off the coast of New England, killing all 129 crew members.[476] Although not entirely conclusive, it was determined that the most likely cause of the sinking was a piping system failure in one of *Thresher's* saltwater systems, probably the engine room, in combination with the incredible amounts of seawater pressure, which subjected the interior to significant flooding that caused the electrical systems to fail, then resulting in the loss of power and sinking of Thresher to 8,400 feet (over 1.5 miles / 2.56 kilometers) beneath the surface.[477]

By 1962, the United States had 26 nuclear submarines operating in its fleet and 30 under construction. Nuclear power had forever altered the course of undersea travel.[478] One particular advantage of nuclear submarines is that non-nuclear submarines would have to approach the target near the surface to avoid draining the battery, submerging only just before the attack (this would allow the submarine's pilots to conserve the battery for their escape attempt). After the attack, the submarine would only have 1-2 hours of battery power for escaping at a speed of 7-10 knots.[479] This was problematic against high-speed aircraft carriers and battleships of the era, which could travel at speeds in excess of 30 knots.[480]

Nuclear submarines' virtually unlimited range allows them to pursue fast surface ships before an attack and affords them the ability to freely evade counterattack after an initial attack. In the only known instance of a nuclear submarine sinking an enemy ship, the British *HMS Conquerer* attacked and sank the fast Argentine cruiser *General Belgrano* after 48 hours of pursuit. Due to the inherent advantages of nuclear submarines, and the additional cost of maintaining parallel diesel and nuclear fleets, the U.S. Navy ceased development of its non-nuclear fleet in 1959.[481]

It is important to note here that strategic submarines (those intended for a nuclear strike) existed before the advent of submarines driven by nuclear propulsion. That is to say, there were nuclear-armed submarines in the 1950s powered by diesel engines. During this era, U.S. submarines were equipped with Regulus cruise missiles, and Soviet ships were equipped with NATO designation SS-N-3 Shaddock and SS-N-4 Sark short-range submarine-launched ballistic missiles (SLBMs).[482] The Regulus 1 was the first U.S. Navy cruise missile. It carried a nuclear warhead and flew at subsonic speeds, with a range of 500 miles (805 km),[483] a speed of 600 miles per hour (970 km/h), and carried a W5 warhead with a yield of 47 kilotons.[484] The SS-N-4 Sark carried a 1-2 megaton warhead, had a range of 350 miles (563 km), and had to be launched from the water's surface.[485] All that said, while submarine-launched ballistic missiles (SLBMs) were in use by the 1950s, because of diesel subs' inherent limitations, these deployed SLBMs couldn't yet be considered a genuinely effective component of the nuclear triad.

The United States *George Washington Class* was the first modern strategic submarine, which became operational in 1959. Three hundred eighty-two feet (116 m) long, with a displacement of 5,900 tons, this class of submarines could carry 16 UGM-27 Polaris missiles.[486] Standing 28 feet tall, with a range of 1,380 miles (2,221 km), the Polaris could carry a 500-kiloton nuclear warhead.[487] At 16 weapons on a single submarine, this would account for a total potential launch yield of 8 megatons from this class of sub.

We should pause for a minute to talk about how these SLBM launches work because it's a bit more complicated than simply launching a missile from a silo. Launching from the water's surface is one thing, but launching a ballistic missile from beneath the ocean is another thing entirely because it affords a submarine with stealth, the element of surprise, and the ability to make a quicker escape. Doing so takes a bit of ingenuity, though, as water is a problem for rocket engines, and early detractors weren't certain SLBMs could ever be launched from under the ocean. As it turns out, the various navies of the world devised a way of doing so. SLBMs sit inside missile tubes on board the nuclear submarines under "hard hatches" that protect them from the ocean water. Just before launch, the tube is pressurized to match the water pressure outside the missile tube, and the hard hatches covering the tube are opened, revealing a thin diaphragm that protects the missile. The diaphragm is blown open using an explosive charge; at this point, the built-up pressure inside the tube propels the SLBM through the water toward the surface. Once the SLBM breaches the water's surface, its engine ignites, and it's sent into the sky, where it behaves similarly to an ICBM.[488]

Of course, the United States and the Soviet Union were not the only powers developing nuclear submarine technology. The British completed their first in 1963 with the *HMS Dreadnought*. France built its first, *Le Redoubtable,* in 1971. China constructed its first in 1987, and India and North Korea eventually followed. So far, only seven nations possess nuclear-powered ballistic missile submarines.[489] Over the years, submarines became larger, faster, more stealthy, employed safer reactors, and fielded more advanced weapons systems.

The largest Soviet sub ever built was the *Typhoon* Class, with the first of its class, *Dmitri Donskoy (TK-208),* entering service in 1981. At a truly menacing displacement of 23,200 tons, with a length of 566 feet (173 m), a width of 76 feet (23 m), and a height of 38 feet (11.5 m), these seaborne monsters were indeed a force to be reckoned with. Each Typhoon could field 20 RSM-52 (NATO codename SS-N-20 Sturgeon) SLBMs.[490] This was the largest SLBM ever produced, with a combined weight (including the gas booster and gas generator in the launch tube) of 90 tons (180,000 pounds/81,647 kg) and carrying *ten* MIRVs. Each MIRV could yield up to 200 kilotons. At yields of that level, a single RSM-52 could carry up to 2 megatons of destructive power up to 5,220 miles (8,400 km) away. With 20 of them on board a Typhoon, up to 200 targets could be hit with a total yield of *40 megatons*. Each of these 200 hits would by itself deliver 13 times the energy of the Hiroshima blast, with a single Typhoon sub capable of delivering approximately

2,666 times the destructive nuclear energy of the bombs dropped on Japan.[491]

The United States' mega subs of the Cold War era were not quite as large as the Typhoon, but were still leviathans in their own right. In 1974, the United States began fielding its largest-ever submarine, the *Ohio* Class. With a top speed of over 20 knots (23 miles per hour / 37 km/h), a length of 560 feet (171 m), and a displacement of 18,750 tons submerged, the Ohio Class can carry a total crew of 155, consisting of 140 enlisted staff and 15 officers. An Ohio submarine has 24 total SLBM tubes, allowing it to field 24 Trident II SLBMs.[492] Each Trident II can carry up to 8 Mk 5 warheads with a yield of 475 kilotons, providing a combined Trident II yield of 3.8 megatons. With 24 on board, the Ohio Class could hypothetically deliver a maximum of over 91 megatons of thermonuclear energy to 192 targets.[493] With a fleet total of 14 submarines,[494] this class was originally capable of hitting a maximum of 2,688 targets with a combined total yield of *1.274 gigatons* of energy.

The submarine/SLBM leg of the nuclear triad has powerful advantages:

1. Because of the dispersion of a nuclear submarine force, its concealment, and its ability to constantly change position, avoiding existing detection technologies, an SLBM force is *invincible* while at sea and *cannot* be eliminated with a first strike, guaranteeing second strike capability for whichever nation has them;
2. Because of the invincibility of the SLBM force as a whole (even if an enemy manages to hit a couple of subs), a nation's leader will not face the same "use them or lose them" dilemma that they might face with the other legs of the Triad; and
3. A 1993 Government Accountability Office (GAO) study determined that no current or long-range

technologies show any promise to detect the locations of submarines in the ocean.[495]

There are certain concerns with regard to the submarine leg that necessitate the maintenance of the other legs of the Triad:

1. There are a limited number of nuclear-armed submarines on patrol at any time with a limited payload (although, as illustrated before, the Ohio Class could carry over one gigaton of payload in its fleet, and previous estimates have determined that a foreign superpower could be wholly neutralized with approximately only 400 megatons of yield). *If* (and this is a big if, for the reasons illustrated above) adversaries determined how to locate and destroy these submarines, they could do so with relatively few target points;

2. At certain times, submarines can be difficult to communicate with; and

3. An attack on an SLBM force may be difficult to identify and attribute to a particular adversary. Confusion around who or what is responsible for the destruction of submarines, even in the event of an accident, could lead to misattributed retaliation and catastrophic escalation.[496]

From 1945 to 1991, the Soviet Union produced 727 submarines, of which 235 were nuclear-powered. The United States built 212, with 169 powered by nuclear reactors. This averaged almost 16 per year for the Soviets and 4.6 for the Americans. Though NATO and the United States ultimately bested the Soviets on many fronts, given the Soviets' heavy focus on naval warfare, they set many records, including the world's fastest submarine, the world's largest submarine, and the deepest-diving combat submarine.[497]

Chapter 18. Tactical Nuclear Weapons: A Measured Solution?

"A limited nuclear war is a contradiction in terms. Any nuclear exchange, once initiated, would swiftly and inexorably escalate to the strategic level." -Shyam Saran, Chairman of India's National Security Advisory Board (2013)[498]

As mentioned before, tactical nuclear weapons are intended to win a battle, not a war, and the hope for their use is the prospect of devastating enemy military assets without escalating to a full strategic conflict involving cities, infrastructure, and the level of devastation that could potentially end society as we know it. Of course, fortunately, tactical nuclear weapons have never been used in a wartime scenario, as leaders have carefully judged the risk of escalating to a full strategic scenario as too high, the costs of using tactical weapons thus far outweighing the immediate benefits for any military. The future of tactical nuclear weapons and their uses is as yet unknown. Still, I will discuss their history here by highlighting some of the more unique or ubiquitous tactical nuclear systems.

The Mark 7 was probably the first of the class of weapons that could be considered tactical nukes. It was designed to be compact, relatively small (for the time), and could be used by all three United States Armed Services branches (Army, Navy, and Air Force). Entering service in 1952, the Mark 7 could also be carried on fighter aircraft in addition to bombers, and could be used for either air or ground burst applications.[499] The weapon had various yields, some smaller than the bombs used on Japan (8 kilotons) and some significantly larger (up to 61 kilotons, or approximately four times as powerful as the Hiroshima detonation).[500]

In 1953, the United States tested what was initially named "Able Annie," and later renamed "Atomic Annie," a nuclear cannon. The shot was fired from the Army's M65 280 mm Motorized Heavy Gun, the most massive mobile artillery piece ever constructed by the United States[501] (weighing 86.4 tons when accounting for the entire assembly of two trucks and one gun). The artillery gun was initially designed at the Watervliet Arsenal outside Albany, New York, and manufactured by the Dravo Corporation in Pittsburgh, Pennsylvania.[502] On a side note, this U.S. artillery weapon was tiny relative to Nazi Germany's Heavy Gustav Gun, which had a length of 150 feet (46 m), a height of 40 feet (12 m), a weight of 1,500 tons, a prep crew of 3,800 personnel, and 250 soldiers to fire it. Gustav was initially intended to devastate French bunkers along the Maginot Line in WWII with five-ton explosive rounds and seven-ton armor-piercing rounds, but was eventually used against the Soviets in the Crimean port city of Sevastopol.[503] Atomic Annie's yield was colossal for an artillery shell: 15 kilotons, roughly equal to the "Little Boy" bomb. The juxtaposition between Gustav and Annie illustrates the brutal efficiency of nuclear weapons vs. conventional weapons: a U.S. artillery weapon weighing 17 times less than the German Gustav was able to deliver destructive force many, many magnitudes more powerful. The nuclear cannon was tested only once, on May 25, 1953, during a test codenamed *Grable*, the 10th of the series known as Operation Upshot-Knothole.[504]

By 1953, the United States had also introduced the MGR-1 "Honest John." The Honest John was essentially a nuclear-armed artillery rocket mounted on a moveable truck with a transporter, erector, launcher (TEL) system. While initially nuclear-armed, the rocket system would later also use conventional and chemical warheads. The MGR-1 was not a smart-bomb or a cruise missile, nor did it have any guidance system. It was a classic artillery rocket, which would have to be carefully aimed with the correct direction and inclination. The rocket was initially armed with a W7 nuclear warhead with a yield of up to 20 kilotons (133% of the Hiroshima yield). Eventually, it could accept the W31 warhead, which featured interchangeable yields of 2, 10, and 30 kilotons. Its minimum range was 3.4 miles (5.5 km), with a maximum range of 15 miles (24.8 km).[505]

In the 1950s and 1960s, the United States developed a number of lightweight nuclear devices for tactical purposes. One such system is the Atomic Demolition Munition (ADM), or "backpack nuke."[506] ADMs are intended as tools to deny adversaries access to specific areas and avenues of approach, or to destroy enemy fortifications. ADMs were intended for use by small operations forces. The smallest of these ADMs, known as the Special Atomic Demolition Munition (SADM), weighed around 150 pounds (68 kg) and used the W-54 warhead. The W54 warhead weighed only slightly over 50 pounds (23 kg) and is the smallest known nuclear warhead. Ultimately, the goal of the SADM project was to enable an individual to parachute in using any type of aircraft to a strategic location that could be accessed by the sea wearing one of these ADM backpack nukes, with a second parachutist dropping in for support purposes. The two-person team would then arm the device, at which point they would be retrieved and whisked away by a boat or submarine.[507]

The Soviets were also in the business of mini-nukes. In 1997, it was revealed that the Russians developed "suitcase nukes" for the Soviet intelligence service, the KGB. These suitcase bombs allegedly featured a yield of one kiloton. While no Tsar Bomba, such a yield could do massive damage in a downtown urban center. Russian national security advisor Alexander Lebed has previously claimed that the Russians lost track of over 100 suitcase nukes. Russia has denied his claims.[508,509]

Back to U.S. weapons – The Davy Crocket system is perhaps the most well-known delivery system for the W54 warhead. The Davy Crocket would launch a W54 warhead from an XM-388 projectile (sort of akin to a rocket-propelled grenade system [RPG]). The projectile would be fired from a 120-millimeter (XM-28) or 155-millimeter (XM-29) "recoilless rifle." The projectile weighed only 76 pounds. Its launchers could hit targets at a maximum range of either 1 mile (2 km) or 2.5 miles (4 km). Yields were as low as 10 tons of TNT, up to 1 kiloton. For perspective, even at its lowest yield of 10 tons of TNT (.01 kilotons), while this is no megaton-range nation-state destroyer by any means, the destructive power would still have been 2-4 times more potent than the truck bomb containing 7,000 pounds (3,175 kg) of ammonium nitrate detonated outside the Alfred P. Murrah federal building in 1995 by the domestic terrorist Timothy McVeigh. That attack destroyed the federal building, killed 168 people, injured hundreds, incinerated dozens of cars, and damaged 300 nearby buildings.[510,511]

The purpose of the Davy Crockett system, which could be operated by as few as three personnel working together, was to provide smaller Army units with the ability to devastate larger Soviet armored forces, primarily those on the frontlines in Europe, during a hypothetical invasion. The rocket was unguided and spin-stabilized. Deployed from 1961 to 1971, over 2,100 were produced.[512]

Perhaps the most versatile and successful of all the tactical nuclear weapons is the United States B61 family of nuclear gravity bombs. By the early 1960s, U.S. officials were seeking bombs that were smaller, lighter, and could be carried externally at high speed (in case they were mounted outside a plane, especially in the case of fighter aircraft). An ability to control the yield of the weapon on delivery ("dial-a-yield") was also favorable to allow adjustment between multiple tactical and/or strategic scenarios and to effectively employ a weapon "against targets near populous areas where collateral effects must be minimized." Also on the list of requirements was a weapon that could initiate earth shock or cratering to devastate hardened bridges, missile silos, or runways.[513]

Over the years, a total of 15 different versions of the original B61 design have been developed by Los Alamos National Laboratory. The B61's delivery systems include free-fall, retarded airburst (physical methods to slow the bomb from freefalling), or a "laydown" mode from aircraft flying as low as 50 feet, using a parachute. The B61 could even be found on aircraft carriers at one point. It also comes in "bunker busting" varieties, allowing it to penetrate the ground and destroy hardened targets like underground silos with a far lower yield than that which would be required of regular thermonuclear weapons. By way of example, a B61-11, with a hardened steel case and nose cone that add 450 pounds (204 kg) of weight to the weapon, can burrow into frozen solid soil up to 20 feet (6 m), then delivering 400 kilotons of yield into its target. So useful is the B61 that it remains the only nuclear bomb deployed outside U.S. borders today (aside from SLBMs on nuclear submarines); tactical versions are still housed in five NATO allied nations – Germany, Italy, Belgium, the Netherlands, and Turkey. The Obama Administration authorized a life extension of the B61 family through the B61-12 iteration, enabling improved accuracy to permit the bomb to hit targets that would have required a higher yield in the past.[514] The B61-12 saw its first production unit in November 2021 and is expected to be produced until around 2026.[515]

The Soviets, too, developed several tactical nuclear weapons. Likely the most well-known is the Scud rocket (SS-1 "Scud"), designed shortly after WWII and based on the V-2 platform. The conventional version of the Scud has proliferated around the world to over 20 countries, such as North Korea, Iraq, Poland, Afghanistan, Syria, and others. It can take warheads of chemical, high explosive, and nuclear varieties, with a range of 118-341 miles (190-550 km). Perhaps best known for Saddam Hussein's launch of conventional warhead-tipped scuds at Israel at the beginning of the Persian Gulf War, the R-11 "Scud A" was earlier equipped with a relatively low-yield warhead in the 50-kiloton range in 1958. Ultimately, conventional scud missiles always suffered from accuracy problems, although accuracy would be far less of a concern with a 50-kiloton nuclear yield.[516,517]

Another well-known Soviet tactical system is the short-range ballistic missile (SRBM) *OTR-21 Tochka (NATO codename SS-21 "Scarab")*. Entering service in 1975, the Scarab was launched from a road-mobile platform and could accept conventional high explosive, chemical, electromagnetic pulse (EMP), or nuclear warheads. The ballistic mode of the weapon would allow increased range and speed for its missiles, while the cruise missile mode afforded higher accuracy and stealth attributes. So successful was the Scarab that Russia is believed to possess 310 nuclear warheads for all versions of the Scarab, even in contemporary times (it once had 1,200). Submunition (features of the warhead) types include anti-tank, anti-personnel, and anti-runway. The nuclear warheads available for the platform are believed to have selectable yields of between 10 to 100 kilotons. The transporter-erector-launcher (TEL) vehicle that carries weapons for the platform is hardened against nuclear, biological, and chemical (NBC) weapons, and has amphibious capabilities, with the option to travel on land at 37 miles per hour (60 km/h), or in water at 5 miles per hour (8 km/h). Its missiles can hit targets up to 75 miles (120 km) away.[518]

Though these have been scaled down in the recent era, tactical nuclear weapons were incredibly ubiquitous during the Cold War. At one point, the United States had 7,000 tactical nuclear weapons deployed in Europe.[519] Even as of very recently, Russia is said to maintain over 5,000 non-strategic nuclear weapons, possibly 2,000 of which are in a state of tactical combat readiness.[520]

Chapter 19. Command and Control: The *Nuclear Football* and Control Systems for Strategic Arms

"This is the voice of world control. I bring you peace. It may be the peace of plenty and content or the peace of unburied death. The choice is yours." -"*Colossus*," Artificial Intelligence Strategic Nuclear Arms Control System from "Colossus, the Forbin Project" (1970)[521,522]

We should briefly discuss the control systems that are used for strategic arms in order to regulate their deployment and prevent rogue deployment/detonation. After all, these weapons and their use is so sensitive that a single detonation could result in uncontrolled escalation to total nuclear apocalypse; to some degree, it's both a combination of luck and adequate fire control systems that have prevented such an outcome over the years, despite how ubiquitous and widespread nuclear weapons now are.

By this point in the material, we know that nuclear weapons are composed of conventional explosives, fissile material, and the necessary detonator to set off the conventional explosives to begin the chain reaction of fission or fission+fusion. The original protection measure to prohibit unintentional or rogue launches was the concept of "separables." The basic idea with separables is that a weapon does not exist until all of its components are assembled. Therefore, in the early years of atomic weaponry, the weapon's fissile materials (core) would often be kept separate from the conventional explosive portion of the weapon until the last moment, such as in the cargo bay of an aircraft en route to a target just before deployment. This system became impractical during the Cold War when seconds and minutes could count; weapons reassembly took time, and in the event of a first or second strike, a nation-state would need to deploy its bombs very quickly. Therefore, separables were no longer practical.[523]

In the 1950s, before portable computers were available, nuclear weapons were protected by safety devices that were integrated into the electrical contacts controlling the weapons. Firing signals were isolated from critical components by an air gap, and safety switches were operated by DC motors driven from various power sources. These systems relied upon certain assumptions about how weapons would operate in certain environments, such as the assumption that electrical faults to ground would "dud" the system or that soldered joints would melt in the event of a fire, thus creating open circuits that would prevent the bomb from exploding. Some of these assumptions were wrong, though (in fires, wire insulation and circuit boards become unpredictable, and electricity relatively easily crosses the terminals of a safety device in this scenario); modern designs include redundant safety measures specifically engineered to prevent accidental detonations and ensure that the weapon will respond reliably, even in unpredictable or unusual scenarios.[524]

Over time, the solution took the form of the "3 Is," which are *The Principles of Enhanced Nuclear Detonation Safety (ENDS):*

1. **Isolation:** Critical components of the nuclear detonator are *isolated* from their surroundings by putting them in a special region enclosed by an energy barrier (this is the *exclusion region*). This barrier blocks all energy levels that could cause a yield greater than four pounds (1.81 kg) of TNT. Perfect isolation, though, would render the weapon ineffective; some energy must be allowed inside the exclusion region to permit

detonation when necessary. So, an energy control element is required– when it's opened, it stops energy transfer to the exclusion region; when it's closed, it prevents energy transfer and stops the weapon from detonating.

2.**Incompatibility:** This concerns the safety device (hardened against the environment) that allows energy transfer sufficient to cause detonation. The safety device is a sort of combination lock that uses a complex pattern of binary pulses to unlock the safety and permit energy transfer into the exclusion zone. These safety patterns are *incompatible* with naturally occurring signals and are known as *unique* signals. Many nuclear weapons have two safety devices installed. The chances that accidental generation could occur in these codes from naturally occurring sources is far less than one in one million.

3.**Inoperability:** This is where "weaklinks," as in "a chain is only as strong as its weakest link" come in. Since the barrier/safety device can only survive up to a certain point (think a jetfuel fire, which could totally destroy the safety device), weaklinks are intentionally built into a weapon to make it inoperable. The failure point of the weaklinks is designed to be below the threshold of the safety device. For example, certain capacitors are required for the detonator in a bomb to function normally; these capacitors are designed to melt before the barrier/safety control device fails, thus rendering a weapon inoperable in an extreme scenario like a fire breaking out inside a silo or a bomber.[525]

Complementing the *weaklinks* in the control system are the *stronglinks*, which are the safety devices mentioned above. The stronglinks/safety devices are hardened against rugged environments and are not readily triggered by vibration, fire, shock, or electromagnetic fields. Stronglinks, even in the modern era, are typically solenoids or motors that look for a 24-event pattern that is not easily generated or expected in abnormal environments. These "single try" devices lock if the correct pattern is not received the first time, to prevent accidental misfires. When possible, two stronglinks are arranged in a series along a path with weaklinks between them, ensuring independence and preventing common mode failures. As new weapons have been authorized over time, the Department of Energy (DOE) Oversight Committee improved the safety of each successive weapon. New weapons designed at both Los Alamos and Lawrence Livermore national laboratories incorporated improved arming, fuzing, firing, and safety systems. Assessment of weapons safety is performed by an independent nuclear safety group that does not work on the weapon's design and a final safety evaluation is performed by military and DOE officials.[526] Ultimately, these complementary weaklinks and stronglinks are designed to ensure that accidental nuclear detonations due to component malfunction will occur with chances of less than 10^{-9} in normal conditions and 10^{-6} for abnormal situations.[527]

Modern nuclear weapons are also equipped with "Environmental Sensing Devices" (ESDs), intended to prevent both accidental and unauthorized detonations. ESDs are designed to detect and react to the typical physical environment expected for a particular weapon. Nuclear warheads on missiles, for example, typically experience rapid acceleration, a period of free fall, and then deceleration. The ESD in such a weapon will not arm a warhead until such conditions are detected. Therefore, if someone stole the warhead, they could not detonate it unless the launch system (for example, a jet bomber) was also stolen.[528]

More modern fire control systems are known as *Permissive Action Links* (PALs), which are powered long-term by radioisotope thermoelectric generators (through alpha decay of plutonium-238).[529] They consist of coded switches installed in nuclear weapons that prevent the weapons from being used by those without the proper codes, such as disgruntled/rogue officers of a nation-state, terrorists, saboteurs, or other entities otherwise acting alone for nefarious purposes. Until the mid-1960s, anybody who got their hands on a nuclear weapon could detonate it because PALs were not yet in wide use. In the early years of PALs, there was intense debate about their installation on nuclear weapons. As U.S. military leaders initially had the ability to deploy nuclear weapons without direct Presidential authorization, some objected to the installation of PALs, arguing that doing so would hamper their ability to respond to a crisis in time, thus gravely endangering America's national security.[530]

PALs are intended to meet the following objectives (each increasingly more technically complex than the last), as outlined in National Security Action Memorandum 160 (June 6, 1962):

1. Safeguarding weapons against actions by an individual psychotic;
2. Meeting the legal and political requirements of U.S. control [of nuclear weapons];
3. Maintaining control against the unauthorized use of [nuclear] weapons by our own or allied military forces under conditions of high tension or actual military combat; and
4. Assuring that weapons could not be used if forcibly seized by an organized group of individuals or by a foreign power.[531]

Dr. Harold Agnew, a member of the Manhattan Project who trained under Enrico Fermi and directed the Los Alamos National Laboratory, is credited for pioneering the idea of installing PALs on nuclear weapons, and for overseeing NATO's installation of the devices on its nuclear weapons.[532] A particular concern of US leaders was that United States' atomic assets were distributed to NATO member countries, and command and control of those weapons was not regarded as entirely secure. After all, political revolutions, military coups, and treason are a staple of every era.

In a security shortfall that likely contributed to Agnew's development of the PAL, an inspection team in Germany noted that there stood on the German runway a German (or Turkish) quick-reaction alert airplane (QRA) loaded with nuclear arms and a foreign pilot in the cockpit. For context, QRA is a state of readiness and modus operandi of air defense maintained at all times by NATO allies, putting crew on standby 24 hours a day, seven days per week, 365 days a year to react to a threat with a moment's notice. The QRA plane in Germany was in a state of readiness, and was prepared to take off at any moment, at the earliest warning. The only United States control mechanism evident at the airstrip was a single, 18-year-old sentry with a carbine standing on the tarmac. When inspectors asked the sentry how he would maintain control of the nuclear arms if the pilot decided to scramble the plane, he replied that he would shoot the pilot. Agnew directed him, instead, to shoot the bomb.[533] The idea behind Agnew's directive is that it would likely be unclear whether the pilot had gone rogue, or if he had been given a directive from the German military to scramble and launch. In the latter scenario, the bomb, not the pilot, would need to be disabled, because the American sentry could be quickly killed, the dead pilot replaced, and control reasserted over the weapon by the rogue foreign power.

PALs and similar technology were not initially included in the submarine leg of the Triad, as there is relatively little risk of capture, no foreign nationals have control of the weapons, and communications with subs can be problematic (meaning a PAL could actually hamper first or second strike capability in a crisis scenario). However, instead of PALs, a complicated set of protocols that use the PA system, different keys, and participation of most of the crew are necessary to actually initiate a launch from a sub. Once on shore, the PAL is used if the weapon needs to be transported. A coded switch system has recently been added to the submarine fleet to provide additional protection against a rogue launch. PALs are used in bombers; the PAL is unlocked before bomber takeoff and can then be activated once the bomber's crew receives the authorization code before launching/dropping a weapon.[534]

There are various categories of control access affixed to nuclear bombs, each successively more advanced than the prior generation. The earlier versions had 3 or 4-digit mechanical locks, and later ones required input from two individuals for added security. They blocked certain components of the weapon to prevent unauthorized use. As the technology advanced, newer PALs accepted longer keys, had limited-try features, and could enable different functionalities like training modes, yield selection, and weapon disabling. The most advanced version used a 12-digit key for added complexity and security.[535,536] Much about PALs remains classified, so the public does not know precisely how they work. This is for a good reason– the more an adversary (especially a sophisticated one, like a nation-state) knows about these security measures, the more likely it is that adversary can defeat them, circumvent them, or reverse engineer them for the aforementioned purposes. What is likely is that access control systems have various levels of verification that ensure the operator of a bomb is receiving a genuine order from the authorized authorities. It is also likely that PALs are so embedded into modern nuclear devices that tampering with them will render the weapons completely unusable, requiring them to be sent back to one of the national laboratories for recommissioning. This disabling feature is critical, because if PALs could be easily circumvented, they'd be useless in stopping rogue actors and hostile agents. PALs have become more sophisticated over time, but even in early electromechanical versions of PALs, it's possible they utilized sophisticated encryption, perhaps similar in some ways to the WWII German *Enigma Machine* system (cracked by British Dr. Alan Turing's *Bombe*),[537] which, even without using digital cryptography, had 158 quintillion possible combinations.[538] For *Enigma,* at the rate of one try every second, assuming a lifetime of 100 years, it would take over 50 billion lifetimes of key tries to find the right combination. One cannot decrypt a system without access to it; combined with tamper-proof system disabling and limited-try functionality, this presumably makes PALs very secure.

No system, no matter how complex, is perfect. In a stark warning about the risks and failure points of our access control systems, the late Dr. Bruce Blair, a nuclear arms expert and former operator at a Minuteman ICBM launch silo, noted that he observed during his time there that Strategic Air Command (SAC) in Omaha had quietly ordered the PAL locks to be set to all zeroes: "00000000." In fact, the launch checklist had directed personnel to verify when entering a code that none of the digits had been accidentally dialed any other number but "0." President Kennedy and the Secretary of Defense had been completely unaware that SAC had circumvented the PAL architecture; this choice by SAC indicated that its sole priority was ensuring a launch could proceed unimpeded in a crisis scenario at all costs, but in so doing, effectively disabled the access block that is intended to prevent rogue actors, terrorists, foreign agents, or other nefarious persons/entities from initiating a launch. This was incredibly irresponsible and fundamentally reckless behavior that could have resulted in a global disaster. Fortunately, the problem was rectified, at least in this particular case, by the late 1970s.[539]

So now that we understand the basics of the PAL, let's talk about "The Big Red Button" on the President's desk. As it turns out, there isn't such a thing in the United States; not physically, at least. In reality, there is a system of command and control that begins with the President's order and is then followed up by a series of verifications and communications that result in a launch order being carried to fruition. Traveling with the President at all times is a military officer with a briefcase, nicknamed the "Nuclear Football," AKA the "Presidential Emergency Satchel," which contains a satellite communications radio and handset, the nuclear codes (known as the "*Gold Codes*"), a list of secure bunkers where the President can be sheltered, and the President's decision book, which helps serve as a guide to the President should they ever make the fateful decision to authorize a strike, as well as instructions for use of the Emergency Broadcast System.[540] A spare Football is stored at the White House, and a third follows the Vice President at all times in the event that the President were to become incapacitated or removed from office via the 25th Amendment.[541] The Nuclear Football has been around since the Kennedy Administration.[542]

The military officer traveling with the President is sometimes (incorrectly) referred to as the *Yankee White*. In reality, this is a special security screening program. Such military officer must obtain clearance through a special access program (SAP) which goes beyond a basic security clearance. Though this is (sort of) a misnomer, movies and television shows often refer to these SAPs as "above top-secret clearances." One such SAP is the Yankee White, a special administrative clearance required for those working around the U.S. President, and it requires a single-scope background investigation (SSBI). This is an incredibly detailed security check that looks into every facet of a person's life, proclivities, associates, and history. In short, the Yankee White ensures that the individual carrying the

Football is loyal to the Office of the President and to the United States itself.[543,544,545]

The officer carrying the Football is always within the President's immediate access. The officer travels on the same elevators, stays on the same hotel floor, and is protected by the same Secret Service agents as the President. But the carrier of the Football does not have separate decision-making power from the President. The President's authority to authorize a nuclear strike rests with the President and the President alone.[546]

Once the order is issued from the White House, a "two-person" access concept is frequently employed down the military chain of command. The "two-person" modus operandi is likely familiar to the reader. TV shows and movies featuring a nuclear launch often show two separate personnel in a silo turning keys on opposite sides of a room to ensure access is authorized before a launch. For example, the Air Force has instructed personnel charged with safeguarding nuclear weapons in Air Force custody to use the "Two-Person Concept" to ensure that one person cannot tamper with a nuclear weapon, system, or component in an improper or unauthorized manner. "No-lone zones" must be identified where personnel cannot enter alone and must enter with at least two personnel together at a time. In the Air Force, persons working with nuclear weapons must be certified via the Personnel Reliability Program; must possess knowledge of the nuclear surety tasks they perform; must have the ability to detect incorrect or unlawful acts or procedures quickly; must complete Air Force nuclear surety training; and must be designated to perform the task.[547]

In Minuteman II silos, launches could take less than five minutes, and the two-person rule was built into the launch hardware itself. The process would play out as follows in one of these silos:

1. The President initiates a strike command. [This could be in response to warning of an attack from early-warning satellites or ground radars];
2. The combat crew in the silo's Launch Control Center hears an alarm, and then a coded message comes through over the loudspeakers issuing the command to launch. The crew then verifies that the message is real and unlocks a small red steel "Emergency War order" safe above the deputy commander's panel. Inside the safe are two launch keys;
3. The crew strap into their chairs; if there's a nuclear strike, they need to stay physically stable to operate the controls. Without a seatbelt, the physical shock of the blast could incapacitate them. They insert the keys into their consoles to begin the final countdown. The commander will call out codes to verify the launch message; the deputy commander will repeat them. The crew members will turn their keys simultaneously. Employing the two-person safeguard, the keys are located 12 feet apart, making it impossible for one individual to trigger a launch. Outside the Minuteman facility, an airborne command officer must concur with the launch order for it to proceed; and finally
4. A "LAUNCH IN PROCESS" display will light up if the launch sequence is initiated correctly. A "MISSILE AWAY" indicator will light up for each missile launched. Missiles will hit their targets within 30 minutes, regardless of their location.[548]

As part of the verification system for the codes, up until 2019, U.S. Strategic Command's Strategic Automated Command and Control System (SACCS) used floppy disks from the 1970s for code authorization in its nuclear systems. While this may seem highly antiquated (and in some senses, it is), the floppy disks exhibited great utility from a security perspective because they were an older technology that was not connected to the internet, and, thus, was unhackable except through physical means. Since then, a shift has begun toward using a "highly secure solid-state digital storage solution." Per defense officials at the time of the switch to newer technology, the system was "made up of technologies and equipment that are at the end of their useful lives... The system is still running on an IBM Series/1 Computer, which is a 1970s computing system, and written in assembly language code... Replacement parts for the system are difficult to find because they are now obsolete."[549,550]

The assumed scenario in which the President would authorize a nuclear strike would be in response to an incoming threat, AKA issuing a "second strike" (although the President is not precluded from authorizing a first strike per U.S. law). Responsible for informing the President and the military chain of command of an incoming strike from a foreign adversary is the "North American Air Defense Command" (NORAD), which was formed in September 1957 and is currently headquartered at Peterson Air Force Base in Colorado Springs, Colorado, as a bi-national command between the United States and Canada. NORAD's purpose is to centralize the missions of aerospace warning and aerospace control for the North American continent. NORAD monitors for artificially-made objects in space and detects, validates, and warns of attacks against North America by aircraft, missiles, or space vehicles. In short, NORAD protects the skies in and around North America, and as of 2006, it expanded its role to include a maritime warning mission. NORAD uses a sophisticated array of satellites, ground-based radar, airborne radar, and fighters to detect, intercept, and, if necessary, engage any threat to North America. If an order to strike ever makes its way from the White House to the silos, bombers, and nuclear submarines, it will likely be NORAD's systems that spurred the President into action.[551,552]

The Soviets also had a strategic command and control system built on the "launch-on-warning" (LOW) principle. The LOW system calls for a retaliatory nuclear weapons strike as soon as satellites and/or other warning systems detect an incoming first strike. The United States utilized it for a time but has since switched to a procedure that emphasizes confirmation of a detonation before launching a retaliatory strike. In the event of a LOW scenario, Russian command is set up to obtain launch authority within 10 minutes from the President, the Defense Minister, or the Chief of the General Staff via the *Cheget* nuclear suitcase. The *Cheget* is similar to the United States Nuclear Football, but wasn't introduced until some time in the 1980s. The Russians have two options for issuing a nuclear strike:[553,554,555]

1. Codes can be sent to individual weapons commanders, who would execute the launch procedures; or
2. The General Staff can direct missile launches directly from command centers in the Moscow vicinity, bypassing the subordinate chain of command and missile launch crews.

Launch-on-Warning has advantages but also alarming disadvantages. The advantage is that it allows a nation that is on the receiving end of a first strike to get its bombers in the air, its SLBMs out of its submarines, and its ICBMs out of its silos before the suspected first strike hits. On the other hand, it does not consider the potential for false alarms, such as the scenario where faulty radar software or human error falsely indicates to military commanders or national leaders that their cities, military installations, and infrastructure assets are about to be obliterated. This can result in a second strike that could potentially end civilization as we know it without provocation, the simple but indescribably costly result of a mistake.[556]

Point two in the Russians' options for a nuclear strike, "The General Staff can direct missile launches directly from command centers in the Moscow vicinity, bypassing the subordinate chain of command and missile launch crews," is an important one, because it concerns a much-discussed but often misunderstood system. This option has been the cause of extreme global concern, as it allegedly created a Soviet system called *"Perimeter,"* also known as *"Dead Hand."* Per media accounts, the system was intended to allow a comprehensive second strike against NATO in the event that Soviet leadership was completely wiped out. Allegedly, Dead Hand is a computer system connected to Russia's strategic nuclear arsenal. In the event that it is activated, it first tries to reach the nation's leadership to confirm that they're alive and communication lines are available; if it is unable to do so, it takes over. If no command hierarchy is detected, Dead Hand then sweeps its inputs for indicia of a nuclear attack – changes in air pressure, extreme light, and radioactivity. If the system determines an attack has taken place, it launches all of the strategic nuclear weapons in its control at targets in the Northern Hemisphere. While the Russians are opaque about the current status of Dead Hand, some officials have indicated as recently as the 2010s that the system remains active today.[557,558,559]

But these media accounts may be inaccurate. Because not much is known publicly about Dead Hand (as is the case with most sensitive nuclear information at all levels and in all nations), it's difficult to definitively characterize its capabilities. But there are some insights we can glean from the late Soviet Colonel Valery E. Yarynich, who allegedly worked on the system and published a 2003 book about the Soviets' nuclear secrets called "C3: Nuclear, Command, Control, Cooperation." According to Yarynich, Dead Hand situates mid-level officials in a super-hardened underground facility in a remote region. This facility, known as the "radio command and control center," is equipped with communication means to the outside world. A launch could only occur in the event the following three conditions were all met. First, the General Staff must release preliminary authorization for the mid-level personnel to gain control of Russia's strategic nuclear arsenal from the bunker. Next, the bunker must experience a complete loss of contact with the national command authorities (NCA). Finally, the system must detect reliable information about nuclear detonations using various types of sensors (visual, seismic, etc.)[560]

      If all three conditions are met, Dead Hand launches communications missiles that transmit launch orders to command posts at every level; these communications missiles also arm nuclear missiles all over Russia. In this way, arguably, Dead Hand *reduces* the probability of a tragic mistake by providing more time for second strike retaliation to be considered (it arguably reduces the immediate need to launch-on-warning when a first strike is detected, because even if Russian leadership were completely decapitated, the nation would still have the option to retaliate via Dead Hand). According to Yarynich, the system is only partially automated, so there is always a human element present in the system directing its actions. Even so, one cannot be blamed for being concerned that the Soviets created a system intended to launch all of its nuclear assets at once if it determined its leadership was neutralized. The facility that the operators would be located in is highly isolated and hardened; what if communications infrastructure failed in a crisis and the humans inside concluded through false sensor readings that (1) they cannot reach their leadership, and therefore leadership is dead (leadership in this hypothetical is actually still alive), and (2) a nuclear attack has occurred based on sensor readings (the readings in this hypothetical are false)?[561] In that scenario, you could have a world-ending false second strike for no reason.

Part IV. The Price of Peace: Radiological Risks and Broken Arrows

Chapter 20. Tasteless, Colorless, Without Warning: Radiation's Effects on the Human Body

"Plutonium's position is frighteningly high on the lethal ladder. A few millionths of a gram (or a few micrograms) distributed through the lungs, liver, or bones may increase the risk for developing cancer in those organs. Airborne, soluble chemical compounds of plutonium are considered so dangerous by the Department of Energy (DOE) that the maximum permissible occupational concentration in air is an infinitesimal 32 trillionths of a gram per cubic meter!"[562]

So far, our primary focus has been on the *potential* costs of nuclear weapons, aside from the actual costs illustrated by Hiroshima and Nagasaki. The ultimate downside of atomic weapons– bombs intended to *deter* war, as opposed to wage it, to *ensure* peace as opposed to shatter it, is that if they are ever used again, the only peace they may wind up ensuring, in the long run, is that of "unburied death," to put it as eloquently as *Colossus*. But *potential* costs are not the only considerations we have to grapple with; we know through experience that there are actual costs. Several incidents have illustrated in horrid detail the dangers of manipulating the atom, many of which are related to the radiation shed by radioactive materials.

As early as innovators became interested in things such as X-rays in the late 1800s, they were experiencing the adverse effects of ionizing radiation. Ionizing radiation includes alpha particles, beta particles, gamma rays, X-rays, neutrons, high-speed electrons, high-speed protons, and other particles that can produce ions. Forms of radiation that are *not* ionizing are radio waves, microwaves, visible radiation, infrared radiation, and ultraviolet light. Ionizing radiation is much more energetic. When it passes through materials (air, water, and living tissue), it leaves enough energy to produce ions by destroying molecular bonds and removing electrons from atoms and molecules. Removing these electrons ("electron displacement") can lead to changes in living cells, which can be harmful, causing many side effects, including, but certainly not limited to, cancer.[563]

Common types of ionizing radiation sources are listed below:

Particle	Emitter	Type
Alpha	Radium, Radon, Uranium, Thorium	Energetic Helium Nuclei
Pure Beta	Strontium-90, Carbon-14, Tritium, Sulfur-35	Energetic Free Electrons
Gamma	Iodine-131, Cesium-137, Cobalt-60, Radium-226, Technetium-99m	High Energy Protons Emitted from an Atomic Nucleus
Neutron	Californium-252, Nuclear Fission, Nuclear Fusion,	Energetic Neutral Particles Emitted or Spallated (When a

	Spallation	Bombarded Nucleus Breaks Up) from Atomic Nuclei

564

Terminology is important for our current purposes, so distinguishing exactly *what* ionizing radiation *is* will be important. Ionizing radiation carries sufficient energy to release one or more electrons from their positions in atoms in matter through which it travels. The electrons and atoms may immediately recombine (and most do), but if they don't, and one or more electrons are lost, the atoms are left with a positive charge, at which point they are "ionized." An ionized atom or molecule may behave differently in matter, which is especially relevant for biological, living tissue.[565]

We discussed radioactive decay (the spontaneous release of energy in the form of various forms of ionizing radiation) in the first chapters of this book, as its discovery led scientists to better understand how atomic physics work. Most atoms on Earth are stable, but some are naturally unstable (radioactive) and subject to radioactive decay, or become radioactive as a result of artificial means. The energy required for radioactive decay is far higher than what's needed for ionization, so almost all products emitted through radioactive decay are ionizing energy. This becomes particularly concerning when it comes to humans and animals, as ionization can *significantly* alter the chemistry of biomolecules– important among these, deoxyribonucleic acid (DNA), the building blocks of organic tissue. If atoms that compose DNA are ionized, one or more single or double-strand DNA breaks can occur. If the cell survives, but the DNA breaks aren't repaired or are repaired incorrectly, this can cause abnormal cell behavior, like cancer growth, which negatively impacts the living tissue.[566]

In the early years of X-rays, after Röntgen's discovery, it was not yet fully understood that radiation exposure had to be carefully controlled and limited. Shortly after his discovery, the technology became popular with the general public and the medical community, which used it for medical imaging. At the time, there were even "X-ray slot machines," which allowed customers to see the bones in their hands. During that period, X-ray components were so easy to obtain that people could even make X-ray images in their own homes. One man used a 10-hour exposure to create a radiograph of his wife's broken hip; she later developed burns, hair loss, and blistering. Thomas Edison, who also became interested in X-rays, had a lab technician, Clarence Dally, who was exposed to so much radiation that he eventually had to have both of his arms amputated; he died of metastatic skin cancer at only 39 years of age.[567] Edison changed his mind about his experiments after he witnessed how the X-rays ravaged Dally, who underwent multiple amputations and skin grafts after his skin began to deteriorate and his hair fell out. Edison would later remark, "Don't talk to me about X-rays... I am afraid of them." He shared the sentiment about other radioactive materials, "I stopped experimenting with [X-rays] two years ago, when I came near to losing my eyesight, and Dally, my assistant, practically lost the use of both of his arms... I am afraid of radium and polonium too, and I don't want to monkey with them."[568]

As previously mentioned, another well-known innovator in the field of atomic science learned the dangers of radioactivity the hard way. By 1920, Marie Curie had developed a double cataract; today, we know radiation can cause that disease. Her vision became so affected that she had to write her lecture notes in massive letters and have her daughters drive her around. She commented on her troubles in November 1920 and wrote to her sister Bronya, "Perhaps radium has something to do with these troubles, but it cannot be affirmed with certainty." By the mid-1930s, she became too sick to work, but specialists suspected tuberculosis. In reality, it was aplastic anemia; she could not produce blood cells because her bone marrow was so damaged by radiation.[569] As mentioned before, her remains were so radioactive that she had to be buried in a lead-lined coffin. Some of her papers were so severely contaminated with radiation that only replicas of them are featured in the *Musée Curie* in Paris. To this day, they remain too radioactive to be near museum guests.[570]

The "Radium Girls" are another harrowing example of the early misunderstanding of radiation's effects on the human body. At the beginning of World War I, several factories were constructed around the United States to assemble watches and military dials. The watch dials were painted with radium because it glows in the dark (due to its radioactivity), and watches that glowed in the dark would provide a tactical advantage for soldiers. Hundreds of women were hired for the job because their hands were small and thus able to work with the small watch components more easily and precisely. At the time, radium was considered safe because it had already been used to treat some forms of cancer (by destroying cancer cells) shortly after it was discovered. The girls became known as "ghost girls" because the radium would make their clothes, hair, and skin glow; some would even paint their teeth with radium or wear their favorite dresses to work to make them shine when they went dancing after work. What's worse, the watch dials were so small that the women were encouraged to use their lips to bring their paint brushes to a fine point; managers assured the women that the radium was safe.[571]

Amelia Maggia was the first of the "Radium Girls" to fall ill. It began with an ache in one of her teeth, which subsequently had to be removed. Then another. Soon, painful, bleeding, pus-filled ulcers replaced the areas where her teeth used to be. Eventually, the sickness spread through her mouth and lower jaw, which then had to be removed. Soon, it spread through the rest of her body; she died on September 12, 1922, a victim of a massive hemorrhage. Similar fates befell more of the women. For two years, their employer, the United States Radium Corp., denied any association between radium and the illnesses; due to the controversy, the company eventually commissioned a study, which found radium to be the culprit. Refusing to accept the findings, the company commissioned more studies that contradicted the first study, and decried the women who had advocated for themselves. Eventually, many of the women sued, knowing that they had only months left to live. It took until 1938, when a dying radium worker sued the Radium Dial Co. over her illness, that the issue was finally resolved in the courts and the public eye.[572]

We know today that radiation can be very hazardous. Ionizing radiation penetrates the skin and, depending on the dose received, can damage tissues and organs. The level of damage that a dose of radiation can inflict depends on both the type of radiation and the sensitivity of the particular region of the body that is exposed to the radiation. The *effective dose* is used to measure ionizing radiation (and its capacity to cause harm). The sievert (Sv) is the international unit of effective dose that considers both the type of radiation and sensitivity of tissues and organs. It is a way of measuring and quantifying radiation in terms of the harm that radiation can do to the human body. As sieverts are very large units, radiation exposures are typically measured in millisieverts (mSv) or microsieverts (μSv). There are 1,000 μSv in one mSv, and 1,000 mSv in one Sv. Doses of radiation are better spread out over time because the body has the opportunity to repair itself and heal. Large doses in a short period are generally more damaging. All that said, even at "low dose" exposure levels, cancer can develop over time.[573] The United States also uses *rem*, another unit, to measure exposure.[574] One rem equals 10 mSv.[575] You may also see radiation exposure levels measured in *gray* (Gy). Gy can be used to measure any type of radiation, but does not take into account the biological effects of radiation (sieverts, instead, do this). For example, 1 Gy of alpha particles has an effect on living tissue that is twenty times more potent than 1 Gy of X-rays.[576,577]

Low doses are typically not an area of major concern, unless they're on the upper end of low doses that are chronically absorbed, such as *overexposure* to X-rays. People encounter radiation constantly in their everyday lives. For example, sleeping next to someone will expose you to .05 μSv. Living within 50 miles of a nuclear power plant for a year will expose you to .09 μSv. While that last sentence might be alarming to some folks in the immediate sense, everything is relative. Consider that eating a single banana exposes you to .1 μSv, more than living next to a power plant for an entire year. Receiving an arm X-ray equates to 1 μSv. The background dose an average person receives over one typical day is 10 μSv. An airplane flight from New York to LA is 40 μSv. To help put this into perspective, the generally accepted annual radiation dose limits for the general public, per the United States Federal Government, are 1,000 μSv, or 1 mSv.[578,579]

Getting into higher doses, exposure to around 100 mSv per year (10 rem) is the lowest level where authorities have detected an evident increase in cancer risk. This is about 1 million bananas worth of radiation. I cite bananas because they have been used to explain to the general public equivalencies to tiny doses of radiation. Some have called it the "Banana Equivalent Dose" (BED).[580] I should note that eating bananas doesn't actually increase the amount of radiation in the body because homeostasis ejects the potassium that carries the radiation since the body attempts to maintain a stable dose; nonetheless, I believe it's a practical comparative reference point.[581] That said, it's believed that a cumulative dose of 1,000 mSv (1 sievert) of radiation causes fatal cancer many years later in 5% (5/100 or 1/20) persons exposed to it. Standard doses used in medical imaging and other intentional exposures are far below the 1-sievert threshold. For example, CT scans used in medical imaging produce far more radiation than X-ray imaging, but are still significantly under this threshold. While a chest X-ray exposes a person to about .02 mSv, a chest CT exposes an individual to 7 mSv, 350 times the radiation of the X-ray. A coronary CT angiogram will expose a person to 16 mSv.[582,583] Whole-body CT scans can be 20 mSv or more.[584]

The United States Centers for Disease Control and Prevention (CDC) breaks up short-term radiation exposures (amounts received within several days) into five categories of risk level:

1. **Category 1:** Radiation levels naturally occurring in the environment that are within range for a geographic area. Food, air, water, and other natural materials emit radiation in low amounts. This type of radiation is called "natural background radiation" and varies by area.
2. **Category 2:** Radiation levels in the environment that are higher than "natural background radiation" for a geographic area. Generally, these radiation levels are still too low to detect any health effects on the human population.
3. **Category 3:** These are elevated radiation doses that are high enough that authorities have been able to detect increases in cancer; for example, leukemia (blood and bone cancers) and thyroid cancers can appear in as little as five years after exposure. The lifetime risk of cancer for the general population due to natural causes is around 40%; these elevated radiation levels can increase that percentage.
4. **Category 4:** This is the category of dangerously high doses that can make people very sick. These doses are not high enough to cause death, but radiation sickness, known as "Acute Radiation Syndrome" (ARS), can occur. Symptoms include nausea, fatigue,

vomiting, and diarrhea. Hair loss and skin burns may appear in weeks.

5. **Category 5:** These doses of radiation are dangerously high and can be lethal. Category 5 doses can cause incredible damage to bodies, both externally and internally. Exposed populations may experience acute bone marrow damage that destroys their ability to produce white blood cells and fight infections. Diarrhea and vomiting are likely. Medical interventions at these levels may be ineffective, and individuals can lose consciousness and die within hours.[585]

Generally, around a 1 Sv dose will be in the range of Category 3. Doses between 1-5 Sv walk the line between category 4 and 5. At the level of 5 Sv, a single dose will kill half of those exposed within one month. At 10 Sv, death is a foregone conclusion; those exposed will die within weeks, but many, sooner.[586]

Children and fetuses are more sensitive to radiation exposure. This is because the cells in this population divide rapidly (since fetuses, babies, and children are in a growth stage). Radiation disrupts these processes and can cause cell damage; with more cell growth, there are more divisions and more chances for faulty DNA to make cell copies that could result in cancers. Children also have a longer lifespan ahead of them than adults, so cancers have more years and decades to develop.[587,588]

While it's hard to know the actual doses taken in by the "Radium Girls," many of whom died as a result of their exposures, some estimates put their ingestion of radium through the oral route at 4,000% of an annual dose; the ingestion of radon and radium dust may have increased this to up to 13,000% the annual recommended dose. Some of the women may have taken in doses exceeding 2 sieverts (somewhere between Category 4 and 5).[589]

We should discuss acute radiation syndrome (ARS). ARS is caused by a high dose of radiation that (1) is received within a short time (usually minutes), (2) is received by most of the body, and (3) penetrates the skin. Initial symptoms will manifest as nausea, headache, vomiting, and diarrhea. For some time after the initial onset of symptoms, an individual may feel and look well, after which they will become sick again with variable symptoms and severity, which will depend upon the dose of radiation the individual received. This stage of ARS may last for hours or up to several months, during which the victim may experience loss of appetite, fatigue, fever, nausea, vomiting, diarrhea, seizures, and coma. External skin damage ("cutaneous radiation injury" [CRI]) can manifest as swelling, itching, redness, blisters, and ulcers. Complete healing, if a patient survives, may take weeks, months, or years. Hair loss may also occur, which in relatively mild cases, can grow back after several weeks. In most cases, the cause of death from ARS is the destruction of bone marrow by radiation, which leads to infections and internal bleeding.[590]

Acute radiation syndrome has four stages:

1. **Prodromal stage (N-V-D stage):** Symptoms include nausea, vomiting, anorexia, and diarrhea. Symptoms occur minutes or days after exposure and may last for minutes or several days.
2. **Latent stage:** The victim looks and feels generally well. This stage lasts several hours or even a few weeks. This is colloquially referred to as the "walking ghost phase."
3. **Manifest illness stage:** Symptoms depend on the specific circumstances of the exposure and will last from hours to several months.

4.**Recovery or death:** The recovery process can last from several weeks up to two years, if the victim survives.[591]

Classic symptoms of ARS are as follows:

1.**Bone marrow syndrome:** Primary cause of death is the destruction of bone marrow, resulting in infection and hemorrhaging.
2.**Gastrointestinal (GI) syndrome:** Survival is extremely unlikely with this syndrome. Destructive and irreparable changes in the GI tract are present, causing infection, dehydration, and electrolyte imbalance. Death will typically occur within two weeks.
3.**Cardiovascular (CV)/Central Nervous System (CNS) syndrome:** Death is caused by collapse of the circulatory system as well as increased pressure in the cranium as the result of fluid buildup caused by edema, vasculitis, and meningitis. Those suffering from CV or CNS syndromes will die within three days.[592]

On another note, we should talk about the different types of radiation and how they affect us. Various isotopes and reactions (both natural and human-made) produce different types of radiation that carry their own risks and considerations. The four types are Alpha, Beta, Gamma/X-Rays, and Neutrons, in order from least to most penetrating:

Type	Penetration	Risks
Alpha (Internal)	Stopped by the skin. Cannot penetrate	Very high ionizing power. Very harmful if inhaled or

	most matter; a piece of paper will stop it.	absorbed in the bloodstream via wounds.
Beta (Internal and External)	Stopped by a layer of clothing.	Intermediate ionizing power. Unlike alpha particles, can penetrate the skin and cause radiation damage and burns. Like alpha particles, extremely dangerous if inhaled or taken internally.
Neutrons (External)	Penetrating. Stopped by several feet of concrete or a material rich in hydrogen (water).	Concern in and around initial blast zones from nuclear weapons. Radiation hazard for the entire body.
Gamma/ X-Ray (External)	Extremely penetrating. Several feet of concrete or inches of lead required to stop it.	Main risk is extreme penetration. These pass completely through the human body, as they are the most penetrating form of radiation. But some percentage is always absorbed by the body, and will cause damage.

593,594

Ultimately, the amount of radioactive contamination following a detonation depends on the isotopes used, the type of explosion (fission vs. fusion), and whether the explosion occurs in the air or at ground level. Nuclear detonations can produce an incredibly complex mix of over 300 different isotopes and scores of elements. Half-lives on these isotopes can range from a sliver of a second to millions of years. Initial irradiation is very high but declines over time. Seven hours following a nuclear detonation, residual radioactivity decreases to around 10% of the amount present at one hour, and after 48 hours, it falls to about one percent. Early fallout settles on the ground during the first 24 hours after the blast, and delayed fallout comes after day one, consisting of microscopic particles carried by wind and rain.[595]

Chapter 21. The Human Cost: Selected Radiological Incidents and Outcomes

"As we sat there shell-shocked and confused, heavily injured burn victims came stumbling into the bomb shelter en masse. Their skin had peeled off their bodies and faces and hung limply down on the ground, in ribbons. Their hair was burnt down to a few measly centimeters from the scalp. Many of the victims collapsed as soon as they reached the bomb shelter entrance, forming a massive pile of contorted bodies. The stench and heat were unbearable." -Shigeko Matsumoto, Nagasaki Bombing Survivor[596]

We can talk ad nauseam about blast yields, fissile materials, gamma rays, or any number of other topics academically. For the citizens of Hiroshima and Nagasaki, there was nothing academic about the Bomb. In Hiroshima, as birds burst into flames in the sky, individuals near the epicenter were turned to black char in a fraction of a second. Of those who survived, many had the outlines of their clothing seared into their skin. As I mentioned before, within half a mile of the blast's epicenter, 90% of all people were killed. Temperatures at ground zero reached 5,432 to 7,232 degrees Fahrenheit (3,000-4,000 °C).[597] The heat was so intense that it telegraphed the silhouettes from physical objects onto the sides of buildings, including human beings. One such infamous shadow is known as the "Human Shadow Etched in Stone" on the steps of the Hiroshima Branch of Somimoto Bank. Exposed at close range to ground zero of the blast, the individual was sitting on the steps, likely waiting for the bank to open. The stone was turned white from the intensity of the explosion; a dark shadow on the steps was all that remained of whoever had been sitting there that fateful day.[598]

Images from shortly after the blasts show victims with horrific burns over vast areas of their bodies, injuries that many would never recover from. Of those who did survive, many fell victim to the effects of radiation poisoning later on. The intense fires that burned around Hiroshima following the blast sent massive amounts of ash and soot into the atmosphere. The ash seeded the clouds, and it began raining an hour or two after the blasts. This "*black rain,*" which had the consistency of tar, was so radioactive that it caused severe radiation burns in many of those who came into contact with it.[599] In 2021, the Hiroshima High Court ruled that 84 people who were exposed to the black rain and who fell ill in the wake of the bombing are now entitled to benefits as bomb survivors.[600]

Common symptoms observed by American authorities were hair loss, bleeding into the skin, hemorrhaging, lesions of the mouth and throat, vomiting, diarrhea, and fever. Many survivors totally lost their hair. Others exhibited bleeding gums, and those suffering worse acute radiation damage would bleed all over their bodies. Hemorrhaging was evident under the skin in many; wounds often only partially healed and bled copiously. Hemorrhaging of the eyes was not uncommon. Coagulation time for the blood was prolonged, and thus bleeding times were increased, presumably due to radiation damage to the blood and bone marrow. While normal white blood cell counts are around 7,000, some victims of extreme ARS showed results of under 1,000. Some patients, totally uninjured by the initial blast, died from radiation sickness about a week after exposure; some took until 3-4 weeks to die. Around 7-8 weeks, the final deaths from ARS were observed.[601]

The Radiation Effects Research Foundation (RERF) is a joint research collaboration between the United States and Japan, the successor to the Atomic Bomb Casualty Commission (ABCC) set up by the U.S. government to study the health outcomes of the atomic bombings in Japan.[602] RERF has concluded through its studies that the survivors of atomic bombings have suffered a number of health effects, including increased cancers in specific organs; increases in non-cancer diseases such as cataracts, benign thyroid tumors, heart disease, and stroke; deterioration of the immune system similar to that caused by aging; and minor inflammatory reactions.[603]

Within the first few months after the bombings, RERF estimates that between 99,000 and 166,000 people died in Hiroshima, while another 60,000 to 80,000 died in Nagasaki. These deaths were caused by force trauma, heat, explosions, and acute radiation syndrome (ARS). Leukemia was the deadliest long-term outcome, appearing two years after the attacks and peaking 4-6 years later. Children were affected most severely. Relative to the general population, survivors in Hiroshima and Nagasaki were 46% more likely to be diagnosed with leukemia. For solid cancers among survivors, the rate increased 10.7%. For those individuals exposed as fetuses ("in utero"), studies have shown increased incidence rates of small head size and mental disability. Doctors and researchers have also documented impairment in physical growth among this population. Fortunately for those who were exposed in utero, they had a lower relative cancer risk increase than those who were children during the bombings.[604]

Los Alamos itself, which developed the bombs used in Japan, saw its own share of incidents shortly after the bombs were detonated in Japan. Harry Daghlian, an American physicist working at Los Alamos, was assigned to the "criticality group," conducting experiments to determine the critical masses of fissile materials. These experiments were especially dangerous, because an error could result in the materials reaching criticality and engaging a fission chain reaction. On August 21, 1945, Daghlian was working with a plutonium core. He'd build a stack of tungsten carbide bricks around the plutonium sphere to create a "neutron reflector," which Los Alamos had hoped would lower the mass of plutonium required for criticality. By itself, the plutonium sphere had too much surface mass to sustain a fission chain reaction, but by using neutron deflection, colloquially referred to as "tickling the dragon's tail," scientists could achieve it. Daghlian did the work alone, with only a guard in the room, constructing four layers of bricks around the core. His monitoring instruments (geiger counters), started chirping like mad, signaling that if he laid the brick in place, it would send the core into a state of criticality. Moving to pull the brick away, he accidentally dropped the brick directly on top of the core, sending it into a state of supercriticality. He knocked it away, but it was too late.[605,606]

His hands soon swelled up, and the skin sloughed off as his organs began to shut down. Daghlian received a dose of radiation from neutrons and gamma rays higher than any ever experienced by a human being before, even those who fell victim to the Nagasaki blast a mere 12 days before. After intense suffering, Daghlian passed away 26 days later, in September 1945, at 24 years old. It was estimated that he received a dose of 5.1 Sv. Unfortunately, he wouldn't be the plutonium core's only victim.[607,608]

On May 21, 1946, a Canadian physicist named Louis Slotin tickled the dragon's tail of the very same plutonium core that had claimed Daghlian's life. Slotin's procedure was a bit different from Daghlian's. He would use a half-shell of beryllium as a tamper, reflecting the neutrons ejected from the plutonium. Slotin would use a screwdriver to stop the tamper from lowering completely over the core. One of Slotin's colleagues, Raemer Schreiber, was in the room at the time, and turned away to work on another project as the slow process unfolded. Unfortunately, Slotin's screwdriver slipped, and the tamper dropped entirely over the core, inducing supercriticality. As Schreiber turned around, he saw a flash of blue light and felt a wave of heat. Though the reaction was about a million times smaller than Hiroshima and Nagasaki, it still equated to about three quadrillion fission reactions instantaneously.[609]

Slotin suffered a period of vomiting at the hospital soon afterward. By the following day, he felt better, and his health seemed improved. However, though initially tingling and numb, his left-hand exhibited pain. Scientists later estimated that this hand received a radiation dose of 150 Sv. His whole-body dose was 21 Sv, many times what is required to kill a human being. His hand eventually turned blue and exhibited significant blistering. Likely understanding what was coming, the U.S. government flew Slotin's parents in from Canada. Five days later, his white blood cell count plummeted. By one week later, he was experiencing abdominal pain, weight loss, and periods of mental confusion. The radiation had penetrated his skin and burned his internal organs, inflicting a "three-dimensional sunburn," as experts put it. Slotin slipped into a coma, and, nine days later, became what would, from that point on, be known as the "Demon Core's" second victim. Other scientists nearby the accident also became ill; in some cases, the exposure may have contributed to their death later in their lives. Eventually, the Demon Core was melted down and distributed amongst several new warheads, which, as we know now, were never used in war.[610,611,612,613]

In Part II of this book, we explored Operation Crossroads, which constituted a series of tests that were the largest of Post-WWII, involving over 40,000 staff, with approximately 17,000 of those in frequent contact with contaminated naval vessels. Staff were exposed to internal alpha, internal and external beta, and external gamma radiation due to the contamination on the ships. Plutonium contamination was evident on the naval vessels that were tested in the bombings, and around 80 plutonium-contaminated ships were at least partially decontaminated by staff; even a microscopic amount of plutonium in the human body can kill. Internal contamination even occurred to these staff as a result of eating meals on board contaminated ships, causing them to ingest radioactive particles.[614]

Eventually, even the support fleet became radiologically contaminated by marine life in the lagoon. Some ships could be decontaminated, but others were so far gone that they had to be scuttled. Many animals on the ships didn't survive the blasts, but the living ones retrieved from the ships in August had mostly died from gamma poisoning by November 1946.[615]

One of the lessons of Operation Crossroads, Hiroshima, Nagasaki, and subsequent testing was that air bursts (when nuclear bombs are detonated high in the sky) maximize blast damage by reflecting the blast wave off the ground, spreading a massive physical impact, but consequently produce less radioactive fallout. Ground bursts (when nuclear bombs are detonated at or near ground level), on the other hand, dig a massive crater and destroy everything in the immediate vicinity, but the destructive blast wave doesn't extend as far. Air bursts are more useful against cities. Ground bursts are more useful against hardened targets like underground missile silos. From a human health perspective, a drawback of ground bursts is that they create more fallout by irradiating more material on the ground and sucking it into the sky; this spreads fallout over massive areas.[616,617]

The problem of radioactive contamination of a large area, as exhibited by Operation Crossroads, is that it isn't trivial and can be very long-lasting. Let's explore the problem of the "half-life."

A half-life, as briefly mentioned before, is the amount of time required for one-half of the atomic nuclei of a radioactive sample to decay (through the emission of particles and energy). In other terms, it's the time required for the number of disintegrations per second of a radioactive material to decrease by one-half.[618] Imagine a ball pit full of 100 balls. This ball pit (the sample) has a radioactive half-life of 10 years. That means that, after 10 years, half of the balls will have "decayed," changing into balls of another color. After another 10 years, half of the remaining balls will have decayed, and so on.

Find below a chart of half-lives for materials commonly used in nuclear weaponry or found as a byproduct of detonations:

Isotope	Half-Life
Uranium-235	704 Million Years
Plutonium-239	24,000 Years
Hyrogen-3 (Tritium)	12 Years
Iodine-131	8 Days
Cesium-137	30 Years
Strontium-90	29 Years

[619,620]

These byproducts and fissile materials have various half-lives and various levels of radioactivity, so the level of concern varies depending on the spread and contamination. Tritium, for example, is used in fusion bombs but is produced by both natural and industrial nuclear processes. While produced for fusion bombs and for civilian energy purposes, tritium is also produced by the Sun through the interaction of cosmic radiation with atmospheric components. Hence, it occurs naturally in the water that we drink. The current standard for tritium in drinking water is set at .04 mSv per year, about equivalent to the background dose of radiation received by an average person over four days, or about two chest X-rays worth.[621,622,623,624]

Plutonium-239, not found in nature and only created using uranium in enrichment facilities for the purpose of building nuclear weapons, is another story entirely. Highly radioactive and with a half-life of 24,000 years, it poses a significant danger to human health. Plutonium emits alpha particles, and so is most dangerous when inhaled. Alpha particles kill lung cells, causing scarring of the lungs, leading to disease and cancer. In the bloodstream, it travels to the kidneys, exposing them to alpha particles. Circulating through the body, it reaches the bones, spleen, and liver, also exposing those organs. Fortunately for the workers consuming contaminated food at Operation Crossroads, plutonium is not ingested easily by the stomach, so it mostly passes through the body via the digestive tract.[625]

Speaking of plutonium, perhaps the first major nuclear accident outside the individual incidents with the Demon Core was at a facility in Ontario, Canada known as *Chalk River Laboratories*, a facility run by Atomic Energy of Canada Limited (AECL) to conduct joint experiments with the United States and Britain, and to produce plutonium for nuclear weapons. The reactor at the facility contained 175 rods inserted vertically; 163 were uranium fuel, and 12 were composed of boron carbide, intended to "mediate" or slow the nuclear reaction. The reactor also used deuterium (heavy water) as a moderator to slow down neutrons, thereby making them more useful for fission. Mistakes by operators in 1952 led to 7 of the 12 rods being pulled from the reactor, which caused the power to double every two seconds; the reactor, rated only for 30 megawatts of power, rose to 60 and then 100 megawatts. The water boiled instead of circulating to carry heat away from the reactor; the uranium rods melted and poisoned the cooling water. The loss of reactor control lasted a mere 62 seconds, but the effects were catastrophic.[626]

Hydrogen formed in the reactor by the melted uranium rods came into contact with air and exploded. 1.2 million gallons (4.5 million liters) of contaminated, radioactive water was dumped into the basement. Alarms went off, signaling that the air was also contaminated with radiation. Cleanup took a year and the coordinated efforts of 800 employees. The contaminated core was so irradiated that truck drivers driving the truck that carried the core had to be swapped out regularly because being within 3 feet (1 m) of the core would provide a lethal dose of radiation in less than one hour. A similar accident in 1958 did not result in a full core meltdown, but required 800 AECL employees and 300 military personnel to clean up. Workers would later come down with several cancers and a cancer rate four times that of the average population. These included anal, chest, face, and sinus cancers that required surgeries that severely disfigured some of the victims.[627]

As a related aside, nuclear accidents are rated through the International Nuclear and Radiological Event Scale (INES), which communicates the safety significance of radiological events to the public. The 1952 incident at chalk river was a five on the scale.[628] The scale is logarithmic, meaning that for each step up in level, the event's severity is 10 times worse than events in the tier below it. Events are considered in terms of their effects on people and the environment, their impact on radiological barriers and control, and the impact on defense in depth (layered, multiple security measures – control systems, physical barriers, operator training, and emergency procedures). Events without safety significance and those that don't have safety relevance regarding radiation/nuclear safety are not rated on the scale. The scale has seven levels, with levels 1-3 constituting incidents; levels 4-7 constitute accidents, with levels 6 and 7 being the most serious.[629]

While reactor meltdowns and incidents have helped us learn many lessons about improving nuclear safety and making enrichment and power production safer over the years, civilian energy production is not the focus of this work, so I can't get into all of them. However, I will mention a couple more because they help us understand and highlight how dangerous radiation can be and how its effects can devastate human life long-term. No accident in the history of nuclear technology evokes more alarm than the Chernobyl incident, a 1986 Level Seven disaster in Ukraine.

During a reactor safety test, the operators at the plant's Reactor 4 attempted to simulate a power outage and assess the reactor's cooling systems. However, due to a flawed reactor design and operator error, the test triggered a runaway chain reaction, causing a massive power surge. The surge in the reactor led to a steam explosion, which ruptured the reactor and produced enough energy to blow off the reactor's 1,000-ton lid. The exposed reactor core caught fire, releasing a plume of highly radioactive materials into the atmosphere. The graphite moderator within the reactor ignited and burned for about ten days. During that time, workers dumped 5,000 tons of dolomite, sand, clay, lead, and boron onto the burning core in an attempt to extinguish the fire and limit further radioactive contamination.[630]

The accident resulted in the most substantial uncontrolled radioactive release into the environment ever recorded for a civilian operation. Massive amounts of radioactive substances were belched into the air by the flames for ten whole days. Iodine-131 and Cesium-137 were spread into the environment, exposing the public to radioactive contamination. Xenon gas and other radioactive materials were released by the accident. Most of the material was spread nearby the plant, but lighter material was carried by winds throughout Ukraine, Belarus, Russia, Scandinavia, and Europe. Many personnel were killed within weeks by the radiation they received at the facility. Twenty-eight are estimated to have died by ARS, six of them firefighters. An additional two were killed in the initial explosion. It's estimated that doses received by the firefighters who died were in the range of 20,000 mGy (20 Sv). [I should note here that Gy and Sv are not directly convertible, so this is a rough estimate of external radiation sources]. [631] Recall that anything over one sievert is very bad news, anything between 1-5 may be fatal, and beyond 10, a human being cannot survive.

Initially, 116,000 people were displaced. In subsequent years, 220,000 had to be resettled into less contaminated areas, and the exclusion zone was extended to cover 1,660 square miles (4,300 km^2). The main effect over the years has been child thyroid cancer fatalities, with nine deaths recorded from 1986-2002.[632] Experts estimate that it will take between 3,000 to 20,000 years for the area around the reactor to drop to levels of radiation that will be considered safe.[633,634]

The latest major disaster to take place was at the Fukushima Daiichi plant in Fukushima, Japan, on March 11, 2011. A powerful (9.0 on the Richter scale) undersea earthquake caused an automatic shutdown of the plant's reactors. With waves reaching up to 49 feet (15 m) in height, the subsequent tsunami overwhelmed the plant's seawall, flooding the reactor buildings and disabling the backup power generators. As a result, the cooling systems for the reactors failed, leading to a loss of coolant and subsequent overheating of the reactor cores. This triggered a partial meltdown in three of the plant's six reactors, releasing a significant amount of radioactive material into the environment.

The disaster resulted in the evacuation of nearby residents and a wide-scale contamination of the surrounding areas. Like Chernobyl, the disaster released radioactive materials, including iodine-131 and cesium isotopes, spread through the air and contaminated soil, water, and food sources. The Fukushima disaster highlighted the vulnerability of nuclear power plants to natural disasters and sparked global debates about the safety of nuclear energy. It led to a reassessment of safety measures, emergency preparedness, and the regulation of nuclear facilities worldwide. Also like Chernobyl, the Fukushima accident was classified as a Level Seven, but miraculously, only one person died– the cause was lung cancer, and it claimed the victim four years after the accident. One hundred thousand people had to be evacuated.[635,636,637] In August 2013, a subsequent accident resulted in 330 tons of irradiated water used in cooling operations in reactors 1, 2, and 3 being discharged around the facility. The leak was classified as a Level Three incident.[638]

Perhaps no accident better describes the horrors of acute radiation syndrome than that which occurred to Hisashi Ouchi, a Japanese technician who was subject to severe radiation poisoning at the Tokaimura nuclear fuel processing plant. On the morning of September 30, 1999, Ouchi and two other workers were purifying uranium oxide to make fuels for a research reactor. As Ouchi was standing at a tank, holding a funnel, while coworker Masato Shinohara poured in a mixture of intermediate-enriched uranium oxide, a characteristic blue flash, like that given off by the Demon Core decades earlier, lit up the room; they had poured too much uranium into the tank, resulting in the materials reaching criticality. Twenty-one such accidents, though none this severe, had occurred from 1953-1997. Ouchi had received up to 25 Sv of radiation, many times that which is required to kill. One of his colleagues received between 6-9 sieverts.

When he initially arrived at the hospital, Ouchi did not appear to be terribly ill, experiencing only a red and swollen face, bloodshot eyes, and pain in his ears and hand. His condition quickly deteriorated; he required oxygen, his abdomen swelled, and six days after the accident, tests indicated that his bone marrow cells had been shattered into pieces. He received a peripheral blood stem cell transplant from his sister a week after the accident. When medical tape was pulled from his chest, his skin came with it. He soon developed blisters, and because the radiation had destroyed the biological systems that would allow his body to regenerate his epidermis, the outer layer of his skin began to slough off, consisting of virtually all of his skin except that which was located on his back. Massive blood and fluid loss through his skin equated to between one-half to one gallon (2-4.5 liters) per day. He experienced heart failure two months after the accident, but was (perhaps tragically) revived. Multiple organ failure (liver, kidneys, and lungs), blood pressure instability, and hemophagocytic syndrome (excessive immune activation) claimed his life on December 21, about three months after the accident.[639,640,641]

I need to qualify that, with this brief description, I've barely touched on the surface of what Ouchi went through. It is difficult to illustrate, in the space available here, how horrific and protracted his death was, but given international attention to the accident, there is a great deal of information available about the way his medical status declined; I would caution the reader that the more specific details are truly harrowing, and it takes a certain mental constitution to read them. His case is unquestionably one of the most terrible deaths ever experienced by a human person, and is one of the starkest highlights of the horrors of radiological dangers.

Of course, radiological contamination is not limited to civilian accidents and laboratory incidents. Indeed, nuclear testing has been responsible for significant environmental contamination, and in some incidents, even deaths outside the context of Hiroshima and Nagasaki.

The Daigo Fukuryū Maru (Lucky Dragon No. 5), was a Japanese fishing boat that unwittingly sailed into the aftermath of the Castle Bravo test that was conducted on March 1, 1954 at Bikini Atoll in the Marshall Islands. The incident left in its wake widespread concern about the dangers of nuclear weapons. The crew of 23, including fishermen and a radio operator, embarked on a routine fishing expedition to the Pacific Ocean, hoping for a bountiful catch. The Castle Bravo test, with a power much greater than anticipated (15 megatons instead of 5 megatons), released an immense cloud of radioactive particles into the atmosphere and contaminated the surrounding waters. Days later, on March 12, 1954, the Lucky Dragon No. 5 encountered mysterious snow-like ash falling from the sky. The crew, unaware of the nuclear test, collected the ash on their hair, faces, necks, hands, and clothing, and even got some in their eyes and mouths, unknowingly exposing themselves to dangerous levels of radiation.

As the crew returned to Japan, they began to suffer from severe ARS. Their symptoms included nausea, vomiting, burns, bleeding gums, and clumps of their hair falling out. The gravity of their condition became apparent when Aikichi Kuboyama, a crew member, fell critically ill and was rushed to the hospital. The news of the crew's radiation exposure spread rapidly, causing significant outrage in Japan, and the ash came to be known as "shi no hai," or "death ash." Despite medical efforts, Aikichi Kuboyama's condition deteriorated, and he tragically passed away on September 23, 1954, six months after exposure. His death further intensified public outcry and fueled international discussions on the dangers of nuclear weapons.

The Lucky Dragon No. 5 incident put pressure on the United States to release a previously secret film about Operation Ivy and its testing. Ultimately, the film would terrify the world, showing footage of a thermonuclear detonation, overlapping the fireball on the skyline of Manhattan, with the narrator declaring, "the fireball alone would engulf about one-quarter of the island of Manhattan." After showing a map of Washington DC, the narrator continues, ". . . with the Capitol (the building in which the U.S. Congress conducts its business) at point zero, there would be complete annihilation" for three miles around the epicenter. With the world now understanding the existential threat posed by nuclear weapons, Japan ordered tests on any fish caught within a 1,553-mile (2,500 km) radius around the test site. Thousands of samples were contaminated with radioactivity. The incident served as the inspiration for the "Godzilla" franchise and also inspired the birth of modern environmentalism.[642,643,644]

The United States' tests at Bikini Atoll also displaced inhabitants of the islands. While the United States promised they could eventually return, they were instead relocated to other islands in the Marshalls. In the 1960s, the Atomic Energy Commission (AEC) declared Bikini Atoll safe for human habitation and allowed some residents to return. This was ultimately an unfortunate decision, as a study a decade later showed that the levels of cesium-137 in the bodies of those who returned to the island had increased by 75 percent. Ultimately, they again had to be relocated to another island, as testing showed that cesium-137, background radiation, and various radioactive isotopes in the soil and ocean sediment were elevated to unacceptable levels.[645] Seventeen of the nineteen children who were younger than age 10 who were on the island of Rongelap the day Bravo was detonated developed thyroid disorders and growths. Cancer cases, deformities, and miscarriages multiplied.[646]

From 1946 to 1958, the United States detonated 67 nuclear bombs on or above the Marshall Islands. These bombs would vaporize entire islands, blast craters into the shallow lagoons, and displace hundreds of people from their homeland. Over 3 million cubic feet of contaminated, radioactive soil and debris, interspersed with plutonium, have been stored in a giant concrete cask on Runit Island known as "The Tomb." Six workers died during the cleanup, and hundreds of others allegedly developed radiation-induced cancers and other illnesses.[647]

Of particular concern with regard to nuclear testing and detonations is strontium-90 (Sr-90), which decays into yttrium 90 (Y-90), which then itself decays by beta radiation. Y-90 can burn the eyes and skin through external exposure. Sr-90 can be inhaled and taken into the body through food and water, where it then acts like calcium and is incorporated into bones and teeth, then wreaking havoc on the body by causing bone, bone marrow, and soft tissue cancers.[648]

In the 1950s, the United States commissioned a classified worldwide study called Project SUNSHINE to determine the uptake of strontium-90 and effects stemming from Sr-90 being put into the environment from nuclear testing. Bones of 1,500-6,000 cadavers, many of them babies, were gathered from half a dozen countries ranging from Europe to Australia, with 26 "bone collection sites" located worldwide, shipping the remains to Chicago and

New York. Children's remains were used, sometimes without their families' permission, because children are particularly susceptible to the uptake of Sr-90 since they have growing bones and bodies. Dr. Willard Libby, a University of Chicago researcher, and Nobel Prize winner, said of the bones required for the study, "I don't know how to get them... But I do say that it is a matter of prime importance to get them and particularly in the young age group. So, human samples are of prime importance and if anybody knows how to do a good job of body snatching, they will really be serving their country." For some parents, this collection effort was traumatic. One British mother of a stillborn baby could not dress her daughter for her funeral because doctors did not want her to find out that they had removed her baby's legs for the study.[649,650]

The findings of Project SUNSHINE and later analyses are perhaps as disturbing as the study's methodology itself, and these findings helped to put an end to nuclear testing worldwide. The RAND Corporation provided an analysis of the data in 1953, in both classified and unclassified forms. RAND notes, in its conclusions, "The bone-retentive and radioactive properties of Sr^{90} endow it with a high carcinogenic capability; a given amount above threshold (which may be zero) fixed in the bone will cause a certain average percentage of the population to die of bone cancer comparable with that observed in victims of radium poisoning." Attempting to find the upper limit of tolerance is difficult because Sr-90 is distributed through various methods and spread across the world. Certain areas are natural hotspots due to soil qualities, weather patterns, and so on, so RAND worked to determine a way of looking at Sr-90 in an average way. Their conclusion was that approximately 25,000 megatons of detonations would be required to hit the upper limit of Sr-90 and effectively poison the whole planet, seriously endangering *all* human life. An earlier estimate developed on a calcium-strontium model indicated an 800-megaton limit. Since Sr-90 has a half-life of 29 years, the accumulation of Sr-90 globally is very long-lasting. To put this into perspective, the total megatonnage of nuclear testing, while it was still going on, was around 510 megatons, over 2% of what would be required to hit the maximum Sr-90 threshold under the RAND model and make our planet uninhabitable.[651,652,653] Had governments continued testing, they may have, in short order, unwittingly put an end to all of our stories.

Chapter 22. Broken Arrows, Major Nuclear Close Calls, and the Cuban Missile Crisis: The Convergence of Human Fault and Uncontrollable Forces of Destruction

"So long as any state has nuclear weapons, others will want them; so long as any such weapons remain, it defies credibility that they will not one day be used, by accident or miscalculation or design; and any such use would be catastrophic for our world as we know it.." -Report of the International Commission on Nuclear Non-Proliferation and Disarmament, 2009.[654]

There are multiple classes of nuclear incidents, per the United States Department of Defense. Although not an exhaustive list, major relevant designations are copied below:

1. **"Bent Spear"** - Incidents that are of significant interest that involve U.S. nuclear weapons.
2. **"Empty Quiver"** - Seizure, theft, or loss of a U.S. nuclear weapon.
3. **"Emergency Disablement"** - Command disablement or non-violent disablement of U.S. nuclear weapons.
4. **"Faded Giant"** - Accidents involving non-weapon reactors.
5. **"Broken Arrow"** - Accidents involving nuclear weapons.
6. **"NUCFLASH"** - Accidents that may lead to war.[655]

There have been multiple close calls and broken arrows over the past several decades since the advent of nuclear weapons. Some of these had the potential to cause mass destruction of civilian populations, and others risked a full-scale nuclear exchange between the United States and the Soviet Union. In this chapter, we'll explore these incidents to understand just how close we've come to disaster, either through military brinkmanship or human incompetence. Since 1950, at least 32 known nuclear weapons accidents have resulted from United States action alone. These broken arrows sometimes result in nuclear weaponry going missing, and to date, there have been six known incidents where nuclear weapons have gone missing and have *never* been recovered.[656]

Central to the discussion of broken arrows and nuclear weapons accidents is Operation Chrome Dome. By the late 1950s, General Thomas S. Power, Commander of the Strategic Air Command (SAC), had decided that at all times, a certain number of strategic bombers should be in the air to ensure second strike capabilities and prevent the Soviets from destroying all U.S. air assets while grounded. Of the program, Power said:

> As long as the Soviets know that, no matter what means they employ to stop it, a sizable percentage of SAC's strike force will be in their air for counterattack, within minutes after they have initiated aggression, they will think twice before undertaking such aggression. For this reason, it is my considered opinion that a combat-ready Alert Force of adequate size is the very backbone of our deterrent posture.[657]

During a single month of continuous operations, SAC B-52 Stratofortress strategic bombers flew 2,088 Chrome Dome flights, totaling over 47,000 hours of flight time, with a 97% effectiveness rate. Mid-air refuelings from other planes were required to keep the bombers in the air over the course of a mission, and over the span of a month, approximately 414 million pounds of fuel would be used for the bombers.[658] Operation Chrome Dome would keep multiple plane teams in the sky at all times for around a decade, from the late 1950s to 1968, when the operation was terminated. By this point, ICBMs and SLBMs had eliminated the necessity for 24/7 bomber patrols, accidents had caused alarm in the United States government, and the costs of maintaining SAC bombers in the sky at all times were becoming prohibitive.[659,660]

There is a problem with constantly keeping so many planes in the air– airplanes, like cars, have accidents. With all those flight hours, even with the strictest of cautions and redundancies, there were bound to be mishaps. And indeed, multiple incidents attributed to Chrome Dome could have led to disaster domestically and internationally.

One such harrowing incident took place on January 23, 1961. A United States B-52 Stratofortress was flying in the skies over Goldsboro, North Carolina with two Mark 39 thermonuclear bombs on board. Each Mark 39 had a yield of between 3-4 megatons, enough to devastate North Carolina, vaporize everything within a 17-mile radius, and inflict unimaginable radiological contamination upon the Eastern Seaboard and its populations. The bombs were fully fused, meaning that when dropped, they would detonate either as a ground burst, an air burst, or a "lay down" using parachutes that would slow the bomb to allow time for the Stratofortress to escape. A mid-air refueling accident caused the bomber to spring a leak; fuel imbalance in the bomber made it unstable and it eventually broke apart while the crew attempted an emergency landing. Some crew members were killed, and as the plane disintegrated above North Carolina, the Mark 39 hydrogen bombs dropped toward Earth...

Some military reports indicate that five out of six steps, or six steps out of seven, required for the bombs to detonate, occurred. One of the bombs buried itself into a muddy field at speeds exceeding 700 miles per hour, driving itself 180 feet deep into the mud. The other's parachute deployed, lodging it in a tree. In the case of the bomb with the deployed parachute, some reports indicate that the trigger mechanisms engaged and a single, low-voltage switch prevented a total detonation of the weapon and what would have been the resultant nuclear fusion. That switch may have been the difference between just one more broken arrow and a catastrophic, major accident whose costs to American communities would be severe, and in the end, caused by their own government's weaponry.[661] Other accounts and analyses, including some from the U.S. military, point out that the failsafe systems built into the bombs worked exactly as intended and that the risk of detonation was minimally north of zero.[662] We'll never know for sure how close we came, but the stakes of such an accident couldn't possibly be higher.

A similar incident occurred above Palomares, Spain, in 1966. An attempted mid-air refueling of a B-52 went wrong while it was carrying four hydrogen bombs. Four of the eleven crew were killed, and the four bombs dropped from the bomber bay. Two of the weapons' high explosive materials exploded on impact with the ground. Luckily, the fail-safes apparently functioned as intended, as the detonation of the bombs' conventional explosives did not result in a fission or fusion reaction. It did, however, blast plutonium dust into the air, spreading radioactive materials all over the ground. Around 1,400 tons of contaminated soil and vegetation had to be removed by the United States and stored at an approved site. One of the bombs fell into the ocean, necessitating a multi-month search to recover it using advanced recovery and deep sea technology.[663]

According to the New York Times (NYT), some of the cleanup crew were deployed without personal protective equipment (PPE) such as radiation suits and respirators, and have been involved in a multi-year conflict with the United States government over their exposure to the radiological hazards at the site. Handheld alpha particle counters detected so much alpha radiation that the contamination exceeded the maximum capacity of the detectors (two million alpha particles per minute). Recall that alpha radiation, given off by plutonium, cannot penetrate the skin but causes severe internal risks if inhaled into the body. Of 40 veterans who assisted in the cleanup effort that the NYT was able to identify, 21 had cancer, and nine had died of cancer. Some may have suffered from acute radiation sickness (ARS), rashes, growths, swelling, pain, headaches, and infertility. In at least one case, skin grafts were required.[664]

A similar incident occurred in Thule, Greenland, in the final year of Operation Chrome Dome. A B-52 flying from Plattsburgh Air Force Base in Northern New York crashed and burned around seven miles from the Thule Air Force Base runway. One of the crew members died in the crash. Four nuclear weapons were on board the plane, and all were destroyed in the fire caused by the crash. The accident occurred on the sea ice; approximately 237,000 cubic feet of contaminated ice, snow, and water had to be removed to a safe storage site over the course of four months. The incident caused a diplomatic row between the United States and Denmark, as Greenland belonged to Denmark, and Denmark had a nuclear-free policy. Once again, fortunately for everyone, the fail-safes in the bombs appear to have operated according to design and prohibited the weapons from undergoing fusion.[665,666]

There were several other crashes, burns, and accidental drops that, perhaps partially through sheer luck and (hopefully) well-designed safety mechanisms, did not result in full-scale detonations. In one particularly disturbing incident, even before Operation Chrome Dome was in effect, an MK-4 "Fat Man" style fission bomb was lost in British Columbia on February 13, 1950. In this particular incident, a Convair B-36 Peacemaker strategic bomber was carrying the bomb when three of its four engines failed. Icy conditions complicated the situation at 12,000 feet in the sky, and the flight could not be maintained. The MK-4 bomb was dropped from 8,000 feet and was never recovered.[667] While lost bombs like this one appear not to have caused the same radiological contamination as the Palomares and Thule incidents, their losses are perhaps even more disturbing as they are scenarios in which hostile actors (even non-state entities) may have subsequently obtained control of the weapons and/or the fissile materials present inside them.

It's not just lower-yield fission bombs that have been lost, either. Fusion superbombs have gone completely missing as well. On February 5, 1958, a B-47 bomber and an F-86 fighter jet collided during a training exercise. The bomber was forced to jettison a 7,600-pound (3,447 kg) Mark 15 hydrogen bomb into the waters near Wassaw Sound off the coast of the U.S. state of Georgia. Despite a search mission spanning a timeframe of over two months and involving 100 Navy personnel, the bomb was never recovered.[668] The MK-15 weapons have a yield between 1.7 to 3.8 megatons.[669]

Broken arrows and other incidents are not limited to strategic bomber aircraft. Incidents on submarines and in silos have also occurred. In one such incident in late May of 1968, the USS *Scorpion* (SSN-589), a nuclear submarine, was reported overdue; she was on a NATO operation in the Atlantic and Mediterranean. Present on the sub were two MK 435 ASTOR torpedoes loaded with tactical nuclear warheads. By October, Navy ships had located the ship's wreckage in more than 10,000 feet of water, around 400 miles southwest of the Azores, near Portugal. The entire crew was lost. The weapons, containing enriched uranium and plutonium, were never recovered, although they are assumed by the U.S. military to be rendered inoperable due to extreme water pressure and ocean seawater contamination.[670,671]

Silo accidents have also endangered the United States' homeland. In one such notorious 1980 accident in the Titan II ICBM silo in Damascus, Arkansas, an Air Force repair crew member performing regular maintenance dropped a heavy wrench socket, which rolled off a work platform and fell into the silo. The socket bounced, striking the missile and causing a leak from a pressurized fuel tank (the Titan II was the last liquid-fueled ICBM built by the United States). A team of specialists was called in from Little Rock Air Force Base to attempt to control the situation. Unfortunately, 8.5 hours after the puncture, fuel vapors within the silo ignited and exploded, injuring 21 Air Force personnel and killing one. Fortunately, the missile's reentry vehicle and nuclear warhead were recovered intact and did not undergo fusion; this would have been a particularly violent and destructive outcome, as the Titan II ICBMs carried *incredibly* powerful W-53 nuclear warheads with a yield of nine megatons, around 600 times more powerful than the bombs dropped on Japan.[672],[673]

Incidents and accidents don't all result from explosions and physical crashes, though. Misunderstandings and misinterpretations between world powers also have the potential to lead to nuclear catastrophe. On the night of October 25, 1962, a guard at a base near Duluth Airport in Minnesota noticed a shadowy figure attempting to climb the facility's perimeter fence. He immediately fired upon the intruder and raised the alert, fearing that the incursion could be part of a wider Soviet attack on the United States (as you'll see later in this chapter, things were extremely tense between the United States and the Soviet Union at this time). Upon being notified, at nearby Volk Field, operators flipped the wrong switch. As a result of the improper switch flip, instead of seeing a standard security warning, pilots heard an emergency siren instructing them to scramble immediately, which they did (as it was their sworn duty). The pilots rushed to get nuclear weapons into the air. The pilots had been previously instructed that there would be no practice runs; if they were taking to the skies, it was to wage all-out nuclear war. As it turns out, it was quickly discovered that the "intruder" was likely a black bear, not a human being. A quick-thinking official put the pedal to the metal and gunned it toward the base in a truck, intercepting the pilots just in time as they were firing their engines on the runway, thus potentially averting World War III.[674]

Multiple false alerts from the North American Air Defense Command (NORAD)'s early warning systems would later result in similar incidents and close calls. In one such incident, on November 9, 1979, at 3 a.m., NORAD's missile warning display screens indicated an incoming attack from 1,400 Soviet ICBMs. The warnings simultaneously appeared at the Pentagon, Strategic Air Command (SAC), other U.S. defense installations, and the World Wide Military Command and Control System (WWMCCS [pronounced Wimex]). WWMCCS is an arrangement of personnel, equipment, facilities, communications infrastructure, and procedures used to plan, direct, coordinate, and control the operational activities of the U.S. armed services. WWMCCS uses computer systems and warning infrastructure to help inform and operationalize the decision-making capacity of the U.S. President and the Secretary of Defense. President Carter's national security advisor received three calls that night; the third came as he prepared to wake Carter and scramble the Strategic Air Command. NORAD notified in the third call that the United States' other early warning systems were reporting nothing that would indicate an attack. As it turns out, a lieutenant colonel at NORAD had accessed the wrong machine and inserted a war games tape into the missile-warning component of WWMCCS. The computer, unable to discern between a drill and an actual launch scenario, told the United States military establishment that the entire country was about to be devastated by a full-scale Soviet first strike. Soviet leadership admonished the United States for its grave security failure, but fortunately, as we know today, it did not lead to war as it was caught in time.[675,676,677]

While the aforementioned incidents and accidents can be attributed to the United States and NATO, the Soviet Union and Warsaw had far from a perfect record. Case in point: Stanislav Petrov. On September 26, 1983, only three weeks after the Soviet military had downed Korean Air Lines Flight 007, duty officer Petrov just may have saved the world. In the early morning, Soviet early-warning systems triggered an alert that the United States had initiated a *massive* first strike against the Soviet Union. Soviet protocol was to launch on warning and retaliate with a second strike to ensure that the United States could not walk away unscathed. As sirens blared and his screens lit up with the word "LAUNCH," Mr. Petrov froze. Alarms continued to trigger; now, the screen indicated several missile launches, and the computer system warned that the reliability of the alert was the highest possible, changing its alerts from "LAUNCH" to "MISSILE STRIKE." Petrov's job wasn't to think; it was to indicate to his superiors that Warsaw would soon be devastated by NATO's nuclear arsenal. But like in the earlier American scenario, while the computers were throwing flags, the other warning systems did not corroborate; the satellite radar operators told Petrov they didn't register any missiles. Radar was only a support, though; the protocol said that the decision must be based on computer intelligence. Petrov's quick thinking, his assessment of the radar information, and his gut instinct (this simply did not seem like enough missiles to be a first strike from the United States) led him to tell his superiors that it was a system malfunction; had he been wrong, the Soviet Union would have been obliterated and its second strike capabilities severely damaged. Fortunately for millions, maybe billions, Petrov was right, and he may have saved the world from a nuclear holocaust. Mr. Petrov was later officially reprimanded by Soviet officials. The incident was kept secret for years.[678,679,680]

Aside from weapons incidents, the Soviets also faced issues at its weapons refinement facilities. The Kyshtym Disaster on September 29, 1957, was one particularly awful event. The Kyshtym industrial complex (also known as "Mayak"), built in the 1940s on the eastern slopes of the central Ural Mountains, featured nuclear reactors and a plutonium processing facility. The plant's operators failed to repair the cooling system for a buried tank storing highly radioactive nuclear waste. Over a year, the contents grew hotter and hotter due to radioactive decay until they reached about 660 degrees Fahrenheit (350 °C). The tank exploded with a force in excess of 70 tons of TNT (by no means rivaling a nuclear weapon, but still incredibly powerful). The tank's three-foot (1 m) concrete lid blew off, spewing cesium-137 and strontium-90 into the atmosphere. Ultimately, 9,000 square miles (23,000 km^2) of land were contaminated, 10,000 were evacuated, and perhaps hundreds died due to exposure to radioactive contamination. The Soviet Union denied any such accident took place until over three decades later.[681]

Because of the Soviets' proclivity for extreme secrecy in any events that could cause embarrassment or diplomatic problems for their government on the world stage, we'll likely never know the true extent of their nuclear close calls, broken arrows, and other incidents. A number of Soviet (and later, Russian Federation) nuclear submarine accidents have been documented over the decades.[682] But risks for broken arrows don't surface only from human or mechanical error; they also present themselves in the form of nation-state collapse. When the Soviet Union collapsed in December 1991, thousands of strategic and tactical Soviet nuclear weapons were still deployed over multiple Soviet successor states, namely Ukraine, Kazakhstan, Belarus, and Russia. Twenty-two thousand tactical nuclear weapons were spread across 15 former Soviet states (this includes those weapons that remained in Russia).[683]

Though the United States and Russia agreed that Russia should be the single, centralized successor state to take control of its strategic and tactical nuclear weapons, there were challenges. Namely, the government of Ukraine, cognizant of its hostile history with the Russians, initially wanted to retain control of its strategic arsenal to deter Russian influence and potential invasion. Unfortunately for the United States, many of these weapons were still aimed at strategic targets in the United States, such as military installations and major cities. In one of the Clinton Administration's most important diplomatic achievements, it used financial incentives, diplomacy, and warnings about potential disasters to convince the Ukrainians it was in everyone's best interests that they turn the nukes back over to Russia. Bilateral cooperation between the United States and Russia, and multilateral cooperation with Soviet successor states, resulted in a total of 14,000 tactical warheads being moved from non-Russian states, many of them subsequently disabled and blended down for nuclear fuel. Perhaps as concerning as the tactical weapons was the risk that nuclear materials from disassembled bombs could be sold on the black market. Despite all the risk, through impressive international cooperation, there are no known incidents of ex-Soviet nuclear weapons or materials being sold on the black market.[684] All that said, the reality remains that a nation-state in disarray, revolution, or a state of disintegration could someday lead to its nuclear arsenal winding up in the hands of actors who mean harm to the world, particularly terrorist entities attempting to obtain nuclear materials on the black market. We've dodged that metaphorical bullet so far, but the future is uncertain.

Of course, no mention of nuclear brinkmanship would be legitimate without the mother of all close calls: the Cuban Missile Crisis. The Lockheed Martin U-2 jet-powered spy plane was one of the most valuable tools for global reconnaissance in the United States' arsenal. Able to cruise at 70,000 feet for several hours, its ability to monitor for nuclear weapons and take aerial photography assisted the missions of both the Central Intelligence Agency (CIA) and the United States Air Force (USAF) immeasurably during the Cold War.[685] One such reconnaissance mission led to a series of events that could have *ended* the human experience forever. On October 16, 1962, a U-2 spy plane flying over Cuba photographed nuclear missile sites being built by the

Soviet Union. Subsequent surveillance revealed by the next day that 16-32 missiles were in the process of being situated in Cuba at multiple launch sites. Given Cuba's proximity to the United States and U.S. leadership's severe concern over the construction of silos in Cuba that could launch medium-range ballistic missiles and intermediate-range ballistic missiles (MRBMs and IRBMs– shorter range than global intercontinental ballistic missiles [ICBMs]) at the U.S. homeland, the Kennedy Administration kept this information secret until it could formulate a plan. As it turns out, the intelligence was accurate, and their concerns were not unfounded– Earlier that fall, Soviet Premier Nikita Kruschev had ordered a nuclear strike force to be constructed in Cuba that could reach most major U.S. cities with strategic warheads within *five minutes*. This would take from the United States its ability to respond under a first strike scenario from the Soviet Union, and the Kennedy Administration interpreted it as an aggressive move that could seriously destabilize the global balance of power.[686,687,688]

Kruschev's decision to put nuclear missiles in Cuba did not come lightly. While the lead-up to the crisis has a long and complicated history, suffice it to say that Cuba was a communist ally state to the Soviet Union, and the Soviets were concerned about American tensions with Cuba. The failed Bay of Pigs invasion in April 1961 ratcheted up the tensions between the capitalist and communist nations of the United States and Cuba, situated only 90 miles away from each other. Kruschev and Castro reached an agreement to situate nuclear missiles in Cuba to deter the United States from further action there. There were three main impetuses for this:

1. **U.S. SM-78 Jupiter Missiles in Turkey:** The United States had medium-range, ground-launched ballistic missiles capable of carrying 1.5 megaton nuclear warheads deployed in Turkey – close to the Soviet

Union – and the Soviets wanted similar strike capability against the U.S.;

2. **Internal Soviet Politics:** Kruschev wanted to appear strong to other Soviet government leadership and his people, demonstrating that he could stand up to NATO; and

3. **Deterrence:** The Soviet Union wanted to prevent further United States attempts to overthrow the communist Castro Regime in Cuba.[689,690]

On October 22, Kennedy established the Executive Committee of the National Security Council (ExComm) to address the crisis. ExComm considered multiple options:

1. **Do Nothing in Response:** This was not seen as a viable option, as it would be domestically untenable, and the Soviets would see it as weakness that would encourage more aggressive maneuvering;

2. **Use the United Nations:** The Soviets would block action in the Security Council, so this option would be of little use. It also may be interpreted by foreign nations as America signaling it could not manage its own affairs;

3. **Ask the Soviets to Remove the Missiles:** The Soviets had already denied the existence at the UN emergency meeting on the matter, so it was unlikely they'd respond to such a polite request;

4. **Blockade Cuba:** This would prevent the Soviets from bringing in more hardware and would allow time for discussion, but would leave the existing Soviet missiles intact; or

5. **Air Attacks on and/or Full-Scale Invasion of Cuba:** This would be extremely taxing on United States

resources and personnel and would lead to an escalation with the Soviets that could result in a full-scale nuclear exchange and global annihilation.[691]

Ultimately, the United States determined that the Soviet missile deployment was too dangerous and *must* be stopped. The Administration ordered a naval blockade around Cuba to prevent the Soviets from bringing in additional military hardware. The United States demanded that Krushchev direct the Soviets to destroy the existing missile sites and remove their nukes from Cuba. On October 22, 1962, President Kennedy addressed the American people, informing them of the unfolding crisis [speech truncated and emphasis added]:

Within the past week, unmistakable evidence has established the fact that a series of offensive missile sites is now in preparation on that imprisoned island. The purpose of these bases can be none other than to provide a nuclear strike capability against the Western Hemisphere…

The characteristics of these new missile sites indicate two distinct types of installations. Several of them include medium range ballistic missiles, capable of carrying a nuclear warhead for a distance of more than 1,000 nautical miles. Each of these missiles, in short, is capable of striking Washington, D.C., the Panama Canal, Cape Canaveral, Mexico City, or any other city in the southeastern part of the United States, in Central America or in the Caribbean area.

Additional sites not yet completed appear to be designed for intermediate range ballistic missiles – capable of traveling more than twice as far -- and thus capable of striking most of the major cities in the Western Hemisphere, ranging as far north as Hudson Bay, Canada, and as far south as Lima, Peru. In addition, jet bombers capable of carrying nuclear weapons are now being uncrated and assembled in Cuba while the necessary air bases are being prepared...

Acting, therefore, in the defense of our own security and of the entire Western Hemisphere, and under the authority entrusted to me by the Constitution as endorsed by the resolution of the Congress, I have directed that the following initial steps be taken immediately:

First: To halt this offensive buildup, a strict quarantine on all offensive military equipment under shipment to Cuba is being initiated. All ships of any kind bound for Cuba from whatever nation and port will, if found to contain cargoes of offensive weapons, be turned back...

Second: I have directed the continued and increased close surveillance of Cuba and its military buildup. The foreign ministers of the OAS, in their communique of October 6, rejected secrecy on such matters in this hemisphere. Should these offensive military preparations continue, thus increasing the threat to the hemisphere, further action will be justified. I have directed the Armed Forces to prepare for any eventualities; and I trust that in the interest of both the Cuban people and the Soviet technicians at the sites, the hazards to all concerned of continuing the threat will be recognized.

Third: It shall be the policy of this nation to regard any nuclear missile launched from Cuba against any nation in the Western Hemisphere as an attack on the United States, requiring a full retaliatory response upon the Soviet Union.

Fourth: As a necessary military precaution, I have reinforced our base at Guantanamo,
evacuated today the dependents of our personnel there, and ordered additional military units to be on a standby alert status.

Fifth: We are calling tonight for an immediate meeting of the Organ of Consultation under the Organization of American States, to consider this threat to hemispheric security and to invoke articles 6 and 8 of the Rio Treaty in support of all necessary action. The United Nations Charter allows for regional security arrangements -- and the nations of this hemisphere decided long ago against the military presence of outside powers. Our other allies around the world have also been alerted.

Sixth: Under the Charter of the United Nations, we are asking tonight that an emergency
meeting of the Security Council be convoked without delay to take action against this latest Soviet threat to world peace. Our resolution will call for the prompt dismantling and withdrawal of all offensive weapons in Cuba, under the supervision of U.N. observers, before the quarantine can be lifted.

Seventh and finally: I call upon Chairman Khrushchev to halt and eliminate this clandestine, reckless, and provocative threat to world peace and to stable relations between our two nations. I call upon him further to abandon this course of world domination, and to join in an historic effort to end the perilous arms race and to transform the history of man. He has an opportunity now to move the world back from the abyss of destruction -- by returning to his government's own words that it had no need to station missiles outside its own territory, and withdrawing these weapons from Cuba -- by refraining from any action which will widen or deepen the present crisis -- and then by participating in a search for peaceful and permanent solutions...[692]

By only the next day, October 23, 1962, Kruschev had rejected the idea of a blockade. He wrote a letter to Kennedy informing the United States that Soviet ships would not be stopping at the blockade and would force their way through; the letter read, in part:

> The statement by the Government of the United States of America can only be regarded as undisguised interference in the internal [affairs] of the Republic of Cuba, the Soviet Union, and other states. The United Nations Charter and international norms give no right to any state to institute in international waters the inspection of vessels bound for the shores of the Republic of Cuba.
>
> And naturally, neither can we recognize the right of the United States to establish control over armaments which are necessary for the Republic of Cuba to strengthen of [sic] its defense capability.
>
> We affirm that the armaments which are in Cuba, regardless of the classification to which they may belong, are intended solely for defensive purposes, in order to secure the Republic of Cuba against the attack of an aggressor.[693]

Ultimately, despite Kruschev's strong words, Soviet freighters en route to Cuba with military hardware stopped before the quarantine, but the oil tanker Bucharest continued toward Cuba. By October 24, 180 United States Ships, submarines, troops, and Air Force were placed on full alert.

Nuclear-armed Soviet submarines moved into the Caribbean, threatening the U.S. quarantine. Tensions remained incredibly high as U.S. leadership, following Kruschev's letter, grappled with the real possibility that U.S. combat assets may be forced to fire upon Soviet ships.[694] Kennedy sent another letter to Kruschev citing the resolution of the Organization of American States, which backed the blockade, and asked the Soviets to comply.[695] Kruschev responded on October 24 [letter copied in part]:

> You, Mr. President, are not declaring a quarantine, but rather are setting forth an ultimatum and threatening that if we do not give in to your demands you will use force. Consider what you are saying! And you want to persuade me to agree to this! What would it mean to agree to these demands? It would mean guiding oneself in one's relations with other countries not by reason, but by submitting to arbitrariness. You are no longer appealing to reason, but wish to intimidate us.[696]

Over the next couple of days, tensions continued to rise as the Americans and Soviets bombastically confronted each other in the halls of the United Nations, disregarding the usual decorum and measured, polite language typically used by allies, "frenemies," or rivals on the international stage. Though the Bucharest was allowed through the blockade because it did not possess military hardware on board, the Americans rejected a cooling-off period proposed by some international actors because missiles would remain in Cuba.[697] By October 26, in a private letter to Kruschev, Fidel Castro urged the Soviets to engage a first strike against the United States if the U.S. were to invade Cuba [copied, in part, and emphasis added]:

> [I] consider an attack to be almost imminent--within the next 24 to 72 hours. There are two possible

variants… an air attack against certain objectives with the limited aim of destroying them; the second, and though less probable, still possible, is a full invasion…

If the second variant takes place and the imperialists invade Cuba with the aim of occupying it, the dangers of their aggressive policy are so great that after such an invasion the Soviet Union must never allow circumstances in which the imperialists could carry out a nuclear first strike against it…

I tell you this because I believe that the imperialists' aggressiveness makes them extremely dangerous, and that if they manage to carry out an invasion of Cuba-- a brutal act in violation of universal and moral law-- **then that would be the moment to eliminate this danger forever, in an act of the most legitimate self-defense. However harsh and terrible the solution, there would be no other.**[698]

On this same day, October 26, during the height of the crisis, the Soviet resolve began to show signs of cracking. Kruschev sent Kennedy a letter offering to pull the missiles out of Cuba in exchange for a lifting of the quarantine and a U.S. commitment not to invade Cuba [letter copied, in part]:

These missiles are a means of extermination and destruction. But one cannot attack with these missiles, even nuclear missiles of a power of 100 megatons because only people, troops, can attack. Without people, any means however powerful cannot be offensive…

All the means located there, and I assure you of this, have a defensive character, are on Cuba solely for the purposes of defense, and we have sent them to Cuba at the request of the Cuban Government. You, however, say that these are offensive means...

I believe that you have no basis to think this way. You can regard us with distrust, but, in any case, you can be calm in this regard, that we are of sound mind and understand perfectly well that if we attack you, you will respond the same way. But you too will receive the same that you hurl against us. And I think that you also understand this....

If assurances were given by the President and the Government of the United States that the USA itself would not participate in an attack on Cuba and would restrain others from actions of this sort, if you would recall your fleet, this would immediately change everything . . . Then, too, the question of armaments would disappear, since, if there is no threat, then armaments are a burden for every people. Then too, the question of the destruction, not only of the armaments which you call offensive, but of all other armaments as well, would look different...

Mr. President, we and you ought not now to pull on the ends of the rope in which you have tied the knot of war, because the more the two of us pull, the tighter that knot will be tied. And a moment may come when that knot will be tied so tight that even he who tied it will not have the strength to untie it, and then it will be necessary to cut that knot, and what that would mean is not for me to explain to you, because you yourself understand perfectly of what terrible forces our countries dispose....

Consequently, if there is no intention to tighten that knot and thereby to doom the world to the catastrophe of thermonuclear war, then let us not only relax the forces pulling on the ends of the rope, let us take measures to untie that knot. We are ready for this.

We welcome all forces which stand on positions of peace...[699]

While October 26 and the Kruschev letter initially may have brought some comfort, it was short-lived. By October 27, the Soviets sent a letter demanding the removal of the United States Jupiter missiles from Turkey. Then, tragedy struck. At this point in the crisis, Soviet SA-2 surface-to-air missile (SAM) sites were now active and armed inside Cuban territory. As Major Rudolph Anderson flew his U-2 spy plane over Cuba, photographing Soviet missile sites, the Soviets' radar arrays began to ping his plane. Concerned the Americans were compromising the secrecy of Soviet tactical nuclear assets, and unable to locate the general who had the authority to authorize a launch that could seriously escalate the conflict into full-scale nuclear war, Soviet Lieutenant General Stepan Grechko took it upon himself to greenlight a shoot down: "Destroy Target Number 33." Anderson was killed as the U-2 spy plane, Target Number 33, plummeted 72,000 feet to the ground.[700,701]

The White House, already on edge after another U-2 plane accidentally entered Soviet territory and was chased down by Soviet MiGs, soon received word of the escalation. Assistant Secretary of Defense Paul Nitze declared, "They've fired the first shot," with President Kennedy observing, "We are now in an entirely new ball game." Attorney General and brother of President Kennedy, Robert F. Kennedy, would later write in his memoir that "There was the feeling that the noose was tightening on all of us, on Americans, on mankind, and that the bridges to escape were crumbling." Military leaders implored Kennedy to respond with force, advising him to launch airstrikes against Cuba's air defenses the following morning.[702]

Air Force troop carrier squadrons were ordered to active duty in case an invasion would be required, but Kennedy did not prefer escalation. Allowing cooler heads to prevail, he pushed back against the generals urging him to strike the Cuban SAM sites. Later in the night of October 27, Robert F. Kennedy met with Ambassador Anatoly Dobrynin, wherein they reached an understanding: The Soviet Union would withdraw its missile assets from Cuba under the supervision of the United Nations. In exchange, the Americans would pledge not to invade Cuba. An additional secret understanding would be worked out to pull America's Jupiter missiles out of Turkey.[703]

By October 28, 1962, the 13-day Cuban Missile Crisis had ended. The Americans and the Soviets had agreed that the presence of missiles both in Turkey and in Cuba would be too dangerous for both sides and would risk fatal escalation. Concessions from both sides guaranteed the continued independence of Cuba and a greater sense of security for both Washington and Moscow. What's often left out of accounts of the Cuban Missile Crisis, though, is perhaps humankind's closest brush with midnight, and it happened quietly, underneath the ocean...

On October 27, 1962, the day before the Cuban Missile Crisis ended and cooler heads prevailed, Vasili Arkhipov may have saved the entire world. As mentioned earlier, the Soviets had deployed a number of submarines around Cuba, including the Soviet diesel submarine B-59, commanded by Captain Second Rank Valentin Savitsky. Unbeknownst to American forces, the submarine was equipped with a 10-kiloton nuclear torpedo; officers aboard the sub had authorization to launch, if necessary, without preapproval from Moscow. During this time, U.S. forces in the area (11 U.S. destroyers and the aircraft carrier USS *Randolph*) were dropping non-lethal depth charges in an attempt to pressure Soviet submarines into surfacing. Washington had notified Moscow that the United States had no lethal intent, but B-59, being incommunicado (out of communication) with the outside world at the time, was unaware of the Americans' intentions. To the staff on the sub, the depth charges appeared to be an attempted bombing of the submarines and the outbreak of war.[704,705]

As they were trapped in the submarine with the air conditioning failing, carbon dioxide levels rising, and unaware that the Americans were not actually attempting to sink them, the crew became alarmed. Savitsky, the submarine's commander, was resolute: "We're gonna blast them now! We will die, but we will sink them all - we will not become the shame of the fleet." One of the signatures of the Soviets' command and control regime for nuclear weapons was that multiple persons would have to assent to the use of nuclear weapons before they could be deployed. On this particular ship, all three senior officers had to agree. Arkhipov refused, calming down the captain and explaining that the depth charges were the only way the Americans could communicate with the sub... Arkhipov's even temper and reason prevailed, very likely averting global nuclear catastrophe.[706,707]

In 2017, Arkhipov became the first person to receive the Future of Life award, issued by the Cambridge-based existential risk nonprofit, the "Future of Life Institute (FLI)." Though the reward was posthumous (Arkhipov passed away in 1998), FLI President Max Tegmark said at the award ceremony that Arkhipov is "arguably the most important person in modern history." Indeed, his heroism and measured response at minimum, saved thousands of lives– even, perhaps *all* known life.[708]

The 1963 U.S.-Soviet Memorandum of Understanding ("The Hotline Agreement") established a direct communications link between the United States and Soviet governments following the Cuban Missile Crisis. The agreement was designed to speed up communications between the Soviets and the Americans to prevent the possibility of accidental nuclear war. The hotline had two terminal points, Washington and Moscow, with a full-time duplex wire telegraph circuit with teletype equipment between the two points. In the event of a circuit interrupt, messages could be translated over radio. The hotline has been used multiple times, including in the wake of President Kennedy's assassination. It has since been updated to use more modern technologies, such as satellite and fax machines. Bilateral hotline agreements have been established between multiple nuclear powers, such as the Russia-China Nuclear Hotline, the India-China Hotline, the India-Pakistan Nuclear Hotline, the U.S.-China Nuclear Hotline, and others.[709]

Chapter 23. Duck and Cover: Surviving Nuclear Catastrophe

"There was a turtle by the name of Bert, and Bert the turtle was very alert; when danger threatened him he never got hurt, he knew just what to do..." -Duck and Cover, from the United States 1952 Civil Defense Film[710]

Note: I must preface this section by saying that nothing contained herein should be construed as medical advice or that you should rely solely on this information in an emergency scenario. In such a situation, your local and national authorities and medical professionals are the best sources of information. Further, as science evolves, new treatments and preventative options become available, and collective social knowledge of nuclear effects increases, so when you read this information, it may be outdated. Please consult the latest news and information available from medical and state authorities in an emergency scenario. Finally, while this chapter is addressed to the reader, I have included it partially because a central motivation of this book is to encourage leadership responsibility and respect for the gravity of this technology; this chapter contains a series of U.S. federal recommendations that may be used by state and local policymakers to prepare adequate nuclear disaster and contingency plans for population centers and to inform the public of best practices. In short, it's an outline for the basis of a good preemptive planning guide.

While *"Duck and Cover"* is a fun jingle, we can't pretend there will be any hope in a full-scale thermonuclear exchange that most lives can be saved. Those near the blast's epicenter will be in serious physical jeopardy and casualties will be enormous. Take, for example, a weapon with a yield the size of Tsar Bomba (50 megatons) detonated as an air burst over New York City. The zone at which the population would be exposed to about 5 sieverts of radiation, which is a high lethal dose, would be close to 12 square miles (31 km²) and would consume much of the island of Manhattan. The fireball radius would be approximately 26 square miles (67.1 km²), which would stretch across Manhattan, into the outer boroughs and into New Jersey, vaporizing significant portions of these areas in nuclear fire. Heavy blast damage would be experienced for 96 square miles (249 km²) and would level concrete structures in vast parts of the Bronx, Queens, and Brooklyn, spreading mass destruction into Jersey City, Hoboken, Weehawken, and up to Fort Lee. Moderate blast damage would be seen for 522 square miles (1,350 km²), stretching as far north as Yonkers, as far South as Coney Island, as far east as Montclair, NJ, and as far West as Floral Park in Long Island. In the moderate blast zone, most residential buildings will collapse, injuries will be virtually universal, and fatalities will be widespread. Fires will also spread uncontrollably. Light blast damage would be experienced through all of NYC, vast parts of New Jersey, half of Long Island, and as far Northeast as Stamford, Connecticut. Thermal radiation, which can cause third-degree burns, would extend over an even larger radius.[711] Just as a final note on this paragraph– I've used square miles here to encapsulate the total land affected by the blast in this hypothetical scenario, although the radii of these blast zones are smaller when expressed in this measurement (the fireball radius on Tsar Bomba, for example, would be 2.87 miles [4.62 km]). The thermal radiation zone would have a radius of 37.3 miles (60 km).

All told, in the scenario laid out above, it's estimated New York City and surrounding areas would sustain over 7.6 million casualties, with an additional 4.15 million injured. These numbers do not consider the effects of radioactive fallout, as outcomes from fallout will depend on several factors, including the precautions that government entities and the population collectively take.[712] This chapter will explore the pre- and post-detonation precautions that individuals can take to shield themselves from the initial detonation and the effects that take place afterward.

There are multiple sources of risk created by a nuclear detonation. These include:

1. **Bright Flash** - Never look at a nuclear detonation. At a distance of one mile from detonation, the flash can be as bright as 1,000 midday Suns.[713] This can damage the eyes, causing blindness and inhibiting an individual from taking the immediate steps required to avoid fallout.

2. **Blast Wave** - This can cause injury or death, and significant damage to structures several miles away from the blast's epicenter.

3. **Radiation** - This damages the body, its cells, and its DNA. It can cause acute radiation sickness, cancer and death.

4. **Fire and Heat** - This can cause significant, sometimes fatal, burn injuries, and will ignite structures for miles around.

5. **Electromagnetic Pulse (EMP)** - The EMP generated by a nuclear weapon can damage electronics many miles away from the detonations and cause disruptions to vehicles, communications infrastructure, energy infrastructure, and destroy all range of electronics, from phones, to computers, to electric generators.

6. **Fallout** - This is radioactive in nature and can cause visible dirt, dust, and contaminated rain to blanket the areas near the blast, causing acute radiation sickness and cancer to those nearby.[714]

If you are in the United States, prior to an attack, you will likely receive a wireless emergency alert (WEA) on your cell phone. This system has been used more than 78,000 times to warn the public about dangerous weather, missing children, and other critical situations since 2012. Alerts are broadcast to geographic areas affected by emergencies, so it's likely that the federal government and local/state affiliates would work together with cell phone network providers to send out an alert advising citizens to take shelter in the event of a launch. While consumers can block some alerts, the WARN Act, which authorized the WEA system, did not permit an opt-out for National alerts.[715]

If, despite receiving notice, you are caught outside during the detonation, your level of survivability will be primarily determined by your proximity to ground zero. If you're far enough away, there are some measures you can take to increase your chances of survival. Whether inside or outside, lie face down on the ground and protect your exposed skin. Place your hands under your body, and remain flat until the heat and shock waves have passed. Cover your mouth and nose with a clean cloth to filter particulates from the inhaled air. If you're on the move afterward, leave the area by a route perpendicular to the path of incoming fallout.[716]

The first precaution recommended by the Federal Emergency Management Agency (FEMA), assuming one survives an initial detonation, is to get inside. Fallout can take more than 15 minutes to reach the ground outside the immediate blast damage zones, but can arrive sooner or later depending on conditions outside. Either way, this is typically enough time for individuals to mitigate their exposure to radiation. As mentioned before, brick or concrete buildings are best because they are sturdy, less susceptible to fire, and more likely to block neutron and gamma radiation capable of penetrating structures. Upon making it inside, you should take the following steps:

1.**Decontaminate:** If possible, remove and discard clothing that may have come in contact with fallout. Removing outer layers of clothing may reduce your exposure to radioactive material by up to 90%. Place the contaminated clothing in a bag and throw it outside and away from people. Wash or wipe off contaminated skin, preferably with a shower if available. Do not scrub or scratch your skin; this can open up wounds or pores, forcing radioactive material into your body. Shampoo, soap, and water will help remove radioactive materials from your hair, but do not use conditioner, as this can bind the material to your hair. Blow your nose and wipe your eyelids and eyelashes clean with a wet cloth. Wipe your ears as well.

2.**Find the Building's Core:** Go to the basement of the building or to the center of the building if no basement is available; these areas are likely to provide the greatest protection from fallout and radiation. Close all exterior windows and doors so that fallout does not enter the building.

3.**Stay Inside:** Remain inside the building for at least 24 hours unless you receive other instructions from authorities. Do not leave the building to find family or friends. You will only expose yourselves or them to radiation. Keep pets inside, as they will come into contact with radiation and fallout if they go outside, and returning will contaminate the building's occupants. If possible, deactivate fans, air conditioners, and other HVAC equipment that could bring air in from the outside. Close fireplace dampers. Any air system that brings in air from the outside could force fallout into the building.

4.**Tune In:** The devastation brought on by the nuclear weapons and the resultant EMP will likely have disrupted communications infrastructure, but some radios may still function. In this scenario, a battery-operated or hand crank radio may be useful to obtain more information.[717,718]

The "Rad Resilient City Initiative" (the "Rad Initiative") by the Center for Health Security at Johns Hopkins University sought to provide recommendations for cities in the event of a nuclear detonation. It notes that federal modeling of a 10-kiloton ground burst (a relatively small explosion smaller than Hiroshima and Nagasaki) demonstrated that if those at risk of exposure sought shelter in a shallow basement or equally protective structure such as a large office building, underground garage, or subway tunnel, 280,000 lives could be saved. Interior areas within a building, such as bathrooms and stairwell cores at the building's center, are more likely to be insulated from fallout. Similarly, brick and cement structures will provide much better protection from penetrating radiation than wood, drywall, or sheet metal (such as would be found in a pole barn or shed). Lead is the most protective, followed by solid steel, concrete, earth, and finally, wood, which is not a good radiation insulator. The Rad Initiative notes that fatalities can be minimized if individuals shelter in place (instead of attempting to run once a detonation has already occurred) and take measures to protect themselves in the time between the blast and when emergency responders arrive.[719]

The Rad Initiative recommends finding shelter *immediately*, but if you cannot immediately find a quality shelter and you're forced to locate inside a wood-frame house without a basement near the blast zone, for example, the time to upgrade shelters is at the one-hour mark. The potential for radiation exposure decreases by 55% after the one-hour point following a detonation. Quickly relocating to a protective shelter such as an underground tunnel is advised, and is worth the outdoor exposure for a brief period for the sake of minimizing long-term radiation exposure. Those in a quality shelter environment, such as a deep basement or a large concrete structure, should remain sheltered in place. After one day, potential radiation exposure will have dropped by around 80%. If information is available from authorities on evacuation routes and procedures after one day, shelter occupants should evacuate. If such information is unavailable, staying put for 2-3 days or more is likely the best approach. In some situations, authorities may recommend that you remain sheltered for up to a month.[720,721] Radiation levels may be reduced by 99% after around two days. The direction and severity of fallout and radiation levels will be determined by the prevailing winds, which will generally push the radioactive fallout into a narrow oval pattern away from ground zero of the blast (the "plume trajectory").[722,723]

When traveling to a new shelter or during the evacuation, if you have a mask available or a cloth to cover your face, wear it, but do not do so if the mask has been contaminated or exposed to radioactive dust/debris. Remember, you cannot see or feel radiation, so you won't know it's there. It's better not to wear a mask if it may have been outdoors and contaminated, as this will increase your chance of contamination, not decrease it. Food that is wrapped in containers/packing or is found in a refrigerator following a blast will be less likely to be contaminated with radioactive fallout. Still, you must ensure that your hands and face are clean before you eat or drink, as this will cause you to take in radioactive contamination internally, which can more readily harm you from the inside.[724]

Shelters should be prepared with several materials and basic necessities of life to sustain those within the shelter for the time necessary. The more space and materials available, the better, as it will afford shelter occupants with the ability to shelter in place for more extended periods of time, if necessary, as society engages its disaster recovery resources. The U.S. Federal Government recommends the following items for shelter facilities:

1. One gallon of water per person daily for both drinking and sanitation purposes.
2. Non-perishable food (such as canned goods, packaged foods that do not require refrigeration, etc.) Any foods that require heating and water will require additional water use or a fuel source.
3. Manual can opener (for opening canned food).
4. Battery-powered or hand-crank radio.
5. Flashlight.
6. First aid kit.
7. Extra batteries for light sources/radios.
8. Whistle (to signal for help).
9. Dust masks (to filter contaminated air and to prevent fallout dust and debris from entering the body).
10. Plastic sheeting and duct tape (to seal off building entrances).
11. Wrench or pliers (to shut off utilities, if necessary; for example– if there is a water leak).
12. Local maps (remember that cell phones and GPS may be disrupted or disabled altogether).
13. Cell phone with chargers and a backup battery (in the event that cell services are maintained and functioning).
14. Soap, hand sanitizer, and disinfecting wipes for surfaces and removing fallout and dust from the body.
15. Prescription medications, and essential over-the-counter drugs such as pain relievers, cough remedies, allergy medications, antacids, laxatives, topical

medications and disinfectants, and antiemetics (these may be helpful for nausea and sickness in the event of radiation exposure).

16. Prescription eyeglasses and contact lens solution.

17. Baby formula, bottles, wipes, diapers, and diaper rash cream.

18. If you have pets – pet food and extra water.

19. Important family documents, insurance policies, identification, bank records, cash, and traveler's checks.

20. Sleeping bag or a warm blanket for each person, given that heat and utilities may be disrupted.

21. Complete changes of clothing appropriate to the climate, and sturdy shoes.

22. Fire extinguisher.

23. Matches in waterproof containers for fire-starting, if necessary.

24. Feminine supplies and other personal hygiene items.

25. Paper cups, plates, paper towels, and plasticware (you do not want to waste water cleaning non-disposable dishes).

26. Papers and pencils for writing and communication, if necessary.

27. Books, games, puzzles, and other activities. You cannot count on the use of electronic devices or phones, so these should be available for mental health purposes and passing the time.[725,726]

An additional note on medications– a few drugs are available that may help mitigate some of the effects of radiological contamination. The most well-known of these is potassium iodide (KI). The body's thyroid gland cannot determine between non-radioactive and radioactive iodine, and will readily absorb both kinds. The thyroid controls multiple bodily processes, including blood pressure, body temperature, heart rate, and childhood growth and development, and the highest risk is to children and infants if the thyroid is affected. In a nuclear detonation, a radioactive isotope of iodine, iodine-131, is released through fallout. Taking potassium iodide pills or liquid in the event of a detonation will effectively flood the thyroid with iodine and prevent it from intaking iodine-131 for the next 24 hours. This will protect the thyroid and could potentially prevent future damage or thyroid cancers. Table salt and dietary supplements are not a substitute for medical-grade potassium iodide. You should only take potassium iodide on the advice of a medical doctor, public health, or emergency management official, as taking too much or taking it when it isn't recommended can cause substantial health risks. Contrary to unfortunate popular belief, KI is not a cure-all for exposure to fallout; it protects the thyroid and only the thyroid. The rest of the body and other organs/vital systems will still be totally exposed to the effects of any fallout encountered, regardless of the intake of KI.[727,728,729]

Prussian blue is another drug that can be used in a radiation emergency. As noted before, cesium-137 is a common radioactive isotope that causes contamination in the event of a nuclear detonation. Prussian blue can be taken in pill form. It can help remove radioactive isotopes of cesium and thallium from the body by trapping them in the intestines and preventing uptake by the digestive tract. The cesium and thallium will then move through the body and will be excreted in bowel movements. Prussian blue is *highly* effective for removing cesium-137, which can cause burns, acute radiation syndrome, various cancers, and death. Prussian blue may only be taken by prescription, and as with potassium iodide, should only be taken upon recommendation of public health authorities and/or a physician; over-the-counter supplements such as artist's dye are not an adequate replacement for prescribed Prussian blue.[730,731,732]

Diethylenetriamine pentaacetate (DTPA) can also prevent radioactive materials from staying in the body. DTPA binds to americium, curium, and plutonium, decreasing the time it takes to get those elements out of the body. DTPA comes in zinc and calcium forms, and binds those elements to them, passing them through the body in the urine. Calcium DTPA is more effective if administered on the first day after exposure, and zinc-DTPA is equally as effective as the calcium form after day one. DTPA is administered intravenously, so it's unlikely to be available in most shelter environments. It also comes in a mist form that can be taken into the lungs by those who have inhaled radioactive isotopes. As with all these other medications, DTPA should only be taken on doctors' and public health agencies' advice and administered by a licensed health professional.[733]

Finally, in the event that fallout exposure cannot be mitigated or eliminated, there is a new treatment for those who have received high doses of radiation. Filgrastim helps to stimulate the growth of white blood cells, which makes patients less vulnerable to infections. As you'll recall, one of the effects of a high dose of radiation in a short period is bone marrow and blood cell damage. Filgrastim speeds up the process of white blood cell creation, and can help sustain life in those who cannot produce new white blood cells in the short term. Filgrastim is given via injection and should only be administered/taken under the care of a physician and public health authorities.[734]

If you find yourself inside a building or shelter during a detonation, the best place to be is generally as far underground as possible. But even inside a shelter, especially inside an above-ground building (hopefully one made of brick, concrete, and steel), there are safer places to be in a room than others. Recent studies have shown that the safest point inside a room is to be positioned at the corners of the wall that faces the blast, which can shield occupants from the shockwave. The shockwave will fire into the room from the direction of the blast through any opening. Staying away from windows, doors, and corridors is best, as the airspeeds created by the blast's shockwaves can be deadly, even inside reinforced structures. The close quarters of a hallway, in particular, can amplify the force of the shockwave; the air from the blast can reflect off walls, bend around corners, and produce a force equal to 18 times the weight of the human body, enough to crush bones. The best location in a building is the half of the building farthest from the blast, in a room with no windows.[735,736]

Given the modern yields of thermonuclear weapons, there is little any human being can do to avoid catastrophe and death if within close proximity to a blast, even with the most reinforced structures. The best way to avoid harm is to be far away from ground zero of a detonation. As we now know, strategic targets are typically military bases, major cities, infrastructure, and manufacturing/industrial hubs. Being outside these areas makes it less likely that you'll be the victim of a direct hit. Certain places in the world, like Antarctica, will naturally be safer because there are no strategic targets there. Countries in the southern hemisphere, like New Zealand, will likely be safer because the nuclear powers (and their adversaries) are located in the northern hemisphere, and so that's where strategic targets will be found. Regardless of which country you live in, remote, rural areas (albeit ones that are far away from ICBM silos) are typically safest.[737]

While this section serves as a general guide regarding safety and mitigation measures that can be taken in case of a limited detonation scenario, I need to be careful to hedge here as I don't want to create false hope or misconceptions. In the event of a full-scale exchange of multiple air drop bombs, nuclear-tipped cruise missiles, submarine-launched ballistic missiles, and intercontinental ballistic missiles, millions, perhaps billions, will perish. There will be little anybody can do to protect themselves; even for those who survive, the world left afterward will be contaminated and poisoned, and the infrastructure and modern comforts of life will be a mere memory of the "before times." Death and dying will be ubiquitous, and even those not claimed by the bombs may suffer horrible fates later due to starvation, anarchy, and/or exposure. To close this chapter, although no form of the quotation has been 100% verified in the writings or speeches of former Soviet Premier Nikita Kruschev, it's often been said that he surmised of nuclear war, "The survivors would envy the dead."[738]

Part V. The Future: Disarmament, Contemporary Weapons, Hypothetical Doomsday Devices, and Conclusions

Chapter 24. Disarmament: Putting the Toothpaste Back in the Tube

"Today, every inhabitant of this planet must contemplate the day when this planet may no longer be habitable. Every man, woman and child lives under a nuclear sword of Damocles, hanging by the slenderest of threads, capable of being cut at any moment by accident or miscalculation or by madness. The weapons of war must be abolished before they abolish us." -U.S. President John F. Kennedy, Address Before the General Assembly of the United Nations, September 25, 1961[739]

Attempts at disarmament (the elimination of nuclear weapons) began as soon as attempts to actually utilize nuclear weaponry in Japan were being considered. As mentioned in Part I of this book, the Franck Report, sent by several Manhattan Project scientists to the Secretary of War in June 1945, urged the U.S. government to consider abstaining from using the weapons. Signatories to the report, among others, were prominent scientists James Franck,

Eugene Rabinowitch, Glenn T. Seaborg, and Leo Szilard. In their report summary, they noted that:

> Nuclear bombs cannot possibly remain a "secret weapon" at the exclusive disposal of this country for more than a few years. The scientific facts on which construction is based are well known to scientists of other countries. Unless an effective international control of nuclear explosives is instituted, a race for nuclear armaments is certain to ensue following the first revelation of our possession of nuclear weapons to the world. Within ten years other countries may have nuclear bombs, each of which, weighing less than a ton, could destroy an urban area of more than ten square miles...

> We believe that these considerations make the use of nuclear bombs for an early unannounced attack against Japan inadvisable. If the United States were to be the first to release this new means of indiscriminate destruction upon mankind, she would sacrifice public support throughout the world, precipitate the race for armaments and prejudice the possibility of reaching an international agreement on the future control of such weapons...[740]

Eugene Rabinowitch would go on to found the *Bulletin of Atomic Scientists*, and a declassified version of the Franck Report, published in the May 1, 1946 issue of the *Bulletin*, would stand as inspiration for future disarmament efforts.[741] It wasn't just Manhattan Project scientists who would later change their tune on nuclear weapons, though. Einstein, who authored the Einstein-Szilard Letter urging President Roosevelt to build the Bomb, would later advocate against nuclear proliferation. He later said in a Newsweek interview, "had I known that the Germans would not succeed in developing an atomic bomb, I would have done nothing."

Between 1945 and 1955, Einstein was active politically, speaking and writing often about his desire for peace through international cooperation and the eradication of all nuclear weapons.[742,743]

So too, did J. Robert Oppenheimer change his tune following World War II. Though Oppenheimer had spearheaded the creation of the atomic bomb through the Manhattan Project, his opinion was that thermonuclear weaponry was a bridge too far. Thus, he vehemently opposed the development and construction of the hydrogen bomb. His opposition put him into direct conflict with Edward Teller, the mastermind behind fusion weaponry, and got him into trouble with the U.S. government. Ultimately, his opposition to furthering nuclear weapons development found him at the wrong end of a military security report on December 21, 1953. He was accused of associating with communists in his past, of delaying the naming of Soviet agents, and of opposing the development of the hydrogen bomb. The year after, a security hearing determined he should not have access to military secrets; his contract as an advisor to the U.S. Atomic Energy Commission was canceled, and he became a worldwide symbol of those who can fall victim to witch hunts as a result of their own personal moral dilemmas. He would spend his final years grappling with the dilemmas of the nexus between science and society, likely ever cognizant of his words to President Truman, who ordered the atomic bombings on Japan, "Mr. President, I have blood on my hands."[744,745,746]

But it wasn't just the scientists having buyer's remorse about the creation of the Bomb. Governments were also working toward control and disarmament, because they were cognizant that too many bombs in too many hands would inevitably result in disaster, perhaps on a global scale. Many multiparty treaties would be entered into during the 20th Century that would attempt to control nuclear weaponry and reduce the number and types of weapons available. The first such treaty was the **1959 Antarctic Treaty**, a 12-nation agreement to protect the peaceful status of Antarctica. It provided that no military bases could be established on Antarctica, no military maneuvers could be made, no stationing or testing of any type of weapon could occur, no nuclear explosives could be used, and no radioactive waste disposal could occur there. In what would be a feature of later treaties, the treaty provided that each party would have a right to full on-site and aerial inspections of Antarctic installations in order to verify compliance with the treaty provisions.[747]

Even before the Antarctic Treaty, the United Nations General Assembly had established the United Nations Disarmament Commission (UNDC) in 1952, under the auspices of the Security Council. It was charged with preparing proposals for treaties on the regulation, limitation, and balanced reduction of all weapons of mass destruction (including nuclear weaponry). Eventually, its successor bodies would manage future disarmament efforts. These were the Ten-Nation Disarmament Committee (1960), the Eighteen-Nation Disarmament Committee (1962), the Conference of the Committee on Disarmament (1969), and the Conference on Disarmament (1978).[748] By 1955, the Subcommittee of Five (the United States, the Soviet Union, the United Kingdom, France, and Canada) of the UNDC was already in discussion about limiting nuclear testing due to the known effects of fallout and global accumulation of radiological contamination in the environment. Years later, these efforts would eventually culminate in the **Limited Test Ban Treaty**, effective October 10, 1963. This treaty banned weapons tests in the atmosphere, outer space, and underwater. Underground testing was still permitted, so long as it did not cause "radioactive debris to be present outside the territorial limits of the State under whose jurisdiction or control" the tests would be conducted. The purpose of the Limited Test Ban Treaty was to end "the contamination of [the] environment by radioactive substances."[749]

The space race and the technologies that stemmed from it were inexorably tied to the arms race. For example, the rocket technologies used to bring astronauts to space are the same used to deploy intercontinental ballistic missiles (ICBMs). As such, concerns grew among the Americans and Soviets that space itself would be weaponized as the arms and space races heated up in tandem. In 1967, the "**Outer Space Treaty**" went into effect, providing that outer space would be free for exploration by all States, that it would not be subject to national appropriation for claims of sovereignty, that states could not place nuclear weapons or other weapons of mass destruction in orbit or on celestial bodies or station them in outer space by any means, the Moon and other celestial bodies must be used only for peaceful purposes, astronauts are envoys of all humankind, states will be liable for the actions of their citizens and government/non-government organs alike, states will avoid harmful contamination of space and other celestial bodies, and states will be liable for damage caused by space objects.[750]

The Treaty for the Prohibition of Nuclear Weapons in Latin America (the "*Tlatelolco Treaty*") entered into force on April 25, 1969. All 33 states in the region of Latin America and the Caribbean have signed and ratified the treaty. It applies to the entire Latin American and Caribbean region and large areas of the Atlantic and Pacific oceans. The treaty effectively banned all nuclear weapons from Latin America. Cuba was the last state to ratify the treaty on October 23, 2002.[751]

Next came the **Nuclear Non-Proliferation Treaty (NPT)** of 1968. Several nuclear and non-nuclear states were involved, and while it did not ultimately prevent nuclear proliferation, the treaty set a precedent for international cooperation to prevent proliferation. While the United States and the Soviet Union had vast stockpiles of weapons by this point, only a few nations had the ability to deploy weapons effectively. Allowing nuclear weapons technology to spread further would disrupt the delicate balance of deterrence between the global superpowers, and proliferation to developing nations could lead to nuclear weapons being used in regional wars or land disputes. Ultimately, the nuclear powers agreed not to transfer nuclear weapons or their associated technology to other states. Non-nuclear states agreed they wouldn't receive, develop, or otherwise obtain atomic weaponry. Parties also agreed to cooperate in the development of peaceful nuclear technology, such as for power generation. This treaty was not a perfect solution, as India, Pakistan, and likely Israel obtained nuclear weapons despite the treaty. North Korea signed the NPT but developed nuclear weapons anyway.[752,753]

Entering into force on May 18, 1972, **the Seabed Treaty**, an agreement between the United States, the Soviet Union, and the United Kingdom, in addition to 84 other nations, banned the placement of nuclear weapons on the ocean floor beyond a 12-mile coastal zone. Signatories are entitled to observe all seabed activities of other signatories beyond the 12-mile zone to monitor compliance.[754]

The Strategic Arms Limitation Treaty (SALT I) was the first treaty to meaningfully limit the weapons stockpiles of the superpowers. Ironically, it may not have been the ballistic missiles themselves that were the impetus for SALT, but rather *anti-ballistic missile* (ABM) defense systems. In 1967, President Lyndon Johnson announced that the Soviet Union was building an ABM system around Moscow to prevent a first strike or a second strike retaliation by the United States. This caused concern in the West, as it could potentially upset the system of Mutually Assured Destruction (MAD) and nuclear deterrence if one side could protect itself from nuclear annihilation. The two sides reached a solution on May 26, 1972, when the United States and the Soviet Union signed the interim SALT agreement and the Anti-Ballistic Missile (ABM) Treaty. The ABM treaty limited the US and the USSR to 200 interceptors each, allowing each side to construct two missile defense sites–

one to protect their national capital and the other to protect one ICBM field. The interim SALT agreement froze the number of strategic ballistic missiles at 1972 levels and prohibited additional land-based ICBM silos from being built. It allowed submarine-launched ballistic missile (SLBM) capabilities to be increased only if equal reductions were made in ICBM and other SLBM launchers.[755,756]

The Threshold Test Ban Treaty (TTBT), opened for signature on July 3, 1974, between the United States and the Soviet Union, established a 150-kiloton maximum threshold for detonations in underground testing. Disagreements regarding other potential treaties, verification measures, and a protocol for unintended breaches of the 150-kiloton limit delayed its ratification until 1990. However, in the intervening decade and a half, the United States and the Soviet Union effectively agreed on a handshake basis to limit the size of any underground detonations, despite delayed ratification.[757,758]

By November 1974, with an eye toward a more permanent solution on SALT, United States President Gerald Ford and Soviet Premier Leonid I. Brezhnev recognized that action should be taken on a framework to limit each nation's respective stockpiles in the future. The details were not yet available, but in principle, each country would limit their stockpiles to "equal aggregates" in the total number of weapons available in each leg of the nuclear triad (land-based, bomber-based, and submarine-based). At the time, stockpiles were estimated to be 2,600 delivery vehicles for the Russians and 2,200 for the United States. Essentially agreeing to a future framework, **the Vladivostok Summit** between Ford and Brezhnev set the stage for a future agreement on strategic arms.[759,760]

The Underground Nuclear Explosions for Peaceful Purposes (PNE Treaty) would follow in 1976. The PNE Treaty prevented the United States and Soviet Union from carrying out any group of explosions exceeding 1.5 megatons in the aggregate; this was a companion to the Threshold Test

Ban Treaty. Like the TTBT, this Treaty did not go into force until 1990.[761]

The Convention on the Prohibition of Military or Any Other Hostile Use of Environmental Modification Techniques was agreed to in Geneva on May 18, 1977, and went into force on October 5, 1978. Essentially, this was an agreement not to engage in hostile, militarized weather/climate control activities that would be widespread (affecting several hundred square kilometers or more), long-lasting (a period of multiple months or approximately a season), or severe (involving serious or significant disruption or harm to human life, natural, or economic resources or other assets).[762] While this treaty was not specifically nuclear in nature, this kind of prohibited activity could be facilitated by the use of high-yield nuclear weaponry or could lead to an escalation in the atomic sphere if a nation-state detects that it is under assault through environmental manipulation by an adversary state.

Less than one year later, on June 17, 1979, Ford and Brezhnev signed **the Strategic Arms Limitation Treaty (SALT) II**. One specific shortcoming of SALT I is that it didn't prevent the Soviets or the Americans from enlarging their nuclear forces through the use of Multiple Independently Targeted Re-Entry Vehicles (MIRVs) attached to their ICBMs and SLBMs. SALT II created a 2,400 equal aggregate limit for strategic nuclear delivery vehicles (SNDV), including ICBMs, SLBMs, and strategic bombers. MIRVs were limited to an aggregate limit of 1,320. SALT II also provided for a ban on new land-based ICBM launchers and a limit on the deployment of new types of offensive strategic arms. It also limited ICBM silo launchers and the number of MIRVs permitted on missiles. SALT II additionally provided that each party could use their national technical means (NTM) of verification to ensure the other party's compliance, and provided that neither party should interfere with the other party's use of NTM. NTM includes the use of electronic signatures and reconnaissance satellites.

The parties also agreed not to deliberately obscure weapons in a way that would interfere with the other party's use of NTM. A Protocol to the Treaty also limited the deployment and testing of certain cruise missiles. Although SALT II was never ratified, partially because of the Soviet invasion of Afghanistan, the Americans and Soviets generally abided by the terms of the agreement.[763,764]

In 1986, **the Treaty of Rarotonga** became only the second non-Antarctic, continental nuclear weapons-free zone (NWFZ) after the Treaty for the Prohibition of Nuclear Weapons in Latin America. It established a NWFZ for the South Pacific, which had been the subject of extensive nuclear weapons testing, with the United States conducting around 66 atmospheric and underwater tests in the Marshall Islands that displaced populations in the region. France and the United Kingdom have ratified the treaty. Russia and China have ratified most of the Treaty, and ratification by the United States of all three protocols is still pending.[765,766]

The Intermediate-Range Nuclear Forces (INF) Treaty went into force on June 1, 1988. The Treaty required that the United States and Russia eliminate their ground-launched ballistic and cruise missiles capable of traveling between 300 and 3,400 miles (500-5,500 km). The INF Treaty ultimately compelled the destruction of 1,846 Soviet and 846 American missiles. It was the first successful elimination of an entire category of nuclear delivery system. However, Russia began testing and deploying a new intermediate-range missile known as the 9M729/SSC-8 in subsequent years. The intermediate-range missile allows nuclear warheads to reach NATO states' capitals within minutes. Although never officially nuclear-tipped, the delivery system can travel 1,553 miles (2,500 km) and carry a payload of nearly 1,000 pounds (450 kg). In response to Russia's deployment of the system, the United States pulled out of the INF Treaty on August 2, 2019.[767,768,769]

Signed and effective on May 31, 1988, the United States and the Soviet Union agreed to **notify each other of any ICBM or SLBM launches**. The agreement provides that for any launch, notification will be provided at least 24 hours before to the other party with information on the planned date, launch zone, and place of impact for any launch of an SLBM or ICBM. The agreement was intended to reduce the risk that a nuclear war could break out due to a misinterpretation or mistake.[770]

Entering into force in 1994 and expiring in 2009, **the Strategic Arms Reduction Treaty (START I)** was designed to reduce the strategic arsenals of the United States and Russia. Rather than limiting existing and future weapons systems, as SALT did, START I was intended to *reduce* existing stockpiles. START I limited ICBMs, SLBMs, and strategic bombers to 1,600. It limited deployed heavy ICBMs to 154. It placed a limit of 6,000 nuclear warheads on all legs of the nuclear triad, with maximums per delivery system totaling 4,900 on ICBMs or SLBMs, 1,540 on heavy ICBMs, and 1,100 on mobile ICBMs. It also limited the lifting power of ballistic missiles, placed bans on certain new types of heavy ICBMs and SLBMs, and prevented MIRVs of more than ten warheads. Various means of verification were provided, including NTM and transfer between the parties of the locations of strategic deployment sites. START I is considered perhaps the most successful nuclear arms treaty, as it resulted in an 80% reduction in the world's strategic nuclear arms stockpiles.[771,772]

The **Lisbon Protocol** in 1992 effectively made the Soviet Union's successor states (Russia, Belarus, Ukraine, Kazakhstan) all parties to START I. Estimates as of 1991 put the number of strategic warheads in Belarus, Kazakhstan, and Ukraine at 3,410, with at least 3,000 tactical warheads in the Soviet successor states. All three non-Russian successor states agreed to the Lisbon Protocol and committed to accede to the Nuclear Nonproliferation Treaty (NPT) as non-nuclear-weapon states.[773]

The Strategic Arms Reduction Treaty (START) II
sought to ban land-based (but not submarine-launched)
multiple independently targetable reentry vehicles (MIRVs)
and place further restrictions on nuclear weapons systems.
The United States and Russia signed it in 1993 but never
fully ratified or implemented it. Under the treaty, each side
was to reduce its deployed strategic warheads to a maximum
of 3,000-3,500. START II faced significant challenges and
ultimately became obsolete due to changing geopolitical
circumstances and the lack of full ratification due to
American and Russian disagreements.[774,775]

The next nuclear weapons-free zone (NWFZ) came
as a result of **the Treaty of Pelindaba**, signed in 1996. This
particular treaty covers the continent of Africa. Its primary
goal is to prevent the development, acquisition, or stationing
of nuclear weapons within Africa and promote peaceful uses
of nuclear energy. The treaty prohibits the testing,
production, possession, and deployment of nuclear weapons
by African states, and establishes a verification system to
ensure compliance. It also encourages cooperation in the
peaceful use of nuclear energy, including promoting nuclear
safety and developing alternative energy sources. The Treaty
of Pelindaba contributes to regional stability, security, and
non-proliferation efforts in Africa.[776]

The Comprehensive Test Ban Treaty (CTBT),
adopted in 1996, is an international treaty to ban all nuclear
explosions, whether for military or civilian purposes. It
expands on earlier attempts to prevent nuclear weapons
testing. The CTBT seeks to prevent the development and
qualitative improvement of nuclear weapons and to advance
disarmament efforts. Signatory states are prohibited from
conducting any nuclear weapon test explosions or any other
nuclear explosions, including those conducted underground.
It establishes a global monitoring system to detect and verify
compliance with the ban and promotes international
cooperation in the peaceful uses of nuclear energy. While
most countries have signed the CTBT, its entry into force

requires ratification by 44 specific states (all participating members of the Conference on Disarmament, and all of which possess nuclear power and research reactors). Eight of those have not yet ratified: the United States, China, North Korea, India, Pakistan, Israel, Iran, and Egypt. Nevertheless, the treaty has contributed to a de facto global norm against nuclear testing and reinforces nuclear disarmament and non-proliferation efforts. In 2021, the United States affirmed its support for the CTBT but noted that significant challenges are standing in the way of global ratification.[777]

In 2002, the United States and Russia signed the **Strategic Offensive Reductions Treaty (SORT)**. Its goal was to reduce the number of strategic nuclear weapons deployed by both countries. The treaty established a range of 1,700-2,200 warheads for each side by the end of 2012. Unlike previous agreements, SORT did not include hard numerical targets or comprehensive verification measures, relying instead on mutual trust and transparency, providing that the two sides should hold meetings of a Bilateral Implementation Commission at least twice yearly. The absence of an extension or replacement agreement after its expiration in 2012 raised further questions. Notably, though, SORT did not override START I and the United States reached its strategic arms reduction goals under SORT three years ahead of schedule in 2009.[778,779]

Russia and the United States jointly signed **the New START Treaty** in 2010, aimed at further reducing and limiting their strategic nuclear arsenals. Under the treaty, both parties agreed to reduce their deployed strategic warheads to 1,550 and their deployed ICBMs, SLBMs, and strategic heavy bombers equipped for nuclear armaments to 700. To ensure compliance, it also includes robust updated verification measures based on the START I treaty, such as on-site inspections and data exchanges.

Finally, **the Treaty on the Prohibition of Nuclear Weapons**, adopted in 2017 and entered into force in 2021, is an international treaty that aims to ban nuclear weapons altogether. It prohibits participating states from developing, testing, producing, acquiring, possessing, stockpiling, using, or threatening to use nuclear weapons. It also prohibits the transfer of atomic weapons and the stationing or hosting of these weapons on a state's territory. The treaty establishes obligations for disarmament, including the destruction of existing nuclear weapons and programs, as well as assistance to individuals affected by the use or testing of nuclear weapons. It emphasizes the humanitarian consequences of nuclear weapons and seeks to promote a total shift toward disarmament. All nine states possessing nuclear weapons boycotted the open-ended working group (OEWG) on nuclear disarmament within the United Nations General Assembly.[780]

That very last sentence should be illuminating with regard to where the world is headed on nuclear weapons. While there is a desire among many entities and nation-states to rid the world of nuclear weapons totally, the world powers that already have them are not at all interested in giving up their stockpiles. Nuclear weapons give a state strong tactical and strategic advantages in the domains of geopolitics and international affairs. While global nuclear powers such as the United States and Russia are willing to reduce their stockpiles in the interest of saving money, increasing safety, and reducing possible broken arrow incidents, will they ever agree to give them up completely? To do so, a nation-state would abandon an indescribably powerful and asymmetrical advantage over all other nations that do not possess nuclear weapons. Simultaneously, it would diminish that nation-state's relative defense position against states that continue to possess nuclear weapons. It's one thing to reduce a national stockpile to a level that still maintains the ability to threaten one's adversaries with total erasure; it's another thing entirely to bring that stockpile to levels below that threshold.

Disarmament and nonproliferation agreements, whether ratified treaties or handshake agreements between heads of state (primarily between the United States and

Russia), have shown some effectiveness in reducing stockpiles. In 1986, the world had a combined stockpile of almost 65,000 warheads. Today, it has less than 10,000. One hundred seventy-eight nuclear tests took place in 1962, contaminating the world with long-lasting and highly toxic radioisotopes. By the year 2000, testing had almost been completely eliminated, but for errant, sporadic testing by North Korea.[781]

However, the modern stockpiles of the global nuclear powers, while only fractions of their Cold War height, are still enough to destroy life as we know it and result in mutually assured destruction (MAD) in the event that the world ever reaches the tipping point. As of 2007, the *deployed* megatonnage of the U.S. arsenal (the "Enduring Stockpile") was just short of around 1.43 gigatons, with the total active arsenal clocking in at about 2.33 gigatons, only 11.4% of 1960 levels.[782] Currently, the US Department of Defense alone has a stockpile of 3,708 nuclear warheads ready for upload onto land- and sea-based ballistic missiles, and the Bulletin of the Atomic Scientists estimates that approximately 1,770 warheads are currently deployed. Additional intact warheads are slated for retirement, making the total US inventory around 5,244 warheads.[783] The United States' current largest warhead, the B-83, had a production run of about 650 warheads, with a yield of up to 1.2 megatons each. And while there are ongoing attempts to retire the B83, assuming the warheads are all still in service, the total megatonnage would be around 780 *megatons* for just the stock of B83s, about double the total megatonnage that was previously calculated to be adequate to destroy the entire Soviet Union. That number, of course, does not count all of the other strategic and tactical warheads available to the United States. Suffice it to say that existing global stockpiles are more than adequate to bring about annihilation on a scale we cannot even fathom.[784,785]

Chapter 25. Hypersonics, Rail Guns, and Lasers: Contemporary Weapons and Delivery Systems

"Today, we have a unique situation in our new and recent history. They try to catch up with us. Not a single country possesses hypersonic weapons, let alone continental-range hypersonic weapons. We already have Kinzhal (Dagger) hypersonic missile systems in the field, and Peresvet laser combat systems have already been deployed with the troops as well..." -Russian President Vladimir Putin, 2019[786]

The Cold War was about size and quantity. Nuclear arsenals were built as big as possible to enable destruction on a wide range of territory. Today, world powers focus advanced technology on stealth weaponry, supersonic and hypersonic speeds inside the atmosphere, and guided systems that are highly accurate. Increases in accuracy that allow pinpoint targeting and the ability to deliver payloads effectively deep within enemy territory no longer necessitate multi-megaton weapons, which are heavy, dangerous, and expensive to maintain. Anti-ballistic missile (ABM) systems of the modern era have also caused nation-states to develop delivery systems, such as new hypersonic weapons, to counter those ABM systems.

The United States allowed the Anti-Ballistic Missile Treaty with Russia to expire in 2002, announcing before its expiration that the United States would leave it behind, citing concerns about its ability to counter growing missile threats worldwide. The ABM Treaty had been a significant barrier (a "cornerstone of strategic stability") to a never-ending cascade between the superpowers of ever-more advanced ballistic missiles and systems designed to counter those advances, which could have led to a more protracted and intense global arms race. In the over 20 years since then, ABM systems (and, not surprisingly, the delivery systems designed to counter those ABM systems) have advanced significantly.[787]

Costing around $53 Billion, the United States' signature anti-ICBM system is developed by Boeing, and is known as the Ground-Based Midcourse Defense (GMD) system. It is designed to intercept and destroy incoming enemy ballistic missiles outside Earth's atmosphere during the midcourse phase of their flight. The GMD system consists of a network of radars, sensors, command centers, and interceptors. Based in underground silos in the United States (currently, California and Alaska), the interceptors are launched into space to intercept and destroy incoming missiles using a kinetic kill vehicle. A third location on the East Coast at Fort Drum, NY has also been considered. The GMD system aims to provide a layered defense capability, enhancing the overall ballistic missile defense posture of the United States. It is no secret that this system would have no chance in its existing form of countering a full-scale barrage of nuclear ICBMs from China and Russia (thousands more interceptors would be required, and even then, America's adversaries could simply build more missiles to counter the additional interceptors). Additionally, the system's efficacy has been questioned, with the interceptors killing their targets in only 55% of testing scenarios. Shorter-range systems, such as the Patriot system and THAAD systems, have a much higher efficacy rate, but those are intended for defense against short- and medium-range missiles and cannot serve as a continental defense against long-range ballistic missiles.[788,789,790]

China and Russia also have kinetic ABM systems; China's current systems are the HQ9 and HQ19. Initially intended for short- and medium-range missiles, China announced in April 2023 that its ABM system had successfully intercepted a nuclear-capable ICBM in its mid-course. This would potentially put China's ABM system on par with the United States' GMD system. Russia's current system is the A-135, which went operational in 1990. The system is designed to shoot down short-range and long-range ballistic missiles, and has discrimination capabilities to discern actual re-entry vehicles from decoys and fake warheads. The system has been updated with new PRS-1 hypersonic interceptors that can reach a speed of Mach 15-16 (up to 12,276 mph or 19,757 km/h). Russia has recently tested the new A-235 Nudol, which has silo-based and mobile launchers, and the proven ability also to shoot down satellites.[791,792]

Traditional ABM systems have long been likened to "shooting a bullet with a bullet." Needless to say, doing so is no easy task. Nuclear powers are exploring alternatives to interceptor missiles that hit other ballistic missiles. One such system is the "rail gun," long a feature of science fiction stories set far in the future. The idea of an electromagnetic rail gun is to use electric energy to launch a projectile (typically non-explosive) at such a speed that its kinetic force impacts on target with devastating power. The speed at which these projectiles travel makes them uniquely suited for interceptor purposes. The idea for the rail gun was first conceived in France during World War I, and the United States recently spent half a billion dollars and 15 years developing rail guns for its naval destroyers. Due to the project's cost, the rail gun barrels' propensity to be destroyed by the force of their projectile ejections, and the power requirements to operate rail guns, the project was scrapped in 2021. While no known declassified systems are yet operational except in test environments, research continues on the concept globally. Japan is particularly interested in collaborating with the United States to invest in the technology.[793]

Unlike rail guns, other nonconventional ABM systems have demonstrated proven success in the field. One such ABM system is the *Iron Beam* developed by Israeli scientists, intended to augment Israel's "Iron Dome" air defense system. The 100-kilowatt directed energy system (directed energy weapons use lasers, microwave weapons, and/or particle beams) is designed to neutralize multiple incoming targets, including artillery, rockets, mortar rounds, and uncrewed aerial systems. At a 6-mile (10-km) range, the weapon can be focused to the diameter of a coin. In March 2022, the system was demonstrated through a series of successful live-fire tests, where the laser intercepted uncrewed aerial vehicles (UAVs), rockets, anti-tank missiles, and mortars. The advantages of the system, versus kinetic interceptors (those that use physical objects instead of lasers), are that it has a low cost-per-shot and results in minimal collateral damage.[794] Iron Beam is a short-range system, though, and not powerful enough to serve as an effective long-range anti-ICBM system. While it demonstrates that the technology works in principle, research continues on more advanced laser-based systems for ICBMs.

No matter how fast a kinetic object is, a laser is faster. While missiles can travel several thousands of miles per *hour*, a laser travels at the speed of light, approximately 186,411 miles (300,000 kilometers) per *second*. While Israel's Iron Beam uses a 100-kilowatt system, Lockheed Martin recently delivered to the Pentagon's High Energy Laser Scaling Initiative (HELSI) a new laser three times as powerful, in the 300-kilowatt range. The more powerful a laser is, the further away and more effectively it can engage targets, with the eventual goal being the ability to neutralize ICBMs. HELSI aims to scale these lasers up to 500 kilowatts by 2024, and up to one megawatt by the end of the 2020s. While current advanced U.S. ABM technology shoots down missiles in their mid-course phase, a one-megawatt laser could have the capability to shoot down an ICBM even earlier, in its boost phase.[795,796] In 2022, Lockheed Martin announced that its Layered Laser Defense (LLD) system effectively shot down two cruise missiles.[797]

Speaking of speed, hypersonic weapons are the newest technology being developed and deployed by major nuclear powers. A common misconception about these new hypersonic missiles is that they travel faster than all previously existing missiles, which is untrue. For example, the LGM-30G Minuteman III silo-based ICBM travels at up to Mach 23, or 15,000 miles per hour (24,140 km/h)/4 miles (6.4 km) per second. All ICBMs, which have been around for decades, are hypersonic. However, what newer hypersonic technologies *can* do, though, is travel at more than five times the speed of sound (Mach-5) in the upper atmosphere (around 3,852 mph or 6,200 km/h), and can maneuver away from interceptors or toward a target. ICBMs travel primarily *above* the atmosphere in a set path, not in it, and are likely more vulnerable to ABM systems. Modern hypersonics can also travel for longer distances without being detected by radar. This capability of changing trajectory/direction mid-flight, combined with their breakneck speeds, makes them *much* more difficult to shoot down using traditional interceptor technologies. One potential use case being explored for hypersonic weapons is the ability to put them partially into orbit using a fractional orbital bombardment system (FOBS), which could then fire the missile from partial orbit at hypersonic speeds, affording nation-state adversaries *very* little time to respond.[798,799]

Hypersonic missiles are also potentially particularly effective against ABM systems, since ABM systems are typically designed for mid-course interception of ballistic missiles. Even with short-range defenses, hypersonic missiles can maneuver at high speeds in ways that cannot be predicted. When fully developed, their most effective use cases are that they will hypothetically be less susceptible to advanced long-range defense systems than ballistic missiles, and will potentially allow better penetration of antiaccess and area-denial (A2/AD) systems. That said, these systems are mostly still in development, and suffer from issues related to cost and extreme heating.[800]

Russia has touted its new hypersonic Kinzhal missiles since 2018 as "invulnerable" to existing missile defense systems, saying that these defense systems simply cannot catch up. In one recent misstep for hypersonic technology, Ukrainian forces fighting Russia in the Russo-Ukrainian War reported using United States-built Patriot surface-to-air missiles to shoot down six Russian hypersonic Kinzhals. Russia subsequently arrested three of its hypersonics research scientists for "high treason."[801] Suffice it to say that there is some work to be done on hypersonic missiles before they can be used for strategic purposes.

That said, Russia has already claimed that some of its hypersonic weapons are capable of carrying nuclear warheads. That statement alone is enough to cause concern among Russia's geopolitical adversaries like the United States, as it would be very difficult for them to tell if a barrage of hypersonic missiles is nuclear-capable or not, which could instigate a second strike (or mistaken first strike) reaction. The United States is particularly concerned that China's Dongfeng-17 hypersonic glide vehicles (wherein a glide vehicle is boosted by a rocket to high altitude and then glides at high speed to its target on a maneuverable, variable path) may be more advanced than the United States' hypersonic glide vehicles.[802,803]

Speaking of contemporary Russian missiles, they currently have in development a super heavy ICBM, the successor to the "Satan" missile. This missile, the RS-28 Sarmat (NATO codename: SS-X-29 or SS-X-30), colloquially referred to as the "Satan II," will be capable of carrying between 10 large warheads or 16 smaller warheads in its MIRV system. It will sport a combined yield of around 50 megatons; in comparison, the U.S. Minuteman III missiles can deliver a yield of about 1.425 megatons.[804]

Another technology with potential strategic ramifications that cannot be overlooked is the *scramjet*. The scramjet engine is an advanced propulsion system designed for high-speed flight in the atmosphere. Unlike a ramjet engine used on supersonic aircraft and cruise missiles, which slows air heading into the engine to subsonic speeds before combustion, a scramjet engine does not restrict the incoming air, and relies on supersonic combustion to achieve the desired thrust. The scramjet operates efficiently at hypersonic speeds, typically above Mach 5, by allowing the air to flow through the engine at supersonic speeds, eliminating the need for moving parts like compressor blades. This design feature allows the scramjet to achieve higher speeds and offers the potential for space access and rapid intercontinental travel. However, it also poses challenges, such as efficient fuel injection and managing the intense heat generated by the supersonic combustion process.[805]

Scramjets will have applications in both strategic bombers and missile technologies when fully deployed in weapons systems. Scramjets will allow cruise missiles to fly faster than existing technologies. They may also enable future bombers to travel at hypersonic speeds before dropping a nuclear payload, increasing survivability and avoiding enemy air-to-air or ground-to-air defenses. Development of this groundbreaking new technology is well underway, with the United States Defense Advanced Research Projects Agency (DARPA) announcing in 2021 that its Hypersonic Air-breathing Weapon Concept (HAWC) was released from an aircraft, at which point its scramjet engine fired and propelled the cruiser to hypersonic speeds above Mach-5.[806] The United States tested a hypersonic scramjet missile in April 2022, launched from a B-52 Stratofortress, further proving the propulsion technology's efficacy.[807]

In the modern era, for better or worse, traditional aspects of the nuclear triad have also continued to advance. One example is ever-improving ICBM technology. The United States is currently preparing to replace its Minuteman III ICBM arsenal with a powerful new silo-based delivery system, the LGM-35A Sentinel. The new Sentinel ICBMs will use a modular design that will allow replacement of aging and outdated components, thus decreasing lifecycle costs and enabling easy upgrades in the future. These future modular components could come in the form of improved guidance systems or may include more advanced countermeasures to allow the ICBMs to penetrate adversaries' ABM systems. The new Sentinel missiles will also have increased security measures, allowing them to be maintained without opening the launcher closure door, which currently exposes the Minuteman III missiles to security vulnerabilities. It will also reduce personnel requirements and will likely be capable of control by fewer Launch Control Centers (LCCs) than the Minuteman missiles. Finally, the Sentinel missiles use advanced composite casings to house the missile propellant, instead of heavy steel casings like the Minuteman III missiles, allowing them to achieve greater throw weights, which could also possibly allow them to carry countermeasures or MIRVs with 2-3 warheads instead of the one warhead the Air Force currently envisions for these missiles.[808]

Speaking of modularity, I mentioned in the chapter on strategic bombers the world's most advanced strategic bomber, the Northrop Grumman B-2 Spirit, designed and controlled by the United States government. On December 2, 2022, the U.S. government announced a new, bleeding-edge strategic bomber based on the B-2's flying wing design, the Northrop Grumman B-21 Raider, named after the Doolittle Raid of World War II. Much of the weapon is, of course, classified, but broad strokes have been provided to the public about the bomber's capabilities. Northrop describes the B-21 as the first sixth-generation aircraft, with unrivaled stealth technology and the ability to penetrate and defeat anti-access and area-denial systems. It will also use an open architecture, featuring built-in hardware flexibility and the ability to undergo agile software upgrades efficiently. The bomber will utilize digital and cloud technologies, and it's speculated that the bomber may also employ drone technology at some point to allow the bomber to deliver thermonuclear weaponry without using a human pilot. While Chinese analysts have raised doubts that the United States can afford enough of the bombers to be a strategic threat to China due to advances in radar technologies, it is likely that the B-21 will be far more cost-effective than the B-2 and will have far more advanced capabilities. While the B-2, which has been in operation for around 30 years, cost over $2 billion per unit, the B-21 will clock in at about $750 million per unit adjusted for inflation.[809,810,811,812,813]

  The takeaway from this chapter is that we are in the midst of a paradigm shift. While the Cold War was about bigger, more destructive weapons, the modern era is about targeting, stealth, speed, range, and defense penetration, while simultaneously building more advanced anti-missile and radar capabilities to provide both offensive and defensive strategic advantage over enemies. The crude methods of the past are history, and instead of using an axe, the global nuclear powers are now using a switchblade. The arms race never ended, and while we're no longer building 50-megaton supernukes, we're finding ways to make smaller thermonuclear weapons far more dangerous and vastly more effective.

Chapter 26. Doomsday Devices and Hypothetical Weaponry: A Bridge Too Far?

"Oh my God... We finally really did it. You Maniacs! You blew it up! Ah, damn you! God damn you all to hell!" - Charlton Heston as George Taylor, Planet of the Apes (1968)[814,815]

As we explored earlier, there is an argument to be made that the entire raison d'être of the Bomb is not merely to have a wartime advantage over one's adversaries but rather to end war altogether. The concept of mutually assured destruction and peace through fear necessitated weapons on a scale that would threaten vast swathes of human civilization, if not the entirety of the species. In the lead-up to the Manhattan Project, world powers sought to build bigger, more powerful, more accurate bombs that would inflict maximum devastation and simultaneously allow governments to destroy some adversaries while deterring others from challenging them.

The Manhattan Project led to a previously inconceivable weapon using advanced physics and manufacturing methods that even today, many decades later, most of the world's population does not have the slightest understanding of. While some architects of the atomic age, like Oppenheimer, drew a line in the sand after they saw the destructive capability of the original fission bombs, others, like Edward Teller, believed we needed to go further. Even after the advent of the thermonuclear bomb, there was a drive to increase the power and intensity of these weapons. Look no further than the Soviet Tsar Bomba for evidence of that.

But while disarmament and nonproliferation efforts have somewhat attenuated the drive for higher yield weapons, the underlying physics, chemistry, and ideas for these weapons are still out there, and are still options for global powers to develop. Some designs are even contemplated as doomsday weaponry– devices that are guaranteed to fuel destruction and death beyond even what modern thermonuclear weaponry is capable of. The RAND Corporation, in its "Glossary of Terms on National Security" defined a Doomsday Machine as "A reliable and securely protected device that is capable of destroying almost all human life and that would be automatically triggered if an enemy committed any one of a designated class of violations."[816]

You'll recall Dr. Szilard from earlier chapters as one of the original architects of the atomic bomb, initially conceptualizing the technology and urging FDR to beat the Nazis to the Bomb. In a 1950 radio address, Szilard discussed the concept of the nascent hydrogen bomb, and further discussed a new idea that could serve the ultimate deterrent purpose: a doomsday bomb. Szilard was not suggesting that this device *should* be built, but rather that it *could* be built (recall that Szilard staunchly opposed using the atomic bomb in Japan). His address was more of a warning to humankind about continuing on the path of ever more destructive weaponry. The conceptual doomsday device Szilard spoke of is now known as the "cobalt bomb" or "salted bomb." While it was primarily a theoretical concept and not an actual weapon likely ever built or deployed, it captured immediate attention due to its potentially world-ending effects.

The idea behind a cobalt doomsday device is to enhance the destructive power of a thermonuclear bomb by using a cobalt-59 isotope. Normally, nuclear weapons produce immediate devastation through the explosion and thermal effects, as well as the initial release of radiation. However, the long-term effects can be relatively short-lived, as the radioactive isotopes decay over time. As we discussed in earlier chapters, radiological risks decrease dramatically, reducing by around 80% within a day or two after a detonation. Not so with a cobalt bomb.

In the case of a cobalt doomsday bomb, the weapon's core would be surrounded by a casing made of cobalt-59. When the nuclear bomb detonates, the intense radiation from the explosion would cause the cobalt-59 to absorb neutrons, transmuting the cobalt-59 and resulting in the creation of cobalt-60, a *highly* radioactive isotope with a half-life of not minutes, days, nor weeks, but rather, over *five years.*[817]

What makes a Cobalt doomsday device particularly alarming is that cobalt-60 emits *highly* penetrating, deadly gamma radiation. These penetrating gamma rays travel long distances and can persist in the environment for several years, posing a significant health hazard to all life. The widespread contamination caused by the cobalt-60 fallout could lead to long-term radiation exposure, rendering affected areas uninhabitable for an extended period.[818] Because the half-life is over five years, areas of widespread contamination might not reach safe levels of radioactivity for several decades. Thus, sheltering for long periods would be impractical to avoid lethal doses; even high-ranking government officials and the ultra-wealthy, with access to the most advanced bunkers and life-sustaining, shelf-stable foods and technologies, would be unable to shelter for a period of decades. To put this into perspective, for the first five years, areas contaminated with cobalt-60 would produce a lethal dose of five sieverts *per hour* to unsheltered individuals. That is to say, within that five year timeframe, being outside for one hour will kill you. Even after 53 years, individuals in contaminated areas would receive a dose of one sievert over a period of only four days, enough to cause severe acute radiation syndrome. At 105 years, the radiation dose from the cobalt-60 would still be 30 times higher than the average natural background radiation.[819]

Professor Harrison Brown, a nuclear chemist at the California Institute of Technology, noted in 1950 that a cobalt bomb using one ton of deuterium, detonated on a north-south line in the Pacific approximately 1,000 miles west of California, would produce the following result:

> The radioactive dust would reach California in about a day, and New York in four or five days, killing most life as it traverses the continent. Similarly the western powers could explode hydrogen-cobalt bombs on a north-south line about the longitude of Prague that would destroy all life within a strip 1,500 miles wide, extending from Leningrad to Odessa and 3,000 miles deep from Prague to the Ural mountains.[820]

Cobalt, if added to a single 20-megaton hydrogen bomb, could eliminate vast swathes of life on an entire continent. Edward Teller, the architect of the hydrogen bomb, also noted, "One limitation to such kind of an attack is the effect of these gases on the attacker himself. The radioactive products will eventually drift over his country too." Thus, the cobalt bomb is regarded as a suicide weapon in that it's not clear that its effects can be reasonably limited or controlled, and the risks of its use are perhaps far too significant, even to the nation-state that would deploy such a weapon.[821] Of course, that consideration is not enough to completely prevent the weapon from being created, as in some senses, the global stockpiles of thermonuclear weaponry that already exist risk serving as a sort of weapon of mass homicide/suicide in themselves. Consider that cobalt doomsday devices could serve the unique purpose as a final guarantee of mutually assured destruction for its adversaries should a nation-state be foolish enough to engage in a full-scale first strike; in this way, they possess a dark deterrent capacity.

Though no full-scale cobalt bomb has ever been tested (for reasons that should be obvious to the reader by this point), the underlying physics of the weapon have been proven by smaller-scale testing. In the 1960s, the Soviets conducted underground tests known as the triple "taiga" nuclear salvo. The three bombs created cobalt-60 due to the composition of the bombs' shells and the resultant neutron radiation they emitted. Forty years later, a full *half* of the gamma dose measurable at the triple "taiga" sites emanated from cobalt-60 generated by the bombs.[822]

All this said, there is fear that a cobalt-60 weapon actually exists or may in the near future. While it's possible the following is Russian disinformation, in 2015, Russian news program cameras, showing Russian military leaders attending a meeting, glimpsed a document over the shoulder of one of the uniformed military officials that read, "Ocean Multipurpose System 'Status-6'." The document said that the planned weapons system would have the capability of creating "extensive zones of radioactive contamination" in adversary coastal zones, making them "unsuitable" for economic and military activities "for a long time." The Russian government newspaper "Rossiskaya Gazeta" (the *"Russian Gazette"*) later reported details of the weapon and speculated that it may be a cobalt weapon. President Vladimir Putin's spokesperson said that Russian State Media NTV and First Channel had erred by slipping "secret data" into their broadcasts, and that authorities would work to ensure it doesn't happen again.[823]

Cobalt is only one of many options for producing salted bombs. Other elements are good candidates as well, including Gold-197 (transmuted to Gold-198), Tantalum-181 (transmuted to Tantalum-182), and Zinc–64 (transmuted to Zinc-65). Each of these has various levels of radiation intensity and various half-lives, resulting in different applications.[824] One particularly dangerous option for short-term, highly destructive fallout is sodium oxide. One gram of neutrons absorbed in sodium, resulting in sodium-24, distributed over one square mile (2.6 km^2), would deliver about 200 sieverts to any person caught in the open, approximately 40 times the lethal dose. A bomb composed of

1,000 tons of heavy water, salted with sodium, would heavily contaminate an area of about 200,000 miles (321,868 km); if only 1/10 of the radioactive sodium reached the ground, the average dose would be 2,000 sieverts, destroying all plant and animal life. Being that the half-life of sodium-24 is so short (approximately 15 hours), this type of bomb could be used by a nuclear power across the world from its adversary without the risk of the radiation drifting to Russian territory (it will have decayed by that point). At 3,000 times more radioactivity than cobalt-60, sodium-24 would penetrate even well-protected basement shelters, extinguishing the lives of even those who manage to avoid ground zero.[825]

While we're on the topic of salted bombs, we should address dirty bombs. Though, to date, these have never been used by any state or non-state entity, and are by no means doomsday weapons, they are much easier to build than fission and fusion bombs. State entities and counterterrorism experts are particularly concerned about the potential for the use of these weapons as they can be relatively easily obtained by unsophisticated, non-state entities, and could inflict mass panic and serious radiological contamination. A dirty bomb is simple by design– a conventional explosive mixed with radioactive material. Thermonuclear bombs are millions of times more powerful than any conventional weapon, including dirty bombs, but a dirty bomb could spread radioactive material for several blocks or miles, depending on the yield of the conventional explosives and the radioactive nature of the materials mixed with the conventional explosives.[826]

In one concerning incident in 2014, it was noted that when the Islamic State seized control of the city of Mosul, Iraq, it obtained control of 88 pounds (40 kg) of uranium. While ISIS would likely not have the nation-state-level of sophistication required to enrich that uranium into weapons-grade fissile material, there were concerns they could use it for a dirty bomb. Nonetheless, it has not been used to date, and of the 200,000 instances of terrorism noted in the

National Consortium for the Study of Terrorism and Responses to Terrorism (START)'s Global Terrorism Database, there are zero instances of nuclear terrorism, and only 13 cases of radiological terrorism, none of which involved a dirty bomb. None of these instances resulted in any human injuries or deaths. Some experts believe chemical, biological, and nuclear (actual fission/fusion) attacks are a far higher concern than dirty bombs. The fact that no such large-scale attacks have happened to date is perhaps a testament to either how difficult nuclear materials are to obtain; the exemplary job that nuclear powers have done maintaining, controlling, and monitoring their nuclear materials; or perhaps both.[827,828,829]

To further mitigate the risks of nuclear terror, most nations joined the International Convention on the Suppression of Acts of Nuclear Terrorism, a global treaty aimed at preventing and combating acts of nuclear terrorism. Adopted by the United Nations General Assembly in 2005 and entering into force on July 7, 2007, the convention seeks to enhance international cooperation and coordination to prevent the illegal acquisition, use, or threat of use of nuclear material or devices for terrorist purposes. It criminalizes various activities related to nuclear terrorism, including the possession, use, and transfer of nuclear material and devices with the intent to cause harm, as well as attempts, threats, and participation in such acts. The convention promotes cooperation among states in areas such as extradition, prosecution, and law enforcement, and emphasizes the importance of international assistance and cooperation in response to nuclear terrorist incidents. By establishing a comprehensive legal framework and fostering international collaboration, the convention aims to deter and suppress acts of nuclear terrorism, safeguard global security, and protect humanity from the potentially catastrophic consequences of nuclear attacks. The Convention addresses both nuclear bombs themselves and the materials used to create them, so dirty bombs are included.[830]

One other technology that should be discussed here, though never deployed in a wartime scenario, is the neutron bomb, earlier mentioned in a description of Israel's nuclear arsenal. This technology is not hypothetical; it has been built and tested in the real world. It has one purpose: to terminate life. While a traditional nuclear bomb, both fission and fusion, are designed to create a large explosion that causes a massive blast and levels everything nearby, a neutron bomb serves not to destroy infrastructure, but living beings. At a fraction of the yield of the Nagasaki bomb, perhaps one kiloton, a neutron bomb fires high-energy neutrons through buildings, armor, and even several feet of earth.[831]

Neutron bombs, also known as enhanced radiation weapons (ERWs), are a type of nuclear weapon designed to release many high-energy neutrons. As discussed before, neutrons are subatomic particles with no electric charge. They can penetrate solid objects much more effectively than other forms of radiation. Neutron weapons use fusion between deuterium and tritium to produce a lethal radius of neutrons and gamma rays. Fusion releases vastly more neutrons than fission, so a one-kiloton neutron bomb will emit a lethal radius similar to a 10-kiloton fission weapon without all of the infrastructure damage associated with the fission weapon. This makes neutron bombs particularly suited for scenarios where enemy soldiers might be situated within tanks that may otherwise be shielded from the heat and radiation effects of a conventional blast. A type of tactical nuclear weapon, neutron bombs would allow vast populations of civilians or soldiers to be effectively destroyed while allowing repopulation or seizure of military targets and infrastructure (which would not be destroyed by a massive blast) very quickly thereafter.[832,833]

The United States built multiple neutron bombs. One such device was the W66 warhead for the Sprint antiballistic missile interceptor. The neutron flux generated by the interceptor was designed to destroy incoming nuclear missiles. The W70 warhead was designed by Lawrence

Livermore National Laboratory and deployed in the 1970s with dial-a-yield functionality from 1-100 kilotons. The W70-3 had neutron bomb capabilities and could be used by NATO in a tactical combat scenario to wipe out columns of Soviet invasion forces moving into Europe. The W79, developed in 1976, was a nuclear artillery shell with a deuterium-tritium mixture and dial-a-yield capabilities to select a yield of between 100 tons and 1.1 kilotons. An optional injection function for an additional deuterium boost turned it into a neutron bomb. More than 550 shells of this type were produced and remained in service until 1992. However, the controversy surrounding the W79's ability to lower the threshold of nuclear war caused a 1985 Congressional order that future W79s be built without the enhanced radiation (neutron bomb) capacity.[834,835,836,837]

China also built its own neutron bomb after it learned of U.S. deployment of enhanced radiation weapons in Europe in the late-1970s. It took them about ten years to develop and miniaturize the technology, but they conducted their first successful demonstration of a neutron bomb on September 29, 1988. It's likely the Chinese government initially developed the technology to prevent Soviet armored divisions from invading them. By 1988, the chances of Russian aggression in China were slim to nil, and the neutron bomb was considered an indecorous technology internationally due to its sole focus on ending human life, but a government can't know its technology works without a test, so test it did.[838]

Finally, the archetypal nuclear "Doomsday" scenario– nuclear winter. Humankind doesn't have to *intend* to destroy the world with nuclear technology to do so; some scientists have posited that it just might happen as a consequence of a full-scale nuclear exchange. Nuclear winter is a hypothetical scenario in which the detonation of numerous nuclear weapons, particularly those with high-yield explosions (multi-megaton thermonuclear bombs and MIRVs), could result in severe global climate disruption and a prolonged decrease in temperatures on Earth. The physics behind nuclear winter involves a chain of events triggered by the massive release of heat, light, and radiation from nuclear explosions.

As we know, when a nuclear weapon detonates, an intense fireball is created, releasing an enormous amount of thermal radiation. This radiation causes widespread fires, leading to the combustion of buildings, forests, and other flammable materials over a vast area.

The immense energy released during the detonation also produces a massive upward column of smoke, soot, and dust particles into the atmosphere. These particles, collectively known as aerosols, are injected into the upper atmosphere, forming a dense cloud layer.[839]

The aerosols in the upper atmosphere can significantly impact Earth's climate. They scatter and absorb sunlight, preventing some of the sun's rays from reaching the surface. As a result, less sunlight is available for heating the earth, decreasing surface temperatures.

The combined effects of reduced sunlight and increased heat-trapping can lead to a significant drop in temperatures globally. Sunlight could ultimately be blocked out for several weeks, leading temperatures to plunge by as much as 20-40 degrees Fahrenheit (11-22 °C). The resulting climate disruption, combined with high doses of radiation, could stop plants' ability to photosynthesize and may destroy most of the planet's vegetation and animal life. A massive, globe-wide death toll could result from exposure, starvation, and disease.

It's important to note that the concept of nuclear winter is based on computer modeling and simulations. While it is supported by scientific evidence and has been the subject of extensive research, the exact extent and severity of a potential nuclear winter scenario would depend on various factors, including the scale and number of nuclear explosions, prevailing weather patterns, and other complex interactions within Earth's climate systems. A number of scientists have disputed the original calculations, and question the degree of damage nuclear winter might cause.[840]

Chapter 27. Conclusion: A Tool of Peace, or a Harbinger of Armageddon– The Choice is Ours

"The atomic age has moved forward at such a pace that every citizen of the world should have some comprehension, at least in comparative terms, of the extent of this development of the utmost significance to every one of us. Clearly, if the people of the world are to conduct an intelligent search for peace, they must be armed with the significant facts of today's existence." -U.S. President Dwight D. Eisenhower, Atoms for Peace[841]

As we learned earlier from historical accounts about the development of nuclear weapons, they do not exist in a void. Political leaders, generals, and scientists didn't just say, "Let's make something that causes a massive explosion," and simply go ahead with it. There was more to it. The intellectual concepts behind the Bomb were and remain major drivers behind its creation and its deployment by world powers. Concepts of mutually assured destruction, deterrence, massive retaliation, existential debates, and a number of philosophical considerations swirl around nuclear weapons and have shaped the way they are developed, used, controlled, or alternatively, curtailed.

Nuclear abolitionists and disarmament advocates throughout the 20th and 21st centuries have gone to great lengths to advocate for eliminating nuclear weapons and defunding the military institutions that enable their maintenance. Their argument is self-evident: nuclear weapons are quite possibly the greatest existential threat to our species, and this threat should be destroyed at all costs.

Given everything we have covered in this book, most readers likely have, at this point, that gut reaction. But it is not a universally-held view.

Some argue that nuclear weapons brought peace to the world after World War II. After all, WWII resulted in the deaths of between 70 million to 100 million people before the atomic bomb was first designed by the Manhattan Project scientists.[842] It has been 78 years since WWII ended, and World War III has yet to occur. Of course, there have been a series of more minor conflicts and proxy wars between the world's great powers. One analysis of the "Nuclear Peace Hypothesis" by Robert Rauchhaus provides some illumination on the idea that nuclear weapons enable peace. Rauchhaus's study is titled, "Evaluating the Nuclear Peace Hypothesis: A Quantitative Approach." Quantitative analyses study and interpret data using mathematical and statistical techniques to understand patterns, relationships, and trends. Rauchhaus notes that multiple quantitative studies have been done on the relationship between nuclear weapons and peace. Ultimately, Rauchhaus' own quantitative study found the following (I've added additional notes added in brackets):[843]

> As the results presented in the previous section indicate, both proliferation optimists [those who believe more nuclear weapons are good, because they contribute to peace/security] and proliferation pessimists [those who believe having more nuclear weapons is bad] find validation for some of their key claims. Kenneth Waltz and other proponents of nuclear deterrence find strong empirical support for their claims that nuclear powers are less likely to fight one another–nuclear weapons may indeed help explain the Long Peace [also known as "Pax Americana," the period of relative peace and stability that has existed post-World War II in the sphere of American influence].[844] Nevertheless, Scott Sagan and other proliferation pessimists find support for

their concerns. At lower levels of escalation, nuclear symmetry does not appear to have a pacifying effect. Worse yet, nuclear asymmetry is generally associated with a higher chance of crises, uses of force, fatalities, and war. On balance, however, these findings support the broader themes of this journal issue. Nuclear weapons do not affect the frequency of conflict, but they do affect the timing, intensity, and outcome of conflict. This study demonstrates that nuclear weapons tend to shift the intensity of disputes toward the lower end of the conflict scale.[845]

Per this analysis, while nuclear weapons don't decrease the *number* of conflicts, and may have other negative effects on conflicts, they likely tend to reduce the *intensity* of those conflicts. To understand the impact of that in real terms, imagine if nuclear weapons did not exist and World War III had been fought as a conventional hot war between NATO and the Warsaw Pact nations; it likely would have made WWII look like child's play, given advances in lethal technologies. Many qualitative and quantitative analyses are floating out there that seek to understand the relationship between nuclear weapons, peace, and war. They draw different conclusions, but chiefly, proliferation optimists will point to the "Long Peace" after World War II, and proliferation pessimists will point to the proxy wars that have taken place, such as the Vietnam War, the Soviet-Afghan War, and the Korean War. Ultimately, the question might never be resolved unless a full-scale nuclear war takes place, in which case the proliferation pessimists will win the argument by default, as humankind will be destroyed by the technology it created. Until and if that happens, or if by some unfathomable oddity, a global hot war breaks out between world powers without the use of nuclear weapons, the proliferation optimists will continue to point to relative peace on a worldwide scale to validate their position.

Entire books can and have been written on the debate about whether more nukes are ultimately *good* or *bad*. For a comprehensive dive into that question, I'd recommend "The Spread of Nuclear Weapons: An Enduring Debate" by Scott D. Sagan and Kenneth N. Waltz. Drawing on diverse perspectives and expert insights, the book delves into the complex issues of nuclear weapons acquisition, deterrence, disarmament, and the challenges posed by emerging nuclear states. It thoroughly analyzes the historical context, strategic considerations, and policy implications of nuclear proliferation.[846]

Beyond considerations of whether nuclear weapons promote peace or threaten to destroy it irrevocably are questions of morality. A long-debated topic is whether the use of nuclear weaponry is ever morally justifiable. This question is most relevant with regard to the only real-world example of the Bomb being used in anger: World War II. Most commonly, this debate turns to a utilitarian trolley problem. Proponents of using the atomic bombs against Japan argue that, absent the bombs, hundreds of thousands or millions more would have been killed as the United States would have been forced to orchestrate a land invasion while the Japanese military and leadership refused to surrender. The trolley problem is a classic moral dilemma that presents a scenario involving a runaway trolley heading toward a group of people tied to the tracks. If the trolley reaches them, they will be killed. The dilemma asks an individual to make a difficult decision: whether to take no action and allow the trolley to continue, resulting in the death of multiple individuals, or to actively intervene by engaging a switch that diverts the trolley onto another track, where a single person is tied, and will be actively killed by the switch throw. Engaging the switch will kill one person and save the lives of several others (thus, resulting in the *"lesser evil"* under pure utilitarian analysis).

In some ways (albeit ones that disregard other moral theories and the complexities of human life and dignity, and do not calculate other possibilities with regard to Japanese decision-making at the end of WWII [perhaps they would have ultimately surrendered anyway]), Hiroshima and Nagasaki can be analyzed through the lens of the trolley problem. This is, of course, an oversimplification, but it's a debate that has played out over close to eight decades since the United States made that fateful decision. For an analysis that critiques the decision to use atomic weapons in Japan and serves as an alternative lens to the utilitarian discussion often raised, look into "Just War Theory." I would also encourage readers to look into first-hand accounts, some of which I've cited, from Japanese civilians in those cities and the stories passed on by their family members; moral debates in academic terms are one thing, but grasping the real impact on human life is critical to understanding the gravity of this technology and the horrible outcomes it can visit upon us.

There are other concepts surrounding nuclear weapons that I recommend readers explore to understand better why nation-states do what they do with their nuclear arsenals. However, these concepts are more broadly applicable to the defense, geopolitical, and foreign affairs spheres, and have entire books dedicated to them, so we won't be able to dive into them meaningfully here. Among these are concepts about *types* of deterrence, such as full-spectrum deterrence (for an example of full-spectrum deterrence, look no further than the U.S. and the Soviets in the Cold War, with massive, unstoppable arsenals) vs. credible deterrence (enough to make your enemy think twice about attacking you, but not necessarily enough weapons to be seen as adopting an aggressive posture that is more likely to provoke your adversary into aggressive action). In recent years, geopolitical adversaries Pakistan and India have grappled with decision-making problems regarding credible vs. full-spectrum deterrence.[847] Even credible deterrence can arguably be problematic from an escalation perspective; for deterrence to be *credible*, an adversary must believe that the state seeking to deter their actions will actually *use* nuclear weapons if necessary. In a sort of self-fulfilling prophecy, some argue that this creates a propensity for nation-states to saber rattle and potentially use nuclear weapons on a small scale (such as a tactical deployment) in order to maintain credibility.

Closely related to deterrence is the security dilemma, fervently debated by foreign policy "realists." Realism is a prominent school of thought in international relations that emphasizes the role of power, self-interest, and the pursuit of national security in shaping the behavior of nation-states. Central to realism is the concept of the security dilemma, which refers to a situation where actions taken by one state to enhance its security are perceived as threatening by other states, leading to a spiral of mistrust and potential conflict. (Sound familiar, after having read this book?)[848]

The security dilemma arises from the inherent uncertainty and lack of trust in international relations. Per the realists, while states have taken actions through various multilateral agreements, concepts of "international law," and international diplomatic bodies such as the United Nations, ultimately, in many ways, the global stage remains a "dog eat dog" atmosphere where nation-states (especially the most powerful among them) continue to exert their will regardless of norms and international consensus. Given this harsh reality, states often seek to increase their security by bolstering their military capabilities, forming alliances, or adopting defensive postures. However, these defensive measures can be misinterpreted by other states as aggressive intentions, prompting these other states to respond with their own security-enhancing actions in a never-ending back-and-forth.

As each state tries to maximize its security, the actions taken by one state to protect itself inadvertently undermine the security of others. This creates a cycle of perceived threats and countermeasures that can escalate tensions and increase the risk of conflict, which is particularly dangerous when you add nuclear weapons to the equation (see the Cuban Missile Crisis). Even when nation-states may not have aggressive intentions, the fear of potential aggression and the need to ensure self-preservation drive other states to adopt defensive measures, inadvertently exacerbating the security dilemma.

Realists argue that the security dilemma is a fundamental characteristic of international politics that cannot be easily overcome. They contend that states are primarily motivated by their own self-interest and security concerns, leading to a constant pursuit of power and strategic advantage. From a realist perspective, managing the security dilemma involves carefully balancing power, maintaining credible deterrence, and establishing stable relationships by balancing interests and power dynamics.

However, critics of realism and the security dilemma argue that it overly simplifies international relations, neglecting the importance of cooperation, shared interests, and the potential for diplomatic resolutions. They advocate for alternative approaches that prioritize trust-building, transparency, and the pursuit of common security interests to overcome the security dilemma and promote peaceful coexistence among states. Regarding the nuclear arms race, both schools of thought have influenced the actions of the world's nuclear powers.

Moving forward, we must all confront the complex reality surrounding nuclear weapons, acknowledging both the inherent dangers they pose and the potential utility they have provided in the pursuit of peace. In this work, we have embarked on a journey through the history of this colossal technology, exploring its immense power and the existential threats it presents to humanity. We should expect our global leaders to deeply contemplate these topics regularly, examining the intricate balance between the risks and potential benefits inherent in the nuclear equation.

Before I close, I want to address the issue of contemplation and make a personal plea to readers about holding their national leaders to account. Over the past decades since the end of the Cold War, we have witnessed statements (and occasionally, actions) by world leaders that have evinced, in some cases, a total misunderstanding of nuclear weapons technology, mutually assured destruction and deterrence, and in others, a cavalier disregard for proliferation and escalation risks. Having read this book, you now possess a better understanding of the stakes involved than most of the world's population (and frankly, of many of your national lawmakers). It is incumbent on you to spread this knowledge and to hold your leaders accountable for the existential responsibility they hold in their hands. Nuclear weaponry is one of the most serious, if not *the* most serious, threats to our species, and it should always be top of mind when assessing the fitness of our leadership. Cool heads, steady hands, and a comprehensive understanding of the stakes involved are the base minimum we should demand of them both when we select them and when we assess their performance during their terms of office.

The pages of history serve as an ever-present reminder of the potential violence that can be visited on all of us by nuclear weapons. From the world-altering bombings of Hiroshima and Nagasaki to the precarious standoffs of the Cold War, the weight of these weapons on global politics and our collective psyche cannot be overstated. They represent an unparalleled destructive force, capable of altering the course of human history in an instant, or alternatively, ending it altogether.

However, even amidst the shadow of uncertainty, we cannot discount the potential benefits of nuclear weapons in shaping world affairs. The concept of nuclear deterrence, for instance, has been credited with preventing massive conventional conflicts between nuclear-armed adversaries, fostering stability, and preventing global-scale wars reminiscent of World War II. The fear of mutually assured destruction has no doubt, to some extent, acted as a deterrent against reckless aggression by all sides.

We must recognize that the pursuit of peace is a delicate dance, and the world is a complex and imperfect stage. In weighing the dangers against potential benefits, we grapple with profound moral and strategic questions. Are the risks of nuclear weapons worth the potential benefits they provide in maintaining peace and stability? Can nation-states strike a balance between disarmament and maintaining a credible deterrent against their adversaries? Will total disarmament ever be possible, given that nation-states are aware that their adversaries will always have the knowledge and the means to build more nuclear weapons, possibly in a clandestine fashion? Even if total disarmament were possible, is it advisable?

In any case, the path forward lies in introspection and thoughtful analysis. We must continue to foster dialogue and diplomacy, seeking multilateral agreements that reduce the risks associated with nuclear weapons. All world powers should vigilantly and assertively pursue arms control measures, transparency, and verification mechanisms to mitigate the inherent dangers of thermonuclear weapons arsenals while retaining a degree of strategic stability and deterrence.

We must also acknowledge the need for continued technological advancements to improve our ability to detect, secure, and safeguard nuclear materials and weapons systems. By investing in research and development, we can strive to make these weapons less vulnerable to proliferation and potential misuse. Our leaders must also pay close attention to encryption systems that protect thermonuclear arsenals through permissive action links and digital security measures. Computing power (and the associated ability to break encryption) evolves on a logarithmic scale, frequently doubling (see: Moore's Law). Quantum computers and artificial general intelligence ("strong AI" that will potentially soon become more intelligent than human beings) will pose unique risks for encryption systems and global stability that could dangerously undermine our ability to secure our nuclear stockpiles. Heads of state should dedicate as much attention and resources toward curbing AI risks and enhancing encryption to protect these weapons as they do to the physical security measures now in place and under development. The human element must also be considered; as digital technologies become more advanced and consequently more dangerous, they will demonstrate greater capacity to manipulate human beings, who as of now, hold the keys to our nuclear arsenals.

Ultimately, the responsibility rests on the shoulders of leaders and policymakers to navigate this moral, political, and strategic tightrope. They must make informed decisions based on comprehensive assessments of the risks and potential benefits, always cognizant of the profound impact these choices can have on the course of human history. And their constituents must do their part in holding them accountable for staying true to this solemn duty.

As we conclude this exploration, let it serve as a call for continued introspection and vigilance. Let us recognize that the world is not black and white, and the complexities surrounding nuclear weapons defy easy answers. By critically examining the dangers and potential benefits, we can strive to chart a course that minimizes the risks while harnessing the potential for a more peaceful and secure future.

In any case, while humankind treads its path through the universe and builds its future, it will, for the foreseeable time ahead, likely do so in the shadow of the proverbial Hand of God. For all our sakes, let's hope we never experience the true extent of its power.

Works Cited

[1] Drollette, Dan. "Meeting Einstein's Challenge: New Thinking about Nuclear Weapons." *Bulletin of the Atomic Scientists*, 3 Apr. 2015, https://thebulletin.org/2015/04/meeting-einsteins-challenge-new-thinking-about-nuclear-weapons/. Accessed 15 Mar. 2023.

[2] "E=mc2: As Famous as the Man Who Wrote It." *American Museum of Natural History*, https://www.amnh.org/exhibitions/einstein/energy/e-mc2. Accessed 23 July 2023.

[3] Perkowitz, Sidney. "E = Mc2." *Encyclopedia Britannica*, 12 Feb. 2010, https://www.britannica.com/science/E-mc2-equation. Accessed 5 Mar. 2023.

[4] "E=mc2: As Famous as the Man Who Wrote It." *American Museum of Natural History*, https://www.amnh.org/exhibitions/einstein/energy/e-mc2. Accessed 23 July 2023.

[5] "E=mc2: As Famous as the Man Who Wrote It." American Museum of Natural History, https://www.amnh.org/exhibitions/einstein/energy/e-mc2. Accessed 5 Mar. 2023.

[6] "The Nobel Prize in Physics 1903." *NobelPrize.Org*, https://www.nobelprize.org/prizes/physics/1903/becquerel/biographical/. Accessed 23 July 2023.

[7] *November 8, 1895: Roentgen's Discovery of x-Rays.* https://www.aps.org/publications/apsnews/200111/history.cfm. Accessed 23 July 2023.

[8] "The Nobel Prize in Physics 1903." *NobelPrize.Org*, https://www.nobelprize.org/prizes/physics/1903/becquerel/biographical/. Accessed 5 Mar. 2023.

[9] "What Is Radiation?" *IAEA*, https://www.iaea.org/newscenter/news/what-is-radiation. Accessed 23 July 2023.

[10] "Marie Curie: Discovering Radium (In Her Own Words)." *BRIEF Exhibit*, https://history.aip.org/exhibits/curie/brief/06_quotes/quotes_07.html. Accessed 23 July 2023.

[11] *Marie Curie and the Science of Radioactivity.* https://history.aip.org/exhibits/curie/resbr1.htm. Accessed 23 July 2023.

[12] "Marie Curie: Radium Therapy (In Her Own Words)." *BRIEF Exhibit*, https://history.aip.org/exhibits/curie/brief/06_quotes/quotes_09.html. Accessed 23 July 2023.

[13] Moul, Dr. Russell. "How Radioactive Is Marie Curie?" *IFLScience*, 24 May 2023, https://www.iflscience.com/marie-curies-body-was-so-radioactive-she-was-buried-in-a-lead-lined-coffin-69080. Accessed 23 July 2023.

[14] Sekiya, Masaru, and Michio Yamasaki. "Antoine Henri Becquerel (1852–

1908): A Scientist Who Endeavored to Discover Natural Radioactivity."
Radiological Physics and Technology, vol. 8, no. 1, Oct. 2014, pp. 1–3,
https://doi.org/10.1007/s12194-014-0292-z.

[15] "Radiation Basics." *US EPA*, 12 Nov. 2014,
https://www.epa.gov/radiation/radiation-basics. Accessed 23 July 2023.

[16] *Ernest Rutherford*.
https://chemed.chem.purdue.edu/genchem/history/rutherford.html. Accessed
23 July 2023.

[17] "Atom." *Encyclopedia Britannica*, 4 May 1999,
https://www.britannica.com/science/atom/Discovery-of-radioactivity.
Accessed 23 July 2023.

[18] "Rutherford's Experiment." *Laradioactivite.Com*,
https://radioactivity.eu.com/phenomenon/rutherford_experiment. Accessed
23 July 2023.

[19] Lea, Robert. "What Is the 'Gold Foil Experiment'? The Geiger-Marsden
Experiments Explained." *Live Science*, 12 Feb. 2022,
https://www.livescience.com/gold-foil-experiment-geiger-marsden. Accessed
23 July 2023.

[20] Cooper, Keith. "Protons: The Essential Building Blocks of Atoms." *Space*,
27 Nov. 2022, https://www.space.com/protons-facts-discovery-charge-mass.
Accessed 23 July 2023.

[21] "The Nobel Prize in Physics 1935." *NobelPrize.Org*,
https://www.nobelprize.org/prizes/physics/1935/chadwick/biographical/.
Accessed 14 Feb. 2023.

[22] *Zosimos_of_Panopolis*. ChemEurope,
https://www.chemeurope.com/en/encyclopedia/Zosimos_of_Panopolis.html.
Accessed 23 July 2023.

[23] "Zosimos of Panopolis." *Encyclopedia.Com*,
https://www.encyclopedia.com/science/dictionaries-thesauruses-pictures-
and-press-releases/zosimos-panopolis. Accessed 23 July 2023.

[24] ChemEurope. *Chrysopoeia*. ChemEurope,
https://www.chemeurope.com/en/encyclopedia/Chrysopoeia.html. Accessed
23 July 2023.

[25] "The Nobel Prize in Chemistry 1921." *NobelPrize.Org*,
https://www.nobelprize.org/prizes/chemistry/1921/soddy/biographical/.
Accessed 23 July 2023.

[26] *Frederick_Soddy*. ChemEurope,
https://www.chemeurope.com/en/encyclopedia/Frederick_Soddy.html.
Accessed 23 July 2023.

[27] "Tritium in Exit Signs." *US EPA*, 26 Nov. 2018,
https://www.epa.gov/radtown/tritium-exit-signs. Accessed 23 July 2023.

[28] "DOE Explains...Deuterium-Tritium Fusion Reactor Fuel." *Energy.Gov*,
https://www.energy.gov/science/doe-explainsdeuterium-tritium-fusion-
reactor-fuel. Accessed 23 July 2023.

[29] "Radioisotopes." *ANSTO*, https://www.ansto.gov.au/education/nuclear-

facts/what-are-radioisotopes. Accessed 23 July 2023.

[30] Wells, H. G. "The World Set Free." *Project Gutenberg*, 1914, https://www.gutenberg.org/cache/epub/1059/pg1059-images.html. Accessed 17 July 2023.

[31] Corneliussen, Steven T. *Science and the Media: 19 - 25 March*. Physics Today, https://pubs.aip.org/physicstoday/Online/17707/Science-and-the-media-19-25-March. Accessed 23 July 2023.

[32] Lanouette, William. *Genius in the Shadows: A Biography of Leo Szilard, the Man Behind the Bomb*. Skyhorse, 2013.

[33] Carlson Caspers. *Leo Szilard*. 30 Mar. 2022, https://www.carlsoncaspers.com/diversity-and-inclusion/diversity-spotlight/leo-szilard/. Accessed 23 July 2023.

[34] "Cyclotrons – What Are They and Where Can You Find Them?" *IAEA*, https://www.iaea.org/newscenter/news/cyclotrons-what-are-they-and-where-can-you-find-them. Accessed 14 Feb. 2023.

[35] News, BBC. "A Point of View: The Man Who Dreamed of the Atom Bomb." *BBC News*, 4 Oct. 2013, https://www.bbc.com/news/magazine-24395740. Accessed 14 Feb. 2023.

[36] "H.G. Wells and the Scientific Imagination." *VQR Online*, https://www.vqronline.org/essay/hg-wells-and-scientific-imagination. Accessed 14 Feb. 2023.

[37] Hassan, Yassin A., and Robin A. Chaplin. *Nuclear Energy Materials And Reactors - Volume I*. EOLSS Publications, 2010.

[38] "Nuclear Fuel." *Nuclear Energy Institute*, https://www.nei.org/fundamentals/nuclear-fuel. Accessed 14 Feb. 2023.

[39] "Prelude to Nuclear Weapons ." *Nuclear Weapons Education Project*, https://nuclearweaponsedproj.mit.edu/history/prelude-nuclear-weapons. Accessed 14 Feb. 2023.

[40] The Editors of Encyclopaedia Britannica. "Frédéric and Irène Joliot-Curie." *Encyclopedia Britannica*, 20 July 1998, https://www.britannica.com/biography/Frederic-and-Irene-Joliot-Curie#ref652104. Accessed 15 Feb. 2023.

[41] Badash, Lawrence. "Enrico Fermi." *Encyclopedia Britannica*, 20 July 1998, https://www.britannica.com/biography/Enrico-Fermi. Accessed 15 Feb. 2023.

[42] Spence, Robert. "Otto Hahn." *Encyclopedia Britannica*, 20 July 1998, https://www.britannica.com/biography/Otto-Hahn.

[43] *Manhattan Project: The Discovery of Fission, 1938-1939*. https://www.osti.gov/opennet/manhattan-project-history/Events/1890s-1939/discovery_fission.htm. Accessed 23 July 2023.

[44] "Enrico Fermi." *Encyclopedia Britannica*, 20 July 1998, https://www.britannica.com/biography/Enrico-Fermi/American-career. Accessed 15 Feb. 2023.

[45] Richmond, C. R. "Population Exposure from the Fuel Cycle: Review and Future Direction." *UNT Digital Library*, 1 Jan. 1987,

https://digital.library.unt.edu/ark:/67531/metadc1086292/.

[46] *Manhattan Project: Einstein's Letter, 1939.*
https://www.osti.gov/opennet/manhattan-project-history/Events/1939-1942/einstein_letter.htm. Accessed 23 July 2023.

[47] Einstein, Albert. "Einstein Letter." *FDR Library*, 2 Aug. 1939, http://www.fdrlibrary.marist.edu/archives/pdfs/docsworldwar.pdf. Accessed 23 July 2023.

[48] *Manhattan Project: Einstein's Letter to Roosevelt.*
https://www.osti.gov/opennet/manhattan-project-history/Resources/einstein_letter_photograph.htm#1. Accessed 15 Feb. 2023.

[49] "FDR's Response to Einstein Letter." *Albuquerque Historical Society*, 2 Jan. 2016, https://www.albuqhistsoc.org/source-documents/fdrs-response-einstein-letter/. Accessed 15 Feb. 2023.

[50] *Manhattan Project: Early Uranium Research, 1939-1941.*
https://www.osti.gov/opennet/manhattan-project-history/Events/1939-1942/uranium_research.htm. Accessed 23 July 2023.

[51] *Manhattan Project: Early Uranium Research, 1939-1941.*
https://www.osti.gov/opennet/manhattan-project-history/Events/1939-1942/uranium_research.htm. Accessed 23 July 2023.

[52] *Manhattan Project: People > Civilian Organizations > NATIONAL DEFENSE RESEARCH COMMITTEE.*
https://www.osti.gov/opennet/manhattan-project-history/People/CivilianOrgs/ndrc.html. Accessed 17 Feb. 2023.

[53] *Manhattan Project: Early Uranium Research, 1939-1941.*
https://www.osti.gov/opennet/manhattan-project-history/Events/1939-1942/uranium_research.htm. Accessed 15 Feb. 2023.

[54] Morss, Lester. "Plutonium." *Encyclopedia Britannica*, 20 July 1998, https://www.britannica.com/science/plutonium. Accessed 17 Feb. 2023.

[55] "Chicago Pile-1." *Nuclear Museum*,
https://ahf.nuclearmuseum.org/ahf/history/chicago-pile-1/. Accessed 17 Feb. 2023.

[56] *Manhattan Project: Piles and Plutonium, 1939-1942.*
https://www.osti.gov/opennet/manhattan-project-history/Events/1939-1942/piles_plutonium.htm. Accessed 23 July 2023.

[57] "Metallurgical Laboratory at the University of Chicago." *Energy.Gov*,
https://www.energy.gov/lm/metallurgical-laboratory-university-chicago. Accessed 23 July 2023.

[58] *Manhattan Project: Places > Metallurgical Laboratory > MAKING OF THE MET LAB.* https://www.osti.gov/opennet/manhattan-project-history/Places/MetLab/making-met.html. Accessed 17 Feb. 2023.

[59] "Chicago Pile-1." *Nuclear Museum*,
https://ahf.nuclearmuseum.org/ahf/history/chicago-pile-1/. Accessed 23 July 2023.

[60] "Chicago Pile 1: A Bold Nuclear Physics Experiment with Enduring Impact." *EurekAlert!*, https://www.eurekalert.org/news-releases/973133.

Accessed 23 July 2023.

[61] *Manhattan Project: CP-1 Goes Critical, Met Lab, December 2, 1942*. https://www.osti.gov/opennet/manhattan-project-history/Events/1942-1944_pu/cp-1_critical.htm. Accessed 23 July 2023.

[62] "Chicago Pile-1." *Nuclear Museum*, https://ahf.nuclearmuseum.org/ahf/history/chicago-pile-1/. Accessed 23 July 2023.

[63] *Manhattan Project: CP-1 Goes Critical, Met Lab, December 2, 1942*. https://www.osti.gov/opennet/manhattan-project-history/Events/1942-1944_pu/cp-1_critical.htm. Accessed 18 Feb. 2023.

[64] "Chicago Pile 1: A Bold Nuclear Physics Experiment with Enduring Impact ." *Argonne National Laboratory*, https://www.anl.gov/article/chicago-pile-1-a-bold-nuclear-physics-experiment-with-enduring-impact.

[65] "Chicago Pile-1." *Nuclear Museum*, https://ahf.nuclearmuseum.org/ahf/history/chicago-pile-1/. Accessed 17 Feb. 2023.

[66] "Glenn T. Seaborg: Quotes." *Encyclopedia Britannica*, https://www.britannica.com/quotes/Glenn-T-Seaborg. Accessed 18 Feb. 2023.

[67] "America's National Churchill Museum." *Fifty Years Hence, Winston Churchill December 1931*, https://www.nationalchurchillmuseum.org/fifty-years-hence.html. Accessed 26 Feb. 2023.

[68] "Frisch-Peierls Memorandum, March 1940." *Historical Documents*, https://www.atomicarchive.com/resources/documents/beginnings/frisch-peierls-2.html. Accessed 26 Feb. 2023.

[69] Aaserud, Finn. "Niels Bohr." *Encyclopedia Britannica*, 20 July 1998, https://www.britannica.com/biography/Niels-Bohr. Accessed 23 July 2023.

[70] *Manhattan Project: The Maud Report, 1941*. https://www.osti.gov/opennet/manhattan-project-history/Events/1939-1942/maud.htm. Accessed 23 July 2023.

[71] "British Nuclear Program." *Nuclear Museum*, https://ahf.nuclearmuseum.org/ahf/history/british-nuclear-program/. Accessed 26 Feb. 2023.

[72] "MAUD Committee Report." *Nuclear Museum*, https://ahf.nuclearmuseum.org/ahf/key-documents/maud-committee-report/. Accessed 26 Feb. 2023.

[73] "British Nuclear Program." *Nuclear Museum*, https://ahf.nuclearmuseum.org/ahf/history/british-nuclear-program/. Accessed 26 Feb. 2023.

[74] "Uranium in Canada Appendix 1: Brief History of Uranium Mining in Canada." *World Nuclear Association*, https://www.world-nuclear.org/information-library/country-profiles/countries-a-f/appendices/uranium-in-canada-appendix-1-brief-history-of-uran.aspx. Accessed 26 Feb. 2023.

[75] Moore, Richard. "Rudolf Peierls's 'Outline of the Development of the

British Tube Alloy Project': A 1945 Account of the Earliest UK Work on Atomic Energy." *Nuclear Technology*, vol. 207, no. sup1, Nov. 2021, pp. S374–79, https://doi.org/10.1080/00295450.2021.1910004.

[76] *Manhattan Project: Enter the Army, 1942.* https://www.osti.gov/opennet/manhattan-project-history/Events/1945-present/../1942/enter_army.htm. Accessed 17 Feb. 2023.

[77] *Manhattan Project: Difficult Choices, 1942.* https://www.osti.gov/opennet/manhattan-project-history/Events/1945-present/../1942/1942.htm. Accessed 17 Feb. 2023.

[78] Johnson, Charles W. "Oak Ridge." *Tennessee Encyclopedia*, 8 Oct. 2017, https://tennesseeencyclopedia.net/entries/oak-ridge/. Accessed 18 Feb. 2023.

[79] "History." *Energy.Gov*, https://www.energy.gov/orem/history. Accessed 18 Feb. 2023.

[80] *Manhattan Project: Working K-25 into the Mix, 1943-1944.* https://www.osti.gov/opennet/manhattan-project-history/Events/1942-1944_ur/k-25_working.htm. Accessed 18 Feb. 2023.

[81] *Manhattan Project: Production Reactor (Pile) Design, Met Lab, 1942.* https://www.osti.gov/opennet/manhattan-project-history/Events/1942-1944_ur/../1942-1944_pu/reactor_design.htm. Accessed 18 Feb. 2023.

[82] *Id.*

[83] "Manhattan Project Spotlight: E.I. DuPont de Nemours & Company." *Nuclear Museum*, https://ahf.nuclearmuseum.org/manhattan-project-spotlight-ei-du-pont-de-nemours-company/. Accessed 18 Feb. 2023.

[84] *Manhattan Project: DuPont and Hanford, Hanford Engineer Works, 1942.* https://www.osti.gov/opennet/manhattan-project-history/Events/1942-1944_ur/../1942-1944_pu/dupont_hanford.htm. Accessed 18 Feb. 2023.

[85] *DuPont and Hanford.* https://www.osti.gov/opennet/manhattan-project-history/Events/1942-1944_pu/dupont_hanford.htm. Accessed 18 Feb. 2023.

[86] "Native Americans and the Manhattan Project." *Nuclear Museum*, https://ahf.nuclearmuseum.org/ahf/history/native-americans-and-manhattan-project/. Accessed 23 July 2023.

[87] *Manhattan Project: Seaborg and Plutonium Chemistry, Met Lab, 1942-1944.* https://www.osti.gov/opennet/manhattan-project-history/Events/1942-1944_pu/seaborg_plutonium.htm. Accessed 18 Feb. 2023.

[88] *Manhattan Project: Final Reactor Design and X-10, 1942-1943.* https://www.osti.gov/opennet/manhattan-project-history/Events/1942-1944_pu/final_reactor_x-10.htm. Accessed 18 Feb. 2023.

[89] *Manhattan Project: Hanford Becomes Operational, 1943-1944.* https://www.osti.gov/opennet/manhattan-project-history/Events/1942-1944_pu/hanford_operational.htm. Accessed 18 Feb. 2023.

[90] "John Wheeler's Interview (1965)." *Nuclear Museum*, https://ahf.nuclearmuseum.org/voices/oral-histories/john-wheelers-interview-1965/. Accessed 18 Feb. 2023.

[91] *"Xenon Poisoning" or Neutron Absorption in Reactors.* http://hyperphysics.phy-astr.gsu.edu/hbase/NucEne/xenon.html. Accessed 18

Feb. 2023.

92 "'The Atomic Bomb Still Haunts Me' - Quotes from the Real-Life Manhattan Project." *Guide*, https://www.sbs.com.au/guide/article/2016/06/09/atomic-bomb-still-haunts-me-quotes-real-life-manhattan-project. Accessed 19 Feb. 2023.

93 History.com Editors. "The Trinity Test." *HISTORY*, 23 Apr. 2010, https://www.history.com/topics/world-war-ii/trinity-test. Accessed 21 Feb. 2023.

94 "Los Alamos National Laboratory." *Encyclopedia Britannica*, 13 Nov. 2013, https://www.britannica.com/topic/Los-Alamos-National-Laboratory. Accessed 19 Feb. 2023.

95 Cassidy, David C. *J. Robert Oppenheimer and the American Century*. Dutton, 2005, pp. 184–86, https://openlibrary.org/books/OL3314534M/J._Robert_Oppenheimer_and_the_American_century. Accessed 19 Feb. 2023.

96 Ouellette, Jennifer. "J. Robert Oppenheimer Cleared 'Black Mark' against His Name after 68 Years." *Ars Technica*, https://arstechnica.com/science/2022/12/j-robert-oppenheimer-cleared-of-being-suspected-soviet-spy-after-68-years/. Accessed 19 Feb. 2023.

97 *Groves and Oppenheimer Statues (U.S. National Park Service)*. https://www.nps.gov/places/000/groves-and-oppenheimer-statues.htm. Accessed 19 Feb. 2023.

98 *Manhattan Project: Establishing Los Alamos, 1942-1943*. https://www.osti.gov/opennet/manhattan-project-history/Events/1942-1945/establishing_los_alamos.htm. Accessed 19 Feb. 2023.

99 *Manhattan Project: Early Bomb Design, Los Alamos: Laboratory, 1943-1944*. https://www.osti.gov/opennet/manhattan-project-history/Events/1942-1945/early_bomb_design.htm. Accessed 19 Feb. 2023.

100 "Subcritical Mass." *NRC Web*, https://www.nrc.gov/reading-rm/basic-ref/glossary/subcritical-mass.html. Accessed 19 Feb. 2023.

101 The Editors of Encyclopaedia Britannica. "Atomic Bomb." *Encyclopedia Britannica*, 20 July 1998, https://www.britannica.com/technology/atomic-bomb#ref62815. Accessed 19 Feb. 2023.

102 *Early Bomb Design*. https://www.osti.gov/opennet/manhattan-project-history/Events/1942-1945/early_bomb_design.htm. Accessed 19 Feb. 2023.

103 *Glossary [Thermal Energy to Yield]*. https://www.atomicarchive.com/resources/glossary/glossary10.html. Accessed 19 Feb. 2023.

104 Cameron Reed, B. "Fission Fizzles: Estimating the Yield of a Predetonated Nuclear Weapon." *American Journal of Physics*, vol. 79, no. 7, July 2011, pp. 769–73, https://doi.org/10.1119/1.3569575.

105 *Definition of Artillery*. https://www.merriam-webster.com/dictionary/artillery. Accessed 19 Feb. 2023.

106 *Manhattan Project: Basic Research at Los Alamos, 1943-1944*. https://www.osti.gov/opennet/manhattan-project-history/Events/1942-

1945/basic_research.htm. Accessed 19 Feb. 2023.

[107] *Manhattan Project: Implosion Becomes a Necessity, Los Alamos, 1944.* https://www.osti.gov/opennet/manhattan-project-history/Events/1942-1945/implosion_necessity.htm. Accessed 19 Feb. 2023.

[108] "B-29." *Encyclopedia Britannica*, 20 July 1998, https://www.britannica.com/technology/B-29. Accessed 19 Feb. 2023.

[109] *Historical Snapshot: B-29 Superfortress.* https://www.boeing.com/history/products/b-29-superfortress.page. Accessed 19 Feb. 2023.

[110] *Implosion Becomes a Necessity.* https://www.osti.gov/opennet/manhattan-project-history/Events/1942-1945/implosion_necessity.htm. Accessed 19 Feb. 2023.

[111] Jones, Vincent C. *Manhattan, the Army and the Atomic Bomb.* U.S. Government Printing Office, 1985, pp. 165–66.

[112] Schrader, Robert. "Eastman Kodak and the Manhattan Project." *Atomic Heritage Foundation*, The Atomic Heritage Foundation, https://ahf.nuclearmuseum.org/eastman-kodak-and-manhattan-project-dr-robert-schrader/. Accessed 19 Feb. 2023.

[113] *Manhattan Project: Oak Ridge and Hanford Come Through, 1944-1945.* https://www.osti.gov/opennet/manhattan-project-history/Events/1942-1945/come_through.htm. Accessed 19 Feb. 2023.

[114] Temperton, James. "Oppenheimer Quotes: The Story behind 'Now I Am Become Death, the Destroyer of Worlds.'" *WIRED UK*, 9 Aug. 2017, https://www.wired.co.uk/article/manhattan-project-robert-oppenheimer. Accessed 17 July 2023.

[115] *Manhattan Project: Debate Over How to Use the Bomb, 1945.* https://www.osti.gov/opennet/manhattan-project-history/Events/1945/debate.htm. Accessed 19 Feb. 2023.

[116] *Manhattan Project: The Trinity Test, July 16, 1945.* https://www.osti.gov/opennet/manhattan-project-history/Events/1945/trinity.htm. Accessed 19 Feb. 2023.

[117] Templeton, Patricia. *Plutonium and Poetry: Where Trinity and Oppenheimer's Reading Habits Met Literary Inspirations, Correcting Misinterpretation of His Famous Quote.* Office of Scientific and Technical Information (OSTI), 1 July 2021, http://dx.doi.org/10.2172/1805707. Accessed 20 Feb. 2023.

[118] "100-Ton TNT Shot." *Nuclear Museum*, https://ahf.nuclearmuseum.org/ahf/history/100-ton-tnt-shot/. Accessed 19 Feb. 2023.

[119] "100 Ton Event Served as Dress Rehearsal for Trinity Site." *Www.Army.Mil*, https://www.army.mil/article/148701/100_ton_event_served_as_dress_rehearsal_for_trinity_site. Accessed 19 Feb. 2023.

[120] Yiu, Yuen. "The Fear of Setting the Planet on Fire with a Nuclear Weapon." *Discover Magazine*, 17 July 2020,

https://www.discovermagazine.com/the-sciences/the-fear-of-setting-the-planet-on-fire-with-a-nuclear-weapon. Accessed 21 Feb. 2023.

[121] *The Trinity Test*. https://www.osti.gov/opennet/manhattan-project-history/Events/1945/trinity.htm. Accessed 21 Feb. 2023.

[122] Eby, Nelson, et al. "Trinitite-the Atomic Rock." *Geology Today*, vol. 26, no. 5, Sept. 2010, pp. 180–85, https://doi.org/10.1111/j.1365-2451.2010.00767.x.

[123] "Trinity: World's First Nuclear Test." *Air Force Nuclear Weapons Center*, https://www.afnwc.af.mil/About-Us/History/Trinity-Nuclear-Test/. Accessed 21 Feb. 2023.

[124] *The Trinity Test*. https://www.osti.gov/opennet/manhattan-project-history/Events/1945/trinity.htm. Accessed 21 Feb. 2023.

[125] "What Is the Bhagavadgita?" *Encyclopedia Britannica*, https://www.britannica.com/question/What-is-the-Bhagavadgita. Accessed 21 Feb. 2023.

[126] Eby, Nelson, et al. "Trinitite-the Atomic Rock." *Geology Today*, vol. 26, no. 5, Sept. 2010, pp. 180–85, https://doi.org/10.1111/j.1365-2451.2010.00767.x.

[127] malloryk. "'Destroyer of Worlds': The Making of an Atomic Bomb." *The National World War II Museum*, 15 July 2020, https://www.nationalww2museum.org/war/articles/making-the-atomic-bomb-trinity-test. Accessed 21 Feb. 2023.

[128] Selby, Hugh D., et al. "A New Yield Assessment for the Trinity Nuclear Test, 75 Years Later." *Nuclear Technology*, vol. 207, no. sup1, Nov. 2021, pp. 321–25, https://doi.org/10.1080/00295450.2021.1932176.

[129] Khan, F. A. "Estimating the Photo-Fission Yield of the Trinity Test." *Scientific Reports*, vol. 10, no. 1, Mar. 2020, https://doi.org/10.1038/s41598-020-61201-0.

[130] "Trinity: World's First Nuclear Test." *Air Force Nuclear Weapons Center*, https://www.afnwc.af.mil/About-Us/History/Trinity-Nuclear-Test/. Accessed 21 Feb. 2023.

[131] "Voice and Silence in the First Nuclear War: Wilfred Burchett and Hiroshima." *The Asia-Pacific Journal: Japan Focus*, https://apjjf.org/-Richard-Tanter/2066/article.html. Accessed 24 Feb. 2023.

[132] "The Costs of the Manhattan Project." *Brookings*, 15 Dec. 2016, https://www.brookings.edu/the-costs-of-the-manhattan-project/. Accessed 22 Feb. 2023.

[133] *Harry Truman's Decision to Use the Atomic Bomb (U.S. National Park Service)*. https://www.nps.gov/articles/trumanatomicbomb.htm. Accessed 22 Feb. 2023.

[134] "Milestones: 1937–1945." *Office of the Historian*, https://history.state.gov/milestones/1937-1945/potsdam-conf. Accessed 23 Feb. 2023.

[135] "Foreign Relations of the United States: Diplomatic Papers, The Conference of Berlin (The Potsdam Conference), 1945, Volume II." *Office of*

the Historian,
https://history.state.gov/historicaldocuments/frus1945Berlinv02/d1382.
Accessed 22 Feb. 2023.

[136] "Little Boy and Fat Man." *Nuclear Museum,*
https://ahf.nuclearmuseum.org/ahf/history/little-boy-and-fat-man/. Accessed 23 Feb. 2023.

[137] "The Most Fearsome Sight: The Atomic Bombing of Hiroshima." *The National World War II Museum,* 5 Aug. 2020,
https://www.nationalww2museum.org/war/articles/atomic-bomb-hiroshima.
Accessed 23 Feb. 2023.

[138] "Featured Document Display: The Atomic Bombing of Hiroshima and Nagasaki." *National Archives Museum,*
https://museum.archives.gov/featured-document-display-atomic-bombing-hiroshima-and-nagasaki. Accessed 23 Feb. 2023.

[139] "The Atomic Bombings of Hiroshima and Nagasaki." *The Atomic Bombings of Hiroshima and Nagasaki,*
https://avalon.law.yale.edu/20th_century/mp07.asp. Accessed 23 Feb. 2023.

[140] "'My God, What Have We Done.'" *Global Zero,*
https://www.globalzero.org/legacy-of-hiroshima-and-nagasaki/. Accessed 21 June 2023.

[141] *Manhattan Project: The Atomic Bombing of Hiroshima, August 6, 1945.*
https://www.osti.gov/opennet/manhattan-project-history/Events/1945/hiroshima.htm. Accessed 23 Feb. 2023.

[142] American Experience. "Announcing the Bombing of Hiroshima." *American Experience,* 22 Jan. 2019,
https://www.pbs.org/wgbh/americanexperience/features/truman-hiroshima/.
Accessed 23 Feb. 2023.

[143] "Leaflets Warning Japanese of Atomic Bomb." *American Experience,* 24 Apr. 2018, https://www.pbs.org/wgbh/americanexperience/features/truman-leaflets/. Accessed 23 Feb. 2023.

[144] *United States Census Website,* Bradbury Science Museum, Los Alamos National Laboratory, https://www.census.gov/history/pdf/fatman-littleboy-losalamosnatllab.pdf. Accessed 23 Feb. 2023.

[145] "The Atomic Bombing of Nagasaki Video." Media Gallery,
https://www.atomicarchive.com/media/videos/nagasaki.html. Accessed 23 Feb. 2023.

[146] *Manhattan Project: The Atomic Bombing of Nagasaki, August 9, 1945.*
https://www.osti.gov/opennet/manhattan-project-history/Events/1945/nagasaki.htm. Accessed 23 Feb. 2023.

[147] *Manhattan Project: Japan Surrenders, August 10-15, 1945.*
https://www.osti.gov/opennet/manhattan-project-history/Events/1945/surrender.htm. Accessed 23 Feb. 2023.

[148] "The Jewel Voice Broadcast." *Nuclear Museum,*
https://ahf.nuclearmuseum.org/ahf/key-documents/jewel-voice-broadcast/.
Accessed 23 Feb. 2023.

[149] "How Many People Died during World War II?" *Encyclopedia Britannica*, https://www.britannica.com/question/How-many-people-died-during-World-War-II. Accessed 23 Feb. 2023.

[150] "Research Starters: Worldwide Deaths in World War II." *The National WWII Museum | New Orleans*, https://www.nationalww2museum.org/students-teachers/student-resources/research-starters/research-starters-worldwide-deaths-world-war. Accessed 23 Feb. 2023.

[151] History.com Editors. "Pearl Harbor." *HISTORY*, 29 Oct. 2009, https://www.history.com/topics/world-war-ii/pearl-harbor#how-many-people-died-in-pearl-harbor. Accessed 23 Feb. 2023.

[152] History.com Editors. "Bombing of Hiroshima and Nagasaki." *HISTORY*, 18 Nov. 2009, https://www.history.com/topics/world-war-ii/bombing-of-hiroshima-and-nagasaki. Accessed 23 Feb. 2023.

[153] "Letter from Richard Abrams to Senator Brien McMahon of Connecticut, February 9, 1946." *U.S. Capitol - Visitor Center*, https://www.visitthecapitol.gov/artifact/letter-richard-abrams-senator-brien-mcmahon-connecticut-february-9-1946. Accessed 7 Mar. 2023.

[154] "Operation Crossroads: A Deadly Illusion." *The National World War II Museum*, 4 July 2021, https://www.nationalww2museum.org/war/articles/operation-crossroads-atomic-bomb-aftermath. Accessed 22 Mar. 2023.

[155] *Operation Crossroads*. https://www.history.navy.mil/browse-by-topic/wars-conflicts-and-operations/cold-war/crossroads.html#1. Accessed 22 Mar. 2023.

[156] "Operation Crossroads: A Deadly Illusion." *The National World War II Museum*, 4 July 2021, https://www.nationalww2museum.org/war/articles/operation-crossroads-atomic-bomb-aftermath. Accessed 22 Mar. 2023.

[157] *Manhattan Project: Operation Crossroads, Bikini Atoll, July 1946*. https://www.osti.gov/opennet/manhattan-project-history/Events/1945-present/crossroads.htm. Accessed 8 Apr. 2023.

[158] "Operation Crossroads: A Deadly Illusion." *The National World War II Museum*, 4 July 2021, https://www.nationalww2museum.org/war/articles/operation-crossroads-atomic-bomb-aftermath. Accessed 8 Apr. 2023.

[159] *Fact Sheet*. https://www.history.navy.mil/content/history/nhhc/research/library/online-reading-room/title-list-alphabetically/o/operation-crossroads/fact-sheet.html. Accessed 8 Apr. 2023.

[160] "Operation Crossroads: A Deadly Illusion." *The National World War II Museum*, 4 July 2021, https://www.nationalww2museum.org/war/articles/operation-crossroads-atomic-bomb-aftermath. Accessed 8 Apr. 2023.

[161] "Henry W. Newson to N. E. Bradbury [Director, Los Alamos Laboratory],

'Possible Difficulties in Naval Tests', 17 December 1945, Secret." *National Security Archive*, https://nsarchive.gwu.edu/document/21878-document-03-henry-w-newson-n-e-bradbury.

[162] *Operation Crossroads: Fact Sheet.* https://www.history.navy.mil/content/history/nhhc/research/library/online-reading-room/title-list-alphabetically/o/operation-crossroads/fact-sheet.html. Accessed 8 Apr. 2023.

[163] "Operation Crossroads: A Deadly Illusion." *The National World War II Museum*, 4 July 2021, https://www.nationalww2museum.org/war/articles/operation-crossroads-atomic-bomb-aftermath. Accessed 8 Apr. 2023.

[164] *Id.*

[165] "Operation Crossroads." *Nuclear Museum*, https://ahf.nuclearmuseum.org/ahf/history/operation-crossroads/. Accessed 8 Apr. 2023.

[166] Weisgall, Jonathan M. *Operation Crossroads: The Atomic Tests at Bikini Atoll.* US Naval Institute Press, 1994, p. 243.

[167] "Operation Crossroads: A Deadly Illusion." *The National World War II Museum*, 4 July 2021, https://www.nationalww2museum.org/war/articles/operation-crossroads-atomic-bomb-aftermath. Accessed 10 Apr. 2023.

[168] Weisgall, Jonathan M. *Operation Crossroads: The Atomic Tests at Bikini Atoll.* US Naval Institute Press, 1994, p. 244.

[169] *Operation Crossroads: The Atomic Tests at Bikini Atoll.* US Naval Institute Press, 1994, pp. 288–90.

[170] *Operation Crossroads: The Atomic Tests at Bikini Atoll.* US Naval Institute Press, 1994, pp. 291–93.

[171] "Newsletter – Spring 2009." *Wilkie Collins Society*, https://wilkiecollinssociety.org/newsletter-spring-2009/. Accessed 24 Feb. 2023.

[172] "Niels Bohr." *Encyclopedia Britannica*, 20 July 1998, https://www.britannica.com/biography/Niels-Bohr/Copenhagen-interpretation-of-quantum-mechanics. Accessed 26 Feb. 2023.

[173] "Edward Teller." *Encyclopedia Britannica*, 20 July 1998, https://www.britannica.com/biography/Edward-Teller. Accessed 26 Feb. 2023.

[174] Eddington. "The Internal Constitution of the Stars." *Zenodo*, 1 Jan. 1988, https://zenodo.org/record/1429642#.Y_tePXbMJD8.

[175] "Nuclear Weapons 101: Back to the Basics." *K=1 Project*, https://k1project.columbia.edu/news/nuclear-weapons-101-back-basics. Accessed 26 Feb. 2023.

[176] *November 1, 1952: Teller and the Hydrogen Bomb.* https://www.aps.org/publications/apsnews/200311/history.cfm. Accessed 26 Feb. 2023.

[177] *Manhattan Project: People > Scientists > Edward Teller.*

https://www.osti.gov/opennet/manhattan-project-history/People/Scientists/edward-teller.html. Accessed 26 Feb. 2023.

[178] Stuewer, Roger H. "Max Planck." *Encyclopedia Britannica*, 20 July 1998, https://www.britannica.com/biography/Max-Planck. Accessed 1 Mar. 2023.

[179] "Klaus Fuchs." *MI5 - The Security Service*, https://www.mi5.gov.uk/klaus-fuchs. Accessed 1 Mar. 2023.

[180] "Fuchs, Klaus." *Encyclopedia.Com*, https://www.encyclopedia.com/people/history/british-and-irish-history-biographies/klaus-fuchs. Accessed 1 Mar. 2023.

[181] reporter, Guardian staff. "Atom Spy Klaus Fuchs Was Motivated by Conscience." *The Guardian*, 20 Aug. 2019, https://www.theguardian.com/world/2019/aug/20/atom-spy-klaus-fuchs-was-motivated-by-conscience. Accessed 1 Mar. 2023.

[182] *"Klaus Fuchs." Nuclear Museum*, https://ahf.nuclearmuseum.org/ahf/profile/klaus-fuchs/. Accessed 1 Mar. 2023.

[183] "National Security Agency." *National Security Agency/Central Security Service*, https://www.nsa.gov/. Accessed 1 Mar. 2023.

[184] "Venona Documents." *National Security Agency/Central Security Service*, https://www.nsa.gov/Helpful-Links/NSA-FOIA/Declassification-Transparency-Initiatives/Historical-Releases/Venona/. Accessed 1 Mar. 2023.

[185] "NOVA Online." *Read Venona Intercepts: February 9, 1944*, https://www.pbs.org/wgbh/nova/venona/inte_19440209.html. Accessed 1 Mar. 2023.

[186] https://www.history.com/news/big-three-allies-wwii-roosevelt-churchill-stalin

[187] "Soviet Atomic Program - 1946." *Nuclear Museum*, https://ahf.nuclearmuseum.org/ahf/history/soviet-atomic-program-1946/. Accessed 1 Mar. 2023.

[188] Norris, Robert S. "Igor Vasilyevich Kurchatov." *Encyclopedia Britannica*, 20 July 1998, https://www.britannica.com/biography/Igor-Vasilyevich-Kurchatov. Accessed 1 Mar. 2023.

[189] "Sarov Closed City. Nuclear Weapon Development and Production Center." *Military-Today.Com*, http://www.military-today.com/bases/arzamas_16.htm. Accessed 5 Mar. 2023.

[190] *The Soviet Atomic Project: How the Soviet Union Obtained the Atomic Bomb*. 2018, p. 539.

[191] Pondrom, Lee G. *The Soviet Atomic Project: How the Soviet Union Obtained the Atomic Bomb*. 2018, p. 577.

[192] *The Soviet Atomic Project: How the Soviet Union Obtained the Atomic Bomb*. 2018, pp. 577–78.

[193] "Statement by the President on Announcing the First Atomic Explosion in the U.S.S.R." *The American Presidency Project*, https://www.presidency.ucsb.edu/documents/statement-the-president-

announcing-the-first-atomic-explosion-the-ussr. Accessed 5 Mar. 2023.
[194] *The Soviet Atomic Project: How the Soviet Union Obtained the Atomic Bomb.* 2018, p. 579.
[195] MIKE, PutinHuiloPutinHuiloAntikomunistsPaul SherrardALBERT. *The First Test of a Soviet Atomic Bomb.*
https://www.weaponews.com/history/65354196-the-first-test-of-a-soviet-atomic-bomb.html. Accessed 5 Mar. 2023.
[196] "Hydrogen Bomb - 1950." *Nuclear Museum,*
https://ahf.nuclearmuseum.org/ahf/history/hydrogen-bomb-1950/. Accessed 11 Apr. 2023.
[197] "General Advisory Committee's Majority and Minority Reports on Building the H-Bomb." *Historical Documents,*
https://www.atomicarchive.com/resources/documents/hydrogen/gac-report.html#Report. Accessed 11 Apr. 2023.
[198] *General Advisory Committee's Majority and Minority Reports on Building the H-Bomb,*
https://www.atomicarchive.com/resources/documents/hydrogen/gac-report.html#Report. Atomic Archive. Accessed 11 Apr. 2023.
[199] "Political Authority or Atomic Celebrity? The Influence of J. Robert Oppenheimer on American Nuclear Policy after the Second World War." *Wilson Center,* https://acrosskarman.wilsoncenter.org/publication/political-authority-or-atomic-celebrity-the-influence-j-robert-oppenheimer-american. Accessed 14 Apr. 2023.
[200] "Statement by the President on the Hydrogen Bomb." *The American Presidency Project,* https://www.presidency.ucsb.edu/documents/statement-the-president-the-hydrogen-bomb. Accessed 14 Apr. 2023.
[201] Hydrogen Bomb - 1950." *Nuclear Museum,*
https://ahf.nuclearmuseum.org/ahf/history/hydrogen-bomb-1950/. Accessed 7 June 2023.
[202] "President Truman Receives NSC-68 Report, Calling for 'Containing' Soviet Expansion." *HISTORY,* 13 Nov. 2009, https://www.history.com/this-day-in-history/president-truman-receives-nsc-68. Accessed 7 June 2023.
[203] "President Truman receives NSC-68 report, calling for "containing" Soviet expansion." *HISTORY,* 13 Nov. 2009, https://www.history.com/this-day-in-history/president-truman-receives-nsc-68. Accessed 7 June 2023.
[204] Race for the Superbomb. "Timeline." *American Experience,* 21 Mar. 2019, https://www.pbs.org/wgbh/americanexperience/features/bomb-timeline/. Accessed 7 June 2023.
[205] "U.S.C. Title 42." *THE PUBLIC HEALTH AND WELFARE,*
https://www.govinfo.gov/content/pkg/USCODE-2010-title42/html/USCODE-2010-title42-chap23-divsnA.htm. Accessed 26 Feb. 2023.
[206] Weinberger. "Born Secret — the Heavy Burden of Bomb Physics." *Nature,* 19 Apr. 2021, https://www.nature.com/articles/d41586-021-01024-9. Accessed 26 Feb. 2023.

[207] McCoy, Alfred W. "How an Article about the H-Bomb Landed Scientific American in the Middle of the Red Scare." *Scientific American*, 1 Sept. 2020, https://www.scientificamerican.com/article/how-an-article-about-the-h-bomb-landed-scientific-american-in-the-middle-of-the-red-scare/. Accessed 26 Feb. 2023.

[208] *4.4 Elements of Thermonuclear Weapon Design.* https://nuclearweaponarchive.org/Nwfaq/Nfaq4-4.html. Accessed 7 June 2023.

[209] Mean Free Path. "The Editors of Encyclopaedia Britannica." *Encyclopedia Britannica*, 20 July 1998, https://www.britannica.com/science/mean-free-path. Accessed 7 June 2023.

[210] *Elements of Thermonuclear Weapon Design.* https://nuclearweaponarchive.org/Nwfaq/Nfaq4-4.html. Accessed 7 June 2023.

[211] *Id.*

[212] *Manhattan Project: Science > Bomb Design and Components > Hydrogen Bomb.* https://www.osti.gov/opennet/manhattan-project-history/Science/BombDesign/hydrogen-bomb.html. Accessed 7 June 2023.

[213] *Elements of Thermonuclear Weapon Design.* https://nuclearweaponarchive.org/Nwfaq/Nfaq4-4.html. Accessed 7 June 2023.

[214] Freudenrich, Craig, and Patrick J. Kiger. "How Nuclear Bombs Work." *HowStuffWorks*, 5 Oct. 2000, https://science.howstuffworks.com/nuclear-bomb.htm. Accessed 7 June 2023.

[215] *Elements of Thermonuclear Weapon Design.* https://nuclearweaponarchive.org/Nwfaq/Nfaq4-4.html. Accessed 7 June 2023.

[216] Siegel, Ethan. "Ask Ethan: How Can A Nuclear Bomb Be Hotter Than The Center Of Our Sun?" *Forbes*, 28 Mar. 2020, https://www.forbes.com/sites/startswithabang/2020/03/28/ask-ethan-how-can-a-nuclear-bomb-be-hotter-than-the-center-of-our-sun/?sh=42ba8a54460b. Accessed 7 June 2023.

[217] Ask Ethan: How Can A Nuclear Bomb Be Hotter Than The Center Of Our Sun?" *Forbes*, 28 Mar. 2020, https://www.forbes.com/sites/startswithabang/2020/03/28/ask-ethan-how-can-a-nuclear-bomb-be-hotter-than-the-center-of-our-sun/?sh=42ba8a54460b. Accessed 7 June 2023.

[218] *The Hydrogen Bomb: The Secret.* https://www.atomicarchive.com/science/fusion/hydrogen-bomb.html. Accessed 7 June 2023.

[219] Race for the Superbomb-. "Timeline." *American Experience*, 21 Mar. 2019, https://www.pbs.org/wgbh/americanexperience/features/bomb-timeline/. Accessed 7 June 2023.

[220] Defense Threat Reduction Agency. *Operation GREENHOUSE.* Sept. 2021, https://www.dtra.mil/Portals/125/Documents/NTPR/newDocs/5-

GREENHOUSE%20-%202021.pdf. Accessed 7 June 2023.
[221] *Id.*
[222] *Id.*
[223] The "George" Test. "U.S. Tests." *American Experience*, 22 Mar. 2019, https://www.pbs.org/wgbh/americanexperience/features/bomb-us-tests/. Accessed 7 June 2023.
[224] *Id.*
[225] "Hiroshima, Nagasaki and Subsequent Weapons Testing." *World Nuclear Association*, https://www.world-nuclear.org/information-library/safety-and-security/non-proliferation/hiroshima,-nagasaki,-and-subsequent-weapons-testin.aspx. Accessed 7 June 2023.
[226] "Hydrogen Bomb - 1950." *Nuclear Museum*, https://ahf.nuclearmuseum.org/ahf/history/hydrogen-bomb-1950/. Accessed 7 June 2023.
[227] The National Archives. "Public Information Films." *Operation Hurricane*, https://www.nationalarchives.gov.uk/films/1951to1964/filmpage_oper_hurr.htm. Accessed 8 June 2023.
[228] Royal Australian Navy. "Semaphore: Operations HURRICANE and MOSAIC." *Royal Australian Navy*, https://www.navy.gov.au/media-room/publications/semaphore-02-16. Accessed 8 June 2023.
[229] Blitz, Matt. "When Kodak Accidentally Discovered A-Bomb Testing." *Popular Mechanics*, 20 June 2016, https://www.popularmechanics.com/science/energy/a21382/how-kodak-accidentally-discovered-radioactive-fallout/. Accessed 23 July 2023.
[230] "Health Imaging." *Kodak*, 29 May 2020, https://www.kodak.com/en/company/page/health-imaging-history. Accessed 23 July 2023.
[231] Blitz, Matt. "When Kodak Accidentally Discovered A-Bomb Testing." *Popular Mechanics*, 20 June 2016, https://www.popularmechanics.com/science/energy/a21382/how-kodak-accidentally-discovered-radioactive-fallout/. Accessed 23 July 2023.
[232] "OPERATION IVY." *1952*, https://www.radiochemistry.org/history/nuke_tests/ivy/index.html. Accessed 8 June 2023.
[233] Fabry, Merrill. "What the First H-Bomb Test Looked Like." *Time*, 2 Nov. 2015, https://time.com/4096424/ivy-mike-history/. Accessed 8 June 2023.
[234] "Operation Ivy." *1952*, https://www.radiochemistry.org/history/nuke_tests/ivy/index.html. Accessed 8 June 2023.
[235] Johnson McMillan, Priscilla. *Science and Secrecy*. New York Times, 2 Oct. 1994, https://archive.nytimes.com/www.nytimes.com/books/98/12/06/specials/holloway-stalin.html. Accessed 17 July 2023.
[236] "Rosenbergs Convicted of Espionage." *HISTORY*, 13 Nov. 2009, https://www.history.com/this-day-in-history/rosenbergs-convicted-of-

espionage. Accessed 8 June 2023.

[237] "The Rosenberg Trial." *Nuclear Museum*, https://ahf.nuclearmuseum.org/ahf/history/rosenberg-trial/. Accessed 8 June 2023.

[238] Jenkins, John Philip. "Julius and Ethel Rosenberg." *Encyclopedia Britannica*, 27 Oct. 1999, https://www.britannica.com/biography/Julius-Rosenberg-and-Ethel-Rosenberg. Accessed 8 June 2023.

[239] "The Rosenberg Trial." *Nuclear Museum*, https://ahf.nuclearmuseum.org/ahf/history/rosenberg-trial/. Accessed 8 June 2023.

[240] *Id.*

[241] "Statement by the President Declining To Intervene on Behalf of Julius and Ethel Rosenberg." *The American Presidency Project*, https://www.presidency.ucsb.edu/documents/statement-the-president-declining-intervene-behalf-julius-and-ethel-rosenberg. Accessed 8 June 2023.

[242] "June 20, 1953: Rosenbergs Executed - The San Diego Union-Tribune." *San Diego Union-Tribune*, 20 June 2018, https://www.sandiegouniontribune.com/news/150-years/sd-me-150-years-june-20-htmlstory.html. Accessed 8 June 2023.

[243] "McCarthyism and the Red Scare." *Miller Center*, 20 Dec. 2017, https://millercenter.org/the-presidency/educational-resources/age-of-eisenhower/mccarthyism-red-scare. Accessed 8 June 2023.

[244] "Andrei D. Sakharov." *Nuclear Museum*, https://ahf.nuclearmuseum.org/ahf/profile/andrei-d-sakharov/. Accessed 8 June 2023.

[245] Malenkov, Georgy. SOVIET NEWS, 8 Aug. 1953, https://michaelharrison.org.uk/wp-content/uploads/2022/03/Malenkov_Speech8August1953.pdf. Accessed 8 June 2023.

[246] "Soviet Hydrogen Bomb Program." *Nuclear Museum*, https://ahf.nuclearmuseum.org/ahf/history/soviet-hydrogen-bomb-program/. Accessed 8 June 2023.

[247] "Andrei D. Sakharov." Nuclear Museum, https://ahf.nuclearmuseum.org/ahf/profile/andrei-d-sakharov/. Accessed 8 June 2023.

[248] Rowberry, Ariana. "Castle Bravo: The Largest U.S. Nuclear Explosion." *Brookings*, 27 Feb. 2014, https://www.brookings.edu/blog/up-front/2014/02/27/castle-bravo-the-largest-u-s-nuclear-explosion/. Accessed 8 June 2023.

[249] "22 November 1955." *This Day in Aviation*, https://www.thisdayinaviation.com/22-november-1955/. Accessed 8 June 2023.

[250] Sakharov, Andrei. "The Danger of Thermonuclear War: An Open Letter to Dr. Sidney Drell." *Foreign Affairs*, vol. 61, no. 5, 1983, pp. 1001–16, https://doi.org/10.2307/20041632.

[251] Harding, Michelle. "Operation Grapple X." *140 Times the Size of the Hiroshima Bomb*, 8 Nov. 2021, https://www.bntva.com/operation-grapple-x. Accessed 11 June 2023.

[252] "Tsar Bomba." *Nuclear Museum*, https://ahf.nuclearmuseum.org/ahf/history/tsar-bomba/. Accessed 8 June 2023.

[253] *Id.*

[254] *Id.*

[255] "Tsar Bomba: The Largest Atomic Test in World History." The National World War II Museum, 28 Aug. 2020, https://www.nationalww2museum.org/war/articles/tsar-bomba-largest-atomic-test-world-history. Accessed 10 June 2023.

[256] "Tsar Bomba." *Nuclear Museum*, https://ahf.nuclearmuseum.org/ahf/history/tsar-bomba/. Accessed 10 June 2023.

[257] *"Tsar Bomba."* https://www.cs.mcgill.ca/~rwest/wikispeedia/wpcd/wp/t/Tsar_Bomba.htm. Accessed 10 June 2023.

[258] Dowling, Stephen. "The Monster Atomic Bomb That Was Too Big to Use." *BBC*, 16 Aug. 2017, https://www.bbc.com/future/article/20170816-the-monster-atomic-bomb-that-was-too-big-to-use. Accessed 10 June 2023.

[259] Gaulkin, Thomas. "The Untold Story of the World's Biggest Nuclear Bomb." *Bulletin of the Atomic Scientists*, 14 Nov. 2021, https://thebulletin.org/2021/11/the-untold-story-of-the-worlds-biggest-nuclear-bomb/. Accessed 10 June 2023.

[260] "Tsar Bomba: The Largest Atomic Test in World History." *The National World War II Museum*, 28 Aug. 2020, https://www.nationalww2museum.org/war/articles/tsar-bomba-largest-atomic-test-world-history. Accessed 10 June 2023.

[261] "Tsar Bomba: The Largest Atomic Test in World History." *The National World War II Museum*, 28 Aug. 2020, https://www.nationalww2museum.org/war/articles/tsar-bomba-largest-atomic-test-world-history. Accessed 10 June 2023.

[262] Tikkanen, Amy. "Tsar Bomba." *Encyclopedia Britannica*, 10 Aug. 2017, https://www.britannica.com/topic/Tsar-Bomba. Accessed 10 June 2023.

[263] "Tsar Bomba." *Encyclopedia Britannica*, 10 Aug. 2017, https://www.britannica.com/topic/Tsar-Bomba. Accessed 11 June 2023.

[264] Sublette, Carey. "4.5 Thermonuclear Weapon Designs and Later Subsections." *Nuclear Weapon Archive*, https://nuclearweaponarchive.org/Nwfaq/Nfaq4-5.html. Accessed 1 July 2023.

[265] Schumann, Anna. "Fact Sheet: China's Nuclear Inventory." *Center for Arms Control and Non-Proliferation*, 2 Apr. 2020, https://armscontrolcenter.org/fact-sheet-chinas-nuclear-arsenal/. Accessed 8 June 2023.

[266] "Sources on China's Nuclear History." *Wilson Center*, https://www.wilsoncenter.org/blog-post/sources-chinas-nuclear-history. Accessed 11 June 2023.

[267] Schumann, Anna. "Fact Sheet: China's Nuclear Inventory." *Center for Arms Control and Non-Proliferation*, 2 Apr. 2020, https://armscontrolcenter.org/fact-sheet-chinas-nuclear-arsenal/. Accessed 8 June 2023.

[268] "Nuclear Weapon." *Encyclopedia Britannica*, 26 July 1999, https://www.britannica.com/technology/nuclear-weapon/France. Accessed 8 June 2023.

[269] "The Atomic Bomb, Statement of the Government of the People's Republic of China, October 16, 1964." *US-China Institute*, https://china.usc.edu/atomic-bomb-statement-government-peoples-republic-china-october-16-1964. Accessed 8 June 2023.

[270] "China Explodes a Hydrogen Bomb." *Current History*, vol. 53, no. 313, 1967, pp. 169–80, https://doi.org/10.2307/45311758.

[271] "Test No. 6." *Media Gallery*, https://www.atomicarchive.com/media/photographs/testing/chinese/test-6-3.html. Accessed 11 June 2023.

[272] Keck, Zachary. "The National Interest." *The National Interest*, 20 Mar. 2021, https://nationalinterest.org/blog/reboot/five-biggest-nuclear-bomb-tests-ever-180708. Accessed 12 June 2023.

[273] "Timeline." *Nuclear Museum*, https://ahf.nuclearmuseum.org/ahf/nuc-history/timeline/. Accessed 12 June 2023.

[274] "Operation Smiling Buddha: The Story of India's First Nuclear Test at Pokhran in 1974." *The Indian Express*, 18 May 2023, https://indianexpress.com/article/explained/explained-history/operation-smiling-buddha-nuclear-first-test-pokhran-history-8616714/. Accessed 12 June 2023.

[275] Mark, Joshua J. "Siddhartha Gautama." *World History Encyclopedia*, https://www.worldhistory.org/Siddhartha_Gautama/. Accessed 12 June 2023.

[276] "Pakistani Nuclear Program." *Nuclear Museum*, https://ahf.nuclearmuseum.org/ahf/history/pakistani-nuclear-program/. Accessed 12 June 2023.

[277] *Id.*

[278] *Id.*

[279] "Revisiting History: North Korea and Nuclear Weapons." *Wilson Center*, https://www.wilsoncenter.org/event/revisiting-history-north-korea-and-nuclear-weapons. Accessed 12 June 2023.

[280] Albert, Eleanor. "North Korean Nuclear Negotiations." *Council on Foreign Relations*, 4 Apr. 2018, https://www.cfr.org/timeline/north-korean-nuclear-negotiations. Accessed 12 June 2023.

[281] "North Korean Nuclear Negotiations." *Council on Foreign Relations*, 4 Apr. 2018, https://www.cfr.org/timeline/north-korean-nuclear-negotiations. Accessed 12 June 2023.

[282] "The US Discovery of Israel's Secret Nuclear Project." *Wilson Center*, https://www.wilsoncenter.org/publication/the-us-discovery-israels-secret-nuclear-project. Accessed 21 June 2023.

[283] *Id.*

[284] "Israeli Nuclear Program." *Nuclear Museum*, https://ahf.nuclearmuseum.org/ahf/history/israeli-nuclear-program/. Accessed 21 June 2023.

[285] "Revisiting the 1979 VELA Mystery: A Report on a Critical Oral History Conference." *Wilson Center*, https://www.wilsoncenter.org/blog-post/revisiting-1979-vela-mystery-report-critical-oral-history-conference. Accessed 21 June 2023.

[286] *Id.*

[287] *Israel's Nuclear Weapons*. https://nuke.fas.org/guide/israel/nuke/farr.htm. Accessed 21 June 2023.

[288] "WarGames (1983)." *IMDb*, Video, https://www.imdb.com/title/tt0086567/characters/nm0939795. Accessed 21 June 2023.

[289] Baker, W., and W. Clarke. *The Letters of Wilkie Collins: Volume 2*. Springer, 1999, p. 344.

[290] Street, Farnam. "Mutually Assured Destruction: When Not to Play." *Farnam Street*, 19 June 2017, https://fs.blog/mutually-assured-destruction/. Accessed 21 June 2023.

[291] "Alfred Nobel's Thoughts about War and Peace." *NobelPrize.Org*, 11 June 2013, https://www.nobelprize.org/alfred-nobel/alfred-nobels-thoughts-about-war-and-peace/. Accessed 21 June 2023.

[292] "Nuclear Strategy." *Encyclopedia Britannica*, 19 Nov. 2007, https://www.britannica.com/topic/nuclear-strategy/Massive-retaliation. Accessed 21 June 2023.

[293] Dulles. "Evolution of Foreign Policy : Text of Speech by John Foster Dulles Secretary of State before the Council on Foreign Relations, New York, N.Y., January 12, ..." *HathiTrust*, https://babel.hathitrust.org/cgi/pt?id=umn.31951d024881358&view=1up&seq=3.

[294] Bauer, Pat. "New Look." *Encyclopedia Britannica*, 9 Mar. 2018, https://www.britannica.com/topic/New-Look-United-States-history. Accessed 21 June 2023.

[295] "Mutual Assured Destruction." Encyclopedia Britannica, 17 July 2020, https://www.britannica.com/topic/mutual-assured-destruction. Accessed 21 June 2023.

[296] "A Brief History of RAND." *RAND*, https://www.rand.org/about/history.html. Accessed 21 June 2023.

[297] Emery, John R. "Moral Choices Without Moral Language: 1950s Political-Military Wargaming at the RAND Corporation." *Texas National Security Review*, 7 Sept. 2021, https://tnsr.org/2021/09/moral-choices-without-moral-language-1950s-political-military-wargaming-at-the-rand-

corporation/. Accessed 21 June 2023.

[298] "Moral Choices Without Moral Language: 1950s Political-Military Wargaming at the RAND Corporation." *Texas National Security Review*, 7 Sept. 2021, https://tnsr.org/2021/09/moral-choices-without-moral-language-1950s-political-military-wargaming-at-the-rand-corporation/. Accessed 22 June 2023.

[299] *Id.*

[300] *Id.*

[301] "How Much Was Enough? Official Estimates of Nuclear Weapons Requirements, 1957-1995." *Brookings*, 6 Dec. 2016, https://www.brookings.edu/how-much-was-enough-official-estimates-of-nuclear-weapons-requirements-1957-1995/. Accessed 22 June 2023.

[302] Castillo, Jasen. "The Cold Comfort of Mutually Assured Destruction." *War on the Rocks*, 16 June 2021, https://warontherocks.com/2021/06/revolutionary-thinking-questioning-the-conventional-wisdom-on-nuclear-deterrence/. Accessed 22 June 2023.

[303] *Key Issues: Nuclear Weapons: History: Cold War: Strategy: Flexible Response.* http://www.nuclearfiles.org/menu/key-issues/nuclear-weapons/history/cold-war/strategy/strategy-flexible-response.htm. Accessed 22 June 2023.

[304] *Nuclear Files: Key Issues: Nuclear Weapons: History: Cold War: Strategy: Counterforce and Countervalue.* http://www.nuclearfiles.org/menu/key-issues/nuclear-weapons/history/cold-war/strategy/strategy-countervalue-force.htm. Accessed 23 July 2023.

[305] "Secure Second Strike." *Encyclopedia Britannica*, 17 July 2017, https://www.britannica.com/topic/second-strike-capability. Accessed 22 June 2023.

[306] "United Nations." *HISTORY*, 4 Apr. 2018, https://www.history.com/topics/stories/united-nations. Accessed 22 June 2023.

[307] "Atlantic Charter - Definition, Purpose & Significance." *HISTORY*, 9 Nov. 2009, https://www.history.com/topics/world-war-ii/atlantic-charter. Accessed 22 June 2023.

[308] "United Nations." *HISTORY*, 4 Apr. 2018, https://www.history.com/topics/stories/united-nations. Accessed 22 June 2023.

[309] United Nations. "United Nations Charter (Full Text)." *United Nations*, https://www.un.org/en/about-us/un-charter/full-text. Accessed 22 June 2023.

[310] United Nations Security Council." *Encyclopedia Britannica*, 20 July 1998, https://www.britannica.com/topic/United-Nations-Security-Council. Accessed 22 June 2023.

[311] "Atoms for Peace." *Eisenhower Presidential Library*, https://www.eisenhowerlibrary.gov/research/online-documents/atoms-peace. Accessed 22 June 2023.

[312] "Remarks at Ceremony Following Ratification of the Statute of the

International Atomic Energy Agency." *The American Presidency Project*, https://www.presidency.ucsb.edu/documents/remarks-ceremony-following-ratification-the-statute-the-international-atomic-energy-agency. Accessed 22 June 2023.

[313] *IAEA*, https://www.iaea.org/about/overview/history. Accessed 22 June 2023.

[314] Haglund, David G. "North Atlantic Treaty Organization." *Encyclopedia Britannica*, 20 July 1998, https://www.britannica.com/topic/North-Atlantic-Treaty-Organization. Accessed 22 June 2023.

[315] NATO. "Collective Defence and Article 5." *NATO*, https://www.nato.int/cps/en/natohq/topics_110496.htm. Accessed 22 June 2023.

[316] Hill, Jonathan. "NATO Review - NATO – Ready for Anything?" *Nato Review*, 24 Jan. 2019, https://www.nato.int/docu/review/articles/2019/01/24/nato-ready-for-anything/index.html. Accessed 22 June 2023.

[317] "Status of World Nuclear Forces." *Federation of American Scientists*, 31 Mar. 2023, https://fas.org/initiative/status-world-nuclear-forces/. Accessed 22 June 2023.

[318] "How Much Was Enough? Official Estimates of Nuclear Weapons Requirements, 1957-1995." *Brookings*, 6 Dec. 2016, https://www.brookings.edu/how-much-was-enough-official-estimates-of-nuclear-weapons-requirements-1957-1995/. Accessed 22 June 2023.

[319] Ruggenthaler, Peter. "The Ongoing Debate." *Journal of Cold War Studies*, vol. 13, no. 4, 2011, pp. 172–212, https://doi.org/10.2307/26924047.

[320] "Molotov's Proposal That the USSR Join NATO, March 1954." *Wilson Center*, https://www.wilsoncenter.org/publication/molotovs-proposal-the-ussr-join-nato-march-1954. Accessed 22 June 2023.

[321] "Milestones: 1953–1960." *Office of the Historian*, https://history.state.gov/milestones/1953-1960/warsaw-treaty. Accessed 22 June 2023.

[322] "Warsaw Pact." *Encyclopedia Britannica*, 20 July 1998, https://www.britannica.com/event/Warsaw-Pact. Accessed 23 June 2023.

[323] Bitzinger, Richard. *Assessing the Conventional Balance in Europe, 1945-1975*. pp. 7–8. Accessed 23 June 2023.

[324] "NATO and Warsaw Pact: Force Comparisons." *NATO*, https://www.nato.int/cps/fr/natohq/declassified_138256.htm. Accessed 23 June 2023.

[325] "News Article." *ETH Zurich*, https://css.ethz.ch/content/specialinterest/gess/cis/center-for-securities-studies/en/services/digital-library/articles/article.html/107840. Accessed 23 June 2023.

[326] Suciu, Peter. "The National Interest." *The National Interest*, 18 Mar. 2021, https://nationalinterest.org/blog/reboot/7-days-how-soviet-union-planned-crush-nato-land-war-180567. Accessed 23 June 2023.

[327] "A Brief History of the Department of Energy." *Energy.Gov*, https://www.energy.gov/lm/brief-history-department-energy. Accessed 23 June 2023.

[328] *Id.*

[329] *Rosatom State Atomic Energy Corporation ROSATOM Global Leader in Nuclear Technologies Nuclear Energy.* https://rosatom.ru/en/press-centre/short-history-of-the-russian-nuclear-industry/. Accessed 23 June 2023.

[330] "Ministry for Atomic Energy (Minatom)." *Russian and Soviet Nuclear Forces*, https://nuke.fas.org/guide/russia/agency/minatom.htm. Accessed 23 June 2023.

[331] Nikitin, Alexander. "Rosatom State Corporation." *Bellona.Org*, 26 Nov. 2007, https://bellona.org/news/nuclear-issues/nuclear-russia/2007-11-rosatom-state-corporation. Accessed 23 June 2023.

[332] *Vice chairman stresses realities of strategic nuclear deterrence.* (2019, April 30). Joint Chiefs of Staff. https://www.jcs.mil/Media/News/News-Display/Article/1831242/vice-chairman-stresses-realities-of-strategic-nuclear-deterrence/

[333] "What Are Tactical Nuclear Weapons?" *Union of Concerned Scientists*, https://www.ucsusa.org/resources/tactical-nuclear-weapons. Accessed 24 June 2023.

[334] *Id.*

[335] *"Tactical Nuclear Weapons."* "The Editors of Encyclopaedia Britannica." *Encyclopedia Britannica*, 3 Jan. 2014, https://www.britannica.com/technology/tactical-nuclear-weapon. Accessed 24 June 2023.

[336] "Strategic Weapons System." *Encyclopedia Britannica*, 20 July 1998, https://www.britannica.com/technology/strategic-weapons-system. Accessed 24 June 2023.

[337] "America's Nuclear Triad." *U.S. Department of Defense*, https://www.defense.gov/Multimedia/Experience/Americas-Nuclear-Triad/. Accessed 24 June 2023.

[338] "Nuclear Triad." *Encyclopedia Britannica*, 3 Dec. 2013, https://www.britannica.com/topic/nuclear-triad. Accessed 24 June 2023.

[339] "History Highlights: The Nuclear Triad." *Defense Logistics Agency*, 1 Oct. 2015, https://www.dla.mil/About-DLA/News/News-Article-View/Article/623030/history-highlights-the-nuclear-triad/. Accessed 24 June 2023.

[340] *Id.*

[341] *The Creation of SIOP-62: More Evidence on the Origins of Overkill.* https://nsarchive2.gwu.edu/NSAEBB/NSAEBB130/press.htm. Accessed 23 July 2023.

[342] Department of Defense. *Single Integrated Operational Plan (Slop) Targeting Philosophy [Memorandum], April 25, 1977.* https://www.archives.gov/files/declassification/iscap/pdf/2010-082-doc1.pdf. Accessed 23 July 2023.

[343] English, Dave. "Dave English." *Great Aviation Quotes*, https://www.aviationquotations.com/airpowerquotes.html. Accessed 25 June 2023.

[344] *History of the Airplane*. https://www.wright-brothers.org/History_Wing/History_of_the_Airplane/History_of_the_Airplane_Intro/History_of_the_Airplane_Intro.htm. Accessed 25 June 2023.

[345] "First Airplane Flies." *HISTORY*, 24 Nov. 2009, https://www.history.com/this-day-in-history/first-airplane-flies. Accessed 25 June 2023.

[346] "Bomber." *Encyclopedia Britannica*, 27 May 1999, https://www.britannica.com/technology/bomber-aircraft. Accessed 25 June 2023.

[347] "Zeppelin Raids." *The National Archives*, 9 Mar. 2014, https://www.nationalarchives.gov.uk/education/resources/zeppelin-raids/. Accessed 25 June 2023.

[348] "Terror from the Air: The German Zeppelin Raids of 1915 - 1918." *Sky HISTORY TV Channel*, https://www.history.co.uk/article/terror-from-the-air-the-german-zeppelin-raids-of-1915-1918. Accessed 25 June 2023.

[349] "Zeppelin Raids." *The National Archives*, 9 Mar. 2014, https://www.nationalarchives.gov.uk/education/resources/zeppelin-raids/. Accessed 25 June 2023.

[350] *Zeppelin-Staaken R-Series*. https://www.militaryfactory.com/aircraft/detail.php?aircraft_id=834. Accessed 25 June 2023.

[351] "Bomber." *Encyclopedia Britannica*, 27 May 1999, https://www.britannica.com/technology/bomber-aircraft. Accessed 25 June 2023.

[352] *"Sikorsky Ilya Mourometz V Specifications."* http://www.theaerodrome.com/aircraft/russia/sim_v.php. Accessed 25 June 2023.

[353] *Bomber Timeline*. http://www.theaerodrome.com/aircraft/bomber_timeline.php. Accessed 25 June 2023.

[354] "Bomber." *Encyclopedia Britannica*, 27 May 1999, https://www.britannica.com/technology/bomber-aircraft. Accessed 25 June 2023.

[355] "Monoplane." The Editors of Encyclopaedia Britannica. *Encyclopedia Britannica*, 20 July 1998, https://www.britannica.com/technology/monoplane. Accessed 25 June 2023.

[356] "Bomber." *Encyclopedia Britannica*, 27 May 1999, https://www.britannica.com/technology/bomber-aircraft. Accessed 25 June 2023.

[357] *"Stuka."* The Editors of Encyclopaedia Britannica. *Encyclopedia Britannica*, 20 July 1998, https://www.britannica.com/technology/Stuka. Accessed 25 June 2023.

[358] "Bomber." *Encyclopedia Britannica*, 27 May 1999, https://www.britannica.com/technology/bomber-aircraft. Accessed 25 June 2023.

[359] *Boeing: Historical Snapshot: B-29 Superfortress.* https://www.boeing.com/history/products/b-29-superfortress.page. Accessed 25 June 2023.

[360] Mall, Scott. "B-29 Superfortress Pulled the Trigger on World War II in the Pacific." *FLYING Magazine*, 2 Sept. 2022, https://www.flyingmag.com/b-29-superfortress-pulled-the-trigger-on-world-war-ii-in-the-pacific/. Accessed 23 July 2023.

[361] *"Historical Snapshot: B-29 Superfortress."* https://www.boeing.com/history/products/b-29-superfortress.page. Accessed 25 June 2023.

[362] *Tupolev Tu-4 (Bull).* https://www.militaryfactory.com/aircraft/detail.php?aircraft_id=701. Accessed 25 June 2023.

[363] *Aircraft Carriers.* https://www.history.navy.mil/browse-by-topic/ships/aircraft-carriers.html. Accessed 25 June 2023.

[364] "How Japan Developed Carrier Aviation." *U.S. Naval Institute*, 1 Apr. 2018, https://www.usni.org/magazines/naval-history-magazine/2018/april/how-japan-developed-carrier-aviation. Accessed 25 June 2023.

[365] *Japanese Forces in the Pearl Harbor Attack.* https://www.history.navy.mil/our-collections/photography/wars-and-events/world-war-ii/pearl-harbor-raid/japanese-forces-in-the-pearl-harbor-attack.html. Accessed 25 June 2023.

[366] Hicks, Gary. "Awakening the Sleeping Giant: The Birth of the Greatest Generation - VA News." *US Department of Veterans Affairs*, 6 Dec. 2013, https://news.va.gov/11713/awakening-the-sleeping-giant-the-birth-of-the-greatest-generation/. Accessed 25 June 2023.

[367] "Convair B-36J Peacemaker." *National Museum of the United States Air Force™*, https://www.nationalmuseum.af.mil/Visit/Museum-Exhibits/Fact-Sheets/Display/Article/197636/convair-b-36j-peacemaker/. Accessed 25 June 2023.

[368] "Washington, D.C. to Moscow Flight Time, Distance, Route Map." *AirportDistanceCalculator.Com*, https://www.airportdistancecalculator.com/Washington--d-c--united-states-to-moscow-russia-flight-time.html. Accessed 25 June 2023.

[369] *Tupolev Tu-85.* https://www.militaryfactory.com/aircraft/detail.php?aircraft_id=1148. Accessed 25 June 2023.

[370] *Tupolev Tu-95 (Bear).* https://www.militaryfactory.com/aircraft/detail.php?aircraft_id=404. Accessed 25 June 2023.

[371] "ODIN." *OE Data Integration Network*,

https://odin.tradoc.army.mil/WEG/Asset/Tu-95_(Bear)_Russian_Strategic_Bomber_Aircraft. Accessed 25 June 2023.

[372] adamgreatflt. "Turboprop vs Jet: Differences, Safety, Pros & Cons." *Great Flight*, 30 Oct. 2019, https://greatflight.com/blog/turboprop-vs-jet/. Accessed 25 June 2023.

[373] Joiner, Stephen. "The Jet That Shocked the West." *Smithsonian Magazine*, 1 Jan. 2014, https://www.smithsonianmag.com/air-space-magazine/the-jet-that-shocked-the-west-180947758/. Accessed 25 June 2023.

[374] B-29." *Encyclopedia Britannica*, 20 July 1998, https://www.britannica.com/technology/B-29. Accessed 25 June 2023.

[375] "Mikoyan – Gurevich MiG-15 Fagot." *Flying Leatherneck Historical Foundation - Aviation Museum | Preserving Marine Aviation and History*, 27 Mar. 2017, https://flyingleathernecks.org/project/mikoyan-gurevich-mig-15-fagot/. Accessed 25 June 2023.

[376] *Arado Ar 234 Blitz*. http://www.aviation-history.com/arado/234.html. Accessed 25 June 2023.

[377] "B-52." *Encyclopedia Britannica*, 20 July 1998, https://www.britannica.com/technology/B-52. Accessed 25 June 2023.

[378] *Boeing: B-52*. https://www.boeing.com/defense/b-52-bomber/. Accessed 25 June 2023.

[379] "B-52H Stratofortress." *Air Force*, https://www.af.mil/About-Us/Fact-Sheets/Display/Article/104465/b-52h-stratofortress/. Accessed 25 June 2023.

[380] *"Myasishchev M-4."* The Editors of Encyclopaedia Britannica. *Encyclopedia Britannica*, 26 Feb. 2009, https://www.britannica.com/technology/Myasishchev-M-4. Accessed 25 June 2023.

[381] Brimelow, Benjamin. "How Navy Aircraft Carriers Have Projected US Military Might All over the World for 86 Years." *Insider*, 5 Oct. 2020, https://www.businessinsider.com/history-of-us-navy-aircraft-carriers-supercarriers-to-ford-class-2020-10. Accessed 25 June 2023.

[382] *CVN-65 USS Enterprise Full History 1965-2012 US Navy*. https://www.seaforces.org/usnships/cvn/CVN-65-USS-Enterprise-history.htm. Accessed 25 June 2023.

[383] "CARRIER. The Ship. Nimitz History." *PBS*, https://www.pbs.org/weta/carrier/the_ship_history.htm. Accessed 25 June 2023.

[384] "Nimitz Class Aircraft Carrier." *Naval Technology*, 6 Jan. 2020, https://www.naval-technology.com/projects/nimitz/. Accessed 25 June 2023.

[385] "Navy to Decommission 2 Carriers in a Row, 2 LCS Set for Foreign Sales, Says Long Range Shipbuilding Plan." *USNI News*, 18 Apr. 2023, https://news.usni.org/2023/04/18/navy-to-decommission-2-carriers-in-a-row-2-lcs-set-for-foreign-sales-says-long-range-shipbuilding-plan. Accessed 25 June 2023.

[386] *Nimitz Class Aircraft Carrier CVN US Navy*. https://www.seaforces.org/usnships/cvn/Nimitz-class.htm. Accessed 25 June

2023.

[387] "Declassified: US Nuclear Weapons At Sea." *Federation of American Scientists*, 3 Feb. 2016, https://fas.org/publication/nuclear-weapons-at-sea/. Accessed 25 June 2023.

[388] "Convair B-58A Hustler." *National Museum of the United States Air Force™*, https://www.nationalmuseum.af.mil/Visit/Museum-Exhibits/Fact-Sheets/Display/Article/196439/convair-b-58a-hustler/. Accessed 25 June 2023.

[389] *CONVAIR B-58 Hustler*. https://www.militaryfactory.com/aircraft/detail.php?aircraft_id=115. Accessed 25 June 2023.

[390] "B-1B Lancer." *Air Force*, https://www.af.mil/About-Us/Fact-Sheets/Display/Article/104500/b-1b-lancer/. Accessed 25 June 2023.

[391] "Tupolev Tu-22M Medium-Range Bomber." *MilitaryToday.Com*, https://www.militarytoday.com/aircraft/tupolev_tu-22m_backfire.htm. Accessed 25 June 2023.

[392] "Tu-22M BACKFIRE (TUPOLEV)." *Russian and Soviet Nuclear Forces*, https://nuke.fas.org/guide/russia/bomber/tu-22m.htm. Accessed 25 June 2023.

[393] *Tupolev Tu-160 (Blackjack)*. https://www.militaryfactory.com/aircraft/detail.php?aircraft_id=289. Accessed 25 June 2023.

[394] "B-2 Spirit." *Air Force*, https://www.af.mil/About-Us/Fact-Sheets/Display/Article/104482/b-2-spirit/. Accessed 25 June 2023.

[395] "10 Cool Facts about the B-2." *Northrop Grumman*, 1 Dec. 2020, https://www.northropgrumman.com/what-we-do/air/b-2-stealth-bomber/10-cool-facts-about-the-b-2/. Accessed 25 June 2023.

[396] *Northrop Grumman B-2 Spirit*. https://www.militaryfactory.com/aircraft/detail.php?aircraft_id=6. Accessed 26 June 2023.

[397] Kindy, David. "Truth Is Stranger Than Fiction With Horten's All-Wing Aircraft Design." *Smithsonian Magazine*, 21 Oct. 2020, https://www.smithsonianmag.com/smithsonian-institution/truth-stranger-fiction-hortens-all-wing-aircraft-design-180976095/. Accessed 26 June 2023.

[398] Pike, John. *Mark 4*. https://www.globalsecurity.org/wmd/systems/mk4.htm. Accessed 26 June 2023.

[399] "Mark 5 Nuclear Weapon." *Media Gallery*, https://www.atomicarchive.com/media/photographs/nuclear-journeys/airforce/nmusaf-21.html. Accessed 26 June 2023.

[400] "Mark 17 Thermonuclear Bomb." *National Museum of the United States Air Force™*, https://www.nationalmuseum.af.mil/Visit/Museum-Exhibits/Fact-Sheets/Display/Article/197628/mark-17-thermonuclear-bomb/. Accessed 26 June 2023.

[401] "Mark 17 Bomb Accident Site." *The Center for Land Use Interpretation*,

https://clui.org/ludb/site/mark-17-bomb-accident-site. Accessed 26 June 2023.

[402] "Mark 17 Thermonuclear Bomb." *National Museum of the United States Air Force™*, https://www.nationalmuseum.af.mil/Visit/Museum-Exhibits/Fact-Sheets/Display/Article/197628/mark-17-thermonuclear-bomb/. Accessed 26 June 2023.

[403] *RDS-3*. https://www.globalsecurity.org/wmd/world/russia/rds-3.htm. Accessed 26 June 2023.

[404] Pike, John. *Kh-20 / AS-3 KANGAROO*. https://www.globalsecurity.org/wmd/world/russia/as-3.htm. Accessed 26 June 2023.

[405] "Kh-20 / AS-3 KANGAROO." *Russian and Soviet Nuclear Forces*, https://nuke.fas.org/guide/russia/bomber/as-3.htm. Accessed 23 July 2023.

[406] *Id.*

[407] *The B-41 Bomb*. https://nuclearweaponarchive.org/Usa/Weapons/B41.html. Accessed 26 June 2023.

[408] *The B83 Bomb*. https://nuclearweaponarchive.org/Usa/Weapons/B83.html. Accessed 26 June 2023.

[409] Dahlgren, Masao Masao Dahlgren. "AGM-86 Air-Launched Cruise Missile (ALCM)." *Missile Threat*, 29 Nov. 2016, https://missilethreat.csis.org/missile/alcm/. Accessed 26 June 2023.

[410] Higginbotham, Adam. "There Are Still Thousands of Tons of Unexploded Bombs in Germany, Left Over From World War II." *Smithsonian Magazine*, 6 Jan. 2016, https://www.smithsonianmag.com/history/seventy-years-world-war-two-thousands-tons-unexploded-bombs-germany-180957680/. Accessed 26 June 2023.

[411] "Pruning the Nuclear Triad? Pros and Cons of Bombers, Missiles, and Submarines." *Center for Arms Control and Non-Proliferation*, 31 Oct. 2011, https://armscontrolcenter.org/pruning-the-nuclear-triad-pros-and-cons-of-bombers-missiles-and-submarines/. Accessed 26 June 2023.

[412] *Id.*

[413] Brodie, Bernard. "The Absolute Weapon: Atomic Power and World Order." *Office of Scientific and Technical Information (OSTI)*, U.S. Department of Energy, 15 Feb. 1946, https://www.osti.gov/opennet/servlets/purl/16380564. Accessed 27 June 2023.

[414] Harvey, Ailsa. "V2 Rocket: Origin, History and Spaceflight Legacy." *Space*, 29 Mar. 2022, https://www.space.com/v2-rocket. Accessed 26 June 2023.

[415] "V-2 Rocket." *Encyclopedia Britannica*, 20 July 1998, https://www.britannica.com/technology/V-2-rocket. Accessed 23 July 2023.

[416] "'Wonder Weapons' and Slave Labor." *National Air and Space Museum*, https://airandspace.si.edu/stories/editorial/wonder-weapons-and-slave-labor. Accessed 23 July 2023

[417] "Missile, Surface-to-Surface, V-2 (A-4)." *National Air and Space Museum*, https://airandspace.si.edu/collection-objects/missile-surface-surface-v-2-4/nasm_A19600342000. Accessed 26 June 2023.

[418] Edwards, Owen. "Wernher von Braun's V-2 Rocket." *Smithsonian Magazine*, 31 July 2011, https://www.smithsonianmag.com/arts-culture/wernher-von-brauns-v-2-rocket-12609128/. Accessed 26 June 2023.

[419] "Poland And Germany Fought Each Other On Horseback... In 1939!" *War History Online*, 10 Dec. 2021, https://www.warhistoryonline.com/war-articles/calvary-battle-1939-poland-and-germany.html?A1c=1. Accessed 26 June 2023.

[420] "Records of the Secretary of Defense (RG 330)." *National Archives*, 15 Aug. 2016, https://www.archives.gov/iwg/declassified-records/rg-330-defense-secretary. Accessed 26 June 2023.

[421] NPR Staff. "The Secret Operation To Bring Nazi Scientists To America." *NPR*, 15 Feb. 2014, https://www.npr.org/2014/02/15/275877755/the-secret-operation-to-bring-nazi-scientists-to-america. Accessed 26 June 2023.

[422] Schumm, Laura. "What Was Operation Paperclip?" *HISTORY*, 2 June 2014, https://www.history.com/news/what-was-operation-paperclip. Accessed 26 June 2023.

[423] Kulik, Rebecca M. "Project Paperclip." *Encyclopedia Britannica*, 4 Nov. 2022, https://www.britannica.com/topic/Project-Paperclip. Accessed 26 June 2023.

[424] "Operation Paperclip at Fort Bliss: 1945-1950." *White Sands Missile Range Museum*, 28 Mar. 2022, https://wsmrmuseum.com/2022/03/28/operation-paperclip-at-fort-bliss-1945-1950/. Accessed 26 June 2023.

[425] "What Was Operation Paperclip?" *HISTORY*, 2 June 2014, https://www.history.com/news/what-was-operation-paperclip. Accessed 26 June 2023.

[426] Harbaugh, Jennifer. "Biography of Wernher Von Braun." *NASA*, 18 Feb. 2016, https://www.nasa.gov/centers/marshall/history/vonbraun/bio.html. Accessed 26 June 2023.

[427] Garza, Alejandro de la. "How Historians Are Reckoning With the Former Nazi Who Launched America's Space Program." *Time*, 18 July 2019, https://time.com/5627637/nasa-nazi-von-braun/. Accessed 26 June 2023.

[428] Project Paperclip." *Encyclopedia Britannica*, 4 Nov. 2022, https://www.britannica.com/topic/Project-Paperclip. Accessed 26 June 2023.

[429] *Operation Paperclip at Forst Bliss: 1945-1950."* White Sands Missile Range Museum*, 28 Mar. 2022, https://wsmrmuseum.com/2022/03/28/operation-paperclip-at-fort-bliss-1945-1950/. Accessed 26 June 2023.

[430] Wall, Mike. "Monkeys in Space: A Brief Spaceflight History." *Space*, 28 Jan. 2013, https://www.space.com/19505-space-monkeys-chimps-history.html. Accessed 26 June 2023.

[431] Tate, Karl. "How Intercontinental Ballistic Missiles Work (Infographic)."

Space, 4 Feb. 2013, https://www.space.com/19601-how-intercontinental-ballistic-missiles-work-infographic.html. Accessed 27 June 2023.

[432] "3 Facts About Suborbital Flights." *EXOS Aerospace Systems & Technologies, Inc.*, 5 Apr. 2016, https://exosaero.com/2016/04/05/about-suborbital-flights/. Accessed 27 June 2023.

[433] *"Operation Paperclip at Forst Bliss: 1945-1950." White Sands Missile Range Museum*, 28 Mar. 2022, https://wsmrmuseum.com/2022/03/28/operation-paperclip-at-fort-bliss-1945-1950/. Accessed 26 June 2023.

[434] "REDSTONE FACT SHEET." *Spaceline*, https://www.spaceline.org/cape-canaveral-rocket-missile-program/redstone-fact-sheet/. Accessed 26 June 2023.

[435] Pike, John. *B39 / Mk.39 / W39.* https://www.globalsecurity.org/wmd/systems/mk39.htm. Accessed 26 June 2023.

[436] "Air Force History of ICBM Development, Safeguarding America." *Air Force Global Strike Command AFSTRAT-AIR*, https://www.afgsc.af.mil/News/Features/Display/Article/455710/air-force-history-of-icbm-development-safeguarding-america/. Accessed 26 June 2023.

[437] *The Atlas Missile (U.S. National Park Service).* https://www.nps.gov/articles/atlas-icbm.htm. Accessed 26 June 2023.

[438] "Missile, Surface-to-Surface, Atlas F." *National Air and Space Museum*, https://airandspace.si.edu/collection-objects/missile-surface-to-surface-atlas-f/nasm_A19750667000. Accessed 26 June 2023.

[439] *Atlas F HGM-16F.* http://www.astronautix.com/a/atlasfhgm-16f.html. Accessed 26 June 2023.

[440] "Missile Bases, Communication Bunkers, & Underground Properties." *Missilebases*, https://www.missilebases.com/atlas-f. Accessed 26 June 2023.

[441] *The Titan Missile (U.S. National Park Service).* https://www.nps.gov/articles/titan-icbm.htm. Accessed 26 June 2023.

[442] "Minuteman III 50th Anniversary Year." *Air Force Nuclear Weapons Center*, https://www.afnwc.af.mil/About-Us/History/Minuteman-III-50th-Anniversary-Year/. Accessed 26 June 2023.

[443] Fields, Dave. *Minuteman Missile Nuclear Warheads.* https://minutemanmissile.com/nuclearwarheads.html. Accessed 26 June 2023.

[444] *MIRVs.* https://nsarchive2.gwu.edu//nsa/NC/mirv/mirv.html. Accessed 26 June 2023.

[445] *"Multiple Independently Targetable Reentry Vehicles (MIRVs)."* https://nsarchive2.gwu.edu//nsa/NC/mirv/mirv.html. Accessed 26 June 2023.

[446] Hartunian, Richard A. "Ballistic Missiles and Reentry Systems: The Critical Years." *Minuteman Missile - A Tribute to the ICBM Program*, https://minutemanmissile.com/documents/ReentryVehicleDesignAndPhysics.pdf. Accessed 26 June 2023.

[447] Leitenberg, Milton. *Studies of Military R&D and Weapons Development: The Origin of MIRV.* Accessed 26 June 2023.

[448] Bin, Li. "China's Thinking on MIRVs." *Carnegie Endowment for International Peace*, https://carnegieendowment.org/2019/10/09/china-s-thinking-on-mirvs-pub-80171. Accessed 26 June 2023.

[449] *"Peacekeeper Missile."* "The Editors of Encyclopaedia Britannica." *Encyclopedia Britannica*, 20 July 1998, https://www.britannica.com/technology/Peacekeeper-missile. Accessed 26 June 2023.

[450] "Decree of the USSR Council of Ministers, 'Questions of Reactive Weaponry.'" *Wilson Center Digital Archive*, https://digitalarchive.wilsoncenter.org/document/decree-ussr-council-ministers-questions-reactive-weaponry. Accessed 26 June 2023.

[451] *R-7 History.* https://www.worldspaceflight.com/addendum/r7/r7_history.php. Accessed 26 June 2023.

[452] "USSR Council of Ministers Decree, 'On the Plan of Scientific-Research Work on Long-Range Missiles in 1953-1955.'" *Wilson Center Digital Archive*, https://digitalarchive.wilsoncenter.org/document/ussr-council-ministers-decree-plan-scientific-research-work-long-range-missiles-1953-1955. Accessed 26 June 2023.

[453] "Memorandum by V. A. Malyshev, et. al., to the Presidium of the CPSU Central Committee on the Creation of a Long-Range Missile with a Nuclear Warhead." *Wilson Center Digital Archive*, https://digitalarchive.wilsoncenter.org/document/memorandum-v-malyshev-et-al-presidium-cpsu-central-committee-creation-long-range-missile. Accessed 26 June 2023.

[454] "Ballistic Missile." *Encyclopedia Britannica*, 10 Mar. 2014, https://www.britannica.com/technology/ballistic-missile. Accessed 26 June 2023.

[455] *R-7 Family of Launchers and ICBMs.* https://www.russianspaceweb.com/r7.html. Accessed 26 June 2023.

[456] *Federation of American Scientists :: R-36M / SS-18 SATAN.* https://programs.fas.org/ssp/nukes/nuclearweapons/russia_nukescurrent/ss18.html. Accessed 26 June 2023.

[457] "RT-2PM Topol (SS-25)." *Missile Threat*, 10 Aug. 2016, https://missilethreat.csis.org/missile/ss-25/. Accessed 26 June 2023.

[458] "RSD-10 Pioneer (SS-20)." *Missile Threat*, 12 Sept. 2017, https://missilethreat.csis.org/missile/ss-20-saber-rsd-10/. Accessed 26 June 2023.

[459] Norris, Robert S., and Hans M. Kristensen. "Nuclear U.S. and Soviet/Russian Intercontinental Ballistic Missiles, 1959-2008." *Bulletin of the Atomic Scientists*, vol. 65, no. 1, Jan. 2009, pp. 62–69, https://doi.org/10.2968/065001008.

[460] "Nuclear U.S. and Soviet/Russian Intercontinental Ballistic Missiles,

1959-2008." *Bulletin of the Atomic Scientists*, vol. 65, no. 1, Jan. 2009, pp. 62–69, https://doi.org/10.2968/065001008.

[461] Pruning the Nuclear Triad? Pros and Cons of Bombers, Missiles, and Submarines." *Center for Arms Control and Non-Proliferation*, 31 Oct. 2011, https://armscontrolcenter.org/pruning-the-nuclear-triad-pros-and-cons-of-bombers-missiles-and-submarines/. Accessed 27 June 2023.

[462] *Id.*

[463] "The Hunt for Red October (1990)." *IMDb*, Video, https://www.imdb.com/title/tt0099810/characters/nm0000125. Accessed 27 June 2023.

[464] *The Hunt for Red October*. Directed by John McTiernan, Paramount Pictures, 1990.

[465] *Submarine Development, A Short History*. https://www.history.navy.mil/content/history/museums/nmusn/education/educational-resources/history-of-submarines/submarine-development--a-short-history.html. Accessed 27 June 2023.

[466] *"Submarine Development: A Short History."* https://www.history.navy.mil/content/history/museums/nmusn/education/educational-resources/history-of-submarines/submarine-development--a-short-history.html. Accessed 27 June 2023.

[467] *Id.*

[468] *Id.*

[469] Klein, Christopher. "How German U-Boats Were Used in WWI—And Perfected in WWII." *HISTORY*, 21 Mar. 2022, https://www.history.com/news/u-boats-world-war-i-germany. Accessed 27 June 2023.

[470] McDermott, Annette. "How the Sinking of Lusitania Changed World War I." *HISTORY*, 17 Apr. 2018, https://www.history.com/news/how-the-sinking-of-lusitania-changed-wwi. Accessed 27 June 2023.

[471] Tenorio, Rich. "How a Jewish Immigrant from the Shtetl Became the Father of America's Nuclear Navy." *The Times of Israel*, 10 June 2022, https://www.timesofisrael.com/how-a-jewish-immigrant-from-the-shtetl-became-the-father-of-americas-nuclear-navy/. Accessed 27 June 2023.

[472] "USS Nautilus Commissioned." *HISTORY*, 9 Feb. 2010, https://www.history.com/this-day-in-history/uss-nautilus-commissioned. Accessed 27 June 2023.

[473] "German U-Boat Crew." *Officers, Petty Officers and Seamen*, http://www.uboataces.com/ref-crew.shtml. Accessed 27 June 2023.

[474] *"USS Nautilus—world's first nuclear submarine—is commissioned."* *HISTORY*, 9 Feb. 2010, https://www.history.com/this-day-in-history/uss-nautilus-commissioned. Accessed 27 June 2023.

[475] *USS Nautilus (SSN-571)*. https://americanhistory.si.edu/subs/history/subsbeforenuc/revolution/nautilus.html. Accessed 27 June 2023.

[476] Military.com. "Military.Com." *Military.Com*, 24 June 2022,

https://www.military.com/history/run-silent-the-birth-of-a-nuclear-navy.html. Accessed 27 June 2023.

[477] "What Killed the Thresher?" *U.S. Naval Institute*, 1 Apr. 2023, https://www.usni.org/magazines/naval-history-magazine/2023/april/what-killed-thresher. Accessed 27 June 2023.

[478] "Nuclear-Powered Ships." *Nuclear Submarines - World Nuclear Association*, https://world-nuclear.org/information-library/non-power-nuclear-applications/transport/nuclear-powered-ships.aspx. Accessed 27 June 2023.

[479] Friedman, Norman. "Nuclear Submarine." *Encyclopedia Britannica*, 23 May 2023, https://www.britannica.com/technology/nuclear-submarine. Accessed 27 June 2023.

[480] *Iowa Class Battleship.* https://www.cs.mcgill.ca/~rwest/wikispeedia/wpcd/wp/i/Iowa_class_battleship.htm. Accessed 27 June 2023.

[481] "Nuclear Submarine." *Encyclopedia Britannica*, 23 May 2023, https://www.britannica.com/technology/nuclear-submarine. Accessed 27 June 2023.

[482] *Id.*

[483] "Missile, Surface-to-Surface, Regulus I, SSM-N-8A." *National Air and Space Museum*, https://airandspace.si.edu/collection-objects/missile-surface-surface-regulus-i-ssm-n-8a/nasm_A19880045000. Accessed 27 June 2023.

[484] *Regulus 1.* http://www.astronautix.com/r/regulus1.html. Accessed 27 June 2023.

[485] "Rocket and Missile System." *Encyclopedia Britannica*, 26 July 1999, https://www.britannica.com/technology/rocket-and-missile-system/Strategic-missiles#ref521030. Accessed 27 June 2023.

[486] "Nuclear Submarine." *Encyclopedia Britannica*, 23 May 2023, https://www.britannica.com/technology/nuclear-submarine. Accessed 27 June 2023.

[487] "POLARIS A1 FACT SHEET." *Spaceline*, https://www.spaceline.org/cape-canaveral-rocket-missile-program/polaris-a1-fact-sheet/. Accessed 27 June 2023.

[488] *Naval & Military.* https://www.processsensing.com/en-us/industries/moisture-in-military-applications.htm. Accessed 27 June 2023.

[489] "All the Nuclear Missile Submarines in the World in One Chart." *Popular Mechanics*, 8 June 2018, https://www.popularmechanics.com/military/navy-ships/a21204892/nuclear-missile-submarines-chart/. Accessed 27 June 2023.

[490] Mizokami, Kyle. "Russia's Monster Submarines Are Even Scarier Than You Imagined." *Popular Mechanics*, 9 May 2022, https://www.popularmechanics.com/military/navy-ships/a35699039/russia-typhoon-class-submarines-true-size/. Accessed 27 June 2023.

[491] "R-39 / SS-N-20 STURGEON SLBM." *Russian / Soviet Nuclear Forces*, https://nuke.fas.org/guide/russia/slbm/r39.htm. Accessed 23 July 2023.

[492] "Ohio Class SSBN." *Submarine Industrial Base Council*, 20 July 2015,

https://submarinesuppliers.org/programs/ssbn/ohio-class-ssbn/. Accessed 27 June 2023.

493 "Trident D5." *Missile Threat*, 19 Sept. 2016, https://missilethreat.csis.org/missile/trident/. Accessed 27 June 2023.

494 "Ballistic Missile Submarines." *Commander, Submarine Force, U.S. Pacific Fleet*, https://www.csp.navy.mil/SUBPAC-Commands/Submarines/Ballistic-Missile-Submarines/. Accessed 27 June 2023.

495 "*Pruning the Nuclear Triad? Pros and Cons of Bombers, Missiles, and Submarines." Center for Arms Control and Non-Proliferation*, 31 Oct. 2011, https://armscontrolcenter.org/pruning-the-nuclear-triad-pros-and-cons-of-bombers-missiles-and-submarines/. Accessed 27 June 2023.

496 *Id.*

497 "The Soviet Navy: How Many Submarines?" *U.S. Naval Institute*, 1 Feb. 1998, https://www.usni.org/magazines/proceedings/1998/february/soviet-navy-how-many-submarines. Accessed 27 June 2023.

498 Thompson, Julia. "Quote of the Week: Strategic vs. Tactical Nuclear Weapons – South Asian Voices." *South Asian Voices*, 8 Oct. 2014, https://southasianvoices.org/quote-of-the-week-strategic-vs-tactical-nuclear-weapons/. Accessed 27 June 2023.

499 "*Mark 7 Nuclear Bomb.*" https://www.globalsecurity.org/wmd/systems/mk7.htm. Accessed 27 June 2023.

500 *List of All U.S. Nuclear Weapons.* https://nuclearweaponarchive.org/Usa/Weapons/Allbombs.html. Accessed 27 June 2023.

501 Pickrell, Ryan. "The US Army Only Ever Fired One Nuclear Artillery Shell from Its 'Atomic Annie' Cannon, and This Is What It Looked Like." *Insider*, 25 May 2022, https://www.businessinsider.com/us-army-atomic-annie-nuclear-artillery-shell-test-2021-5. Accessed 27 June 2023.

502 "The M65 280mm Atomic Cannon." *History*, https://www.theatomiccannon.com/history. Accessed 27 June 2023.

503 Simpson, James. "The National Interest." *The National Interest*, 24 July 2021, https://nationalinterest.org/blog/reboot/german-heavy-gustav-was-largest-gun-ever-built-190368. Accessed 27 June 2023.

504 "The US Army Only Ever Fired One Nuclear Artillery Shell from Its 'Atomic Annie' Cannon, and This Is What It Looked Like." Insider, 25 May 2022, https://www.businessinsider.com/us-army-atomic-annie-nuclear-artillery-shell-test-2021-5. Accessed 27 June 2023.

505 "MGR-1 Honest John." *Weaponsystems.Net*, https://weaponsystems.net/system/908-MGR-1+Honest+John. Accessed 27 June 2023.

506 "The U.S. Military Once Had Nukes That Could Fit in a Backpack." *Popular Mechanics*, 14 Jan. 2019, https://www.popularmechanics.com/military/weapons/a25893776/b-54-cold-

war-backpack-nuke/. Accessed 27 June 2023.

[507] *"W54 Special Atomic Demolition Munition (SADM)."*
https://www.globalsecurity.org/wmd/systems/w54.htm. Accessed 27 June
2023.

[508] "Atomic Suitcase Bombs." *PBS*,
https://www.pbs.org/wgbh/pages/frontline/shows/russia/suitcase/. Accessed
23 July 2023.

[509] Suciu, Peter. "The National Interest." *The National Interest*, 8 Mar. 2021,
https://nationalinterest.org/blog/buzz/russias-atomic-nightmare-100-missing-
suitcase-nuclear-weapons-179577. Accessed 23 July 2023.

[510] *"W54 Special Atomic Demolition Munition (SADM)."*
https://www.globalsecurity.org/wmd/systems/w54.htm. Accessed 27 June
2023.

[511] "The Oklahoma City Bombing." *Federal Bureau of Investigation*,
https://stories.fbi.gov/oklahoma-bombing/. Accessed 27 June 2023.

[512] *"W54 Special Atomic Demolition Munition (SADM)."*
https://www.globalsecurity.org/wmd/systems/w54.htm. Accessed 27 June
2023.

[513] "NATO's European Nuclear Deterrent: The B61 Bomb." *National
Security Archive*, https://nsarchive.gwu.edu/briefing-book/nuclear-
vault/2022-03-28/natos-european-nuclear-deterrent-b61-bomb. Accessed 27
June 2023.

[514] Kristensen, Hans M., and Robert S. Norris. "The B61 Family of Nuclear
Bombs." *Bulletin of the Atomic Scientists*, vol. 70, no. 3, May 2014, pp. 79–
84, https://doi.org/10.1177/0096340214531546.

[515] *B61-12 Life Extension Program*. National Nuclear Security
Administration, Nov. 2021, https://www.energy.gov/nnsa/articles/b61-12-
life-extension-program-lep-fact-sheet. Accessed 27 June 2023.

[516] "SS-1 'Scud.'" *Missile Threat*, 11 Aug. 2016,
https://missilethreat.csis.org/missile/scud/. Accessed 27 June 2023.

[517] "45 Years of Scud; How One of the Most Prolific Ballistic Missiles of All
Time Has Proliferated and Been Deployed Since It First Use in the Yom
Kippur War." *Military Watch Magazine*,
https://militarywatchmagazine.com/article/45-years-of-scud-how-one-of-the-
most-prolific-ballistic-missiles-of-all-time-has-proliferated-and-been-
deployed-since-it-first-use-in-the-yom-kippur-war. Accessed 27 June 2023.

[518] Molenda, Shaan ShaikhWes RumbaughJenevieve. "OTR-21 Tochka (SS-
21)." *Missile Threat*, 11 Aug. 2016, https://missilethreat.csis.org/missile/ss-
21/. Accessed 27 June 2023.

[519] *US Military Options Should Not Include Starting a Nuclear War*. Union of
Concerned Scientists, Dec. 2018,
https://www.ucsusa.org/sites/default/files/attach/2018/12/US-Military-
Options-Should-Not-Include-Starting-a-Nuclear-War.pdf. Accessed 27 June
2023.

[520] *"Soviet / Russian Tactical Nuclear Weapons."*

https://www.globalsecurity.org/wmd/world/russia/tactical.htm. Accessed 27 June 2023.
[521] "Colossus: The Forbin Project (1970)." *IMDb*, Video, https://www.imdb.com/title/tt0064177/characters/nm0293659. Accessed 27 June 2023.
[522] *Colossus: The Forbin Project*. Directed by Joseph Sargent, Universal Pictures, 1970.
[523] Plummer, D. W., and W. H. Greenwood. "The History of Nuclear Weapon Safety Devices." *UNT Digital Library*, 1 June 1998, https://digital.library.unt.edu/ark:/67531/metadc704156/m1/1/.
[524] *Id.*
[525] *Id.*
[526] *Id.*
[527] "Catalog Record: The Report of the Nuclear Weapons Safety Panel : Hearing before the Committee on Armed Services, House of Representatives, One Hundred First Congress, Second Session : Hearing Held December 18, 1990." *HathiTrust Digital Library*, https://catalog.hathitrust.org/Record/007602711.
[528] Plummer, D. W., and W. H. Greenwood. "The History of Nuclear Weapon Safety Devices." *UNT Digital Library*, 1 June 1998, https://digital.library.unt.edu/ark:/67531/metadc704156/m1/1/.
[529] *Id.*
[530] Schlosser, Eric. "Primary Sources: Permissive Action Links and the Threat of Nuclear War." *The New Yorker*, 17 Jan. 2014, https://www.newyorker.com/news/news-desk/primary-sources-permissive-action-links-and-the-threat-of-nuclear-war. Accessed 27 June 2023.
[531] "National Security Action Memorandum 160 to the Secretary of State et al., 'Permissive Links for Nuclear Weapons in NATO,' 6 June 1962, with Memorandum from Jerome Wiesner Attached, 29 May 1962, Secret, Excised Copy." *National Security Archive*, https://nsarchive.gwu.edu/document/28565-document-27-national-security-action-memorandum-160-secretary-state-et-al-permissive. Accessed 28 June 2023.
[532] "Harold Agnew, Head of Atomic Laboratory, Dies at 92." *The Washington Post*, https://www.washingtonpost.com/national/2013/10/02/b4bb81cc-2ba4-11e3-97a3-ff2758228523_story.html. Accessed 27 June 2023.
[533] *Permissive Action Links*. https://www.cs.columbia.edu/~smb/nsam-160/pal.html#A05. Accessed 27 June 2023.
[534] *Id.*
[535] *Id.*
[536] *Principles of Nuclear Weapons Security and Safety*. https://nuclearweaponarchive.org/Usa/Weapons/Pal.html. Accessed 23 July 2023.
[537] *"Permissive Action Links."* https://www.cs.columbia.edu/~smb/nsam-

160/pal.html#D93. Accessed 28 June 2023.

[538] Feldhausen, Russell. *Enigma Machine :: K-State Computational Core.* https://core.cs.ksu.edu/1-cc110/16-cryptography/03-enigma/. Accessed 28 June 2023.

[539] "Keeping Presidents in the Nuclear Dark (Episode #1: The Case of the Missing 'Permissive Action Links')." *Bruce G. Blair, Ph.D.,* http://web.archive.org/web/20120511191600/http:/www.cdi.org/blair/permissive-action-links.cfm. Accessed 28 June 2023.

[540] "The Nuclear Football." *United States Nuclear Forces,* https://web.archive.org/web/20050415200654/http://www.globalsecurity.org/wmd/systems/nuclear-football.htm. Accessed 28 June 2023.

[541] "USATODAY.Com." *Military Aides Still Carry the President's Nuclear "Football,"* http://usatoday30.usatoday.com/news/washington/2005-05-05-nuclear-football_x.htm. Accessed 28 June 2023.

[542] *"The Football."* United States Nuclear Forces, https://nuke.fas.org/guide/usa/c3i/nuclear-football.htm. Accessed 28 June 2023.

[543] Department of Defense. *DoD Instruction Number 5210.87.* Nov. 1998.

[544] Revell, Eric. "Boeing Personnel Working on Air Force One Planes Had Lapsed Security Clearances." *Fox Business,* 23 Mar. 2023, https://www.foxbusiness.com/markets/boeing-personnel-working-air-force-one-planes-lapsed-security-clearances. Accessed 28 June 2023.

[545] "Marine Corps Helicopter Squadron One > About > Join HMX-1." *Marine Corps Helicopter Squadron One,* https://www.hqmc.marines.mil/hmx-1/About/Join-HMX-1/. Accessed 28 June 2023.

[546] Cohen, Zachary, and Brian Todd. "What It Actually Takes to Launch a Nuclear Strike." *CNN,* 3 Jan. 2018, https://www.cnn.com/2018/01/03/politics/trump-nuclear-football-explainer/index.html. Accessed 28 June 2023.

[547] Pedersen. "Two-Person Control: A Brief History and Modern Industry Practices." *OSTI.GOV,* 1 July 2017, https://www.osti.gov/servlets/purl/1374246. Accessed 28 June 2023.

[548] "Launching Missiles." *National Museum of the United States Air Force™,* https://www.nationalmuseum.af.mil/Visit/Museum-Exhibits/Fact-Sheets/Display/Article/197675/launching-missiles/. Accessed 28 June 2023.

[549] Purpose, Task &. "The Military Has Gotten Rid of the Floppy Disks Used to Control US Nuclear Weapons for More than 50 Years." *Insider,* 30 Oct. 2019, https://www.businessinsider.com/military-replaces-floppy-disks-used-to-control-nuclear-weapons-2019-10. Accessed 28 June 2023.

[550] Insinna, Valerie. "The US Nuclear Forces' Dr. Strangelove-Era Messaging System Finally Got Rid of Its Floppy Disks." *C4ISRNet,* 22 Oct. 2019, https://www.c4isrnet.com/air/2019/10/17/the-us-nuclear-forces-dr-strangelove-era-messaging-system-finally-got-rid-of-its-floppy-disks/. Accessed 28 June 2023.

[551] "North American Aerospace Defense Command > About NORAD > NORAD Agreement." *North American Aerospace Defense Command*, https://www.norad.mil/About-NORAD/NORAD-Agreement/. Accessed 28 June 2023.

[552] "North American Aerospace Defense Command > About NORAD > NORAD History." *North American Aerospace Defense Command*, https://www.norad.mil/About-NORAD/NORAD-History/. Accessed 28 June 2023.

[553] "Launch on Warning (LOW)." *Encyclopedia Britannica*, 10 June 2014, https://www.britannica.com/topic/launch-on-warning. Accessed 28 June 2023.

[554] Stewart, Will, and Anthony Blair. "Putin Seen with Nuclear Briefcase on Shock Trip to Ukraine in Closest Ever Visit to Front as Nuke Bombers..." *The US Sun*, 18 Apr. 2023, https://www.the-sun.com/news/7898264/vladimir-putin-ukraine-visit-kherson-nuclear-suitcase/. Accessed 28 June 2023.

[555] "Russian Nuclear Policy and the Status of Detargeting." *Brookings*, https://www.brookings.edu/articles/russian-nuclear-policy-and-the-status-of-detargeting/. Accessed 28 June 2023.

[556] "Launch on Warning (LOW)." *Encyclopedia Britannica*, 10 June 2014, https://www.britannica.com/topic/launch-on-warning. Accessed 28 June 2023.

[557] Bender, Jeremy. "Russia May Still Have An Automated Nuclear Launch System Aimed Across The Northern Hemisphere." *Insider*, 4 Sept. 2014, https://www.businessinsider.com/russias-dead-hand-system-may-still-be-active-2014-9. Accessed 28 June 2023.

[558] *"Russia's 'Dead Hand' Is a Soviet-Built Nuclear Doomsday Device."* "Military.Com." *Military.Com*, 14 Mar. 2022, https://www.military.com/history/russias-dead-hand-soviet-built-nuclear-doomsday-device.html. Accessed 28 June 2023.

[559] Thompson, Nicholas. "Inside the Apocalyptic Soviet Doomsday Machine." *WIRED*, 21 Sept. 2009, https://www.wired.com/2009/09/mf-deadhand/. Accessed 28 June 2023.

[560] Yarynich, Valery E. *C3: Nuclear Command, Control, Cooperation*. 2003. Center for Defense Information, 2003, pp. 156–59, https://books.google.com/books?id=qG6UAQAACAAJ&printsec=frontcover&source=gbs_ge_summary_r&cad=0#v=onepage&q=dead%20hand&f=false . Accessed 28 June 2023.

[561] *Id.*

[562] Voelz, George L. and Ileana G. Buican. "Plutonium and Health. How Great Is the Risk?" *Federation of American Scientists*, Los Alamos Science, 26 Nov. 2000, pp. 74–89, https://sgp.fas.org/othergov/doe/lanl/pubs/00818013.pdf. Accessed 29 June 2023.

[563] "Ionizing Radiation." *NRC Web*, https://www.nrc.gov/reading-rm/basic-

ref/glossary/ionizing-radiation.html. Accessed 28 June 2023.

[564] Department of Energy, Office of Environmental, Health, Safety, and Security. *The DOE Ionizing Radiation Dose Ranges Chart.* Dec. 2017.

[565] *Id.*

[566] *Id.*

[567] Kennedy, Kelsey. "The Existential Horror Created by the First X-Ray Images." *Atlas Obscura*, 5 Oct. 2017, https://www.atlasobscura.com/articles/roentgen-xrays-discovery-radiographs. Accessed 28 June 2023.

[568] King, Gilbert. "Clarence Dally — The Man Who Gave Thomas Edison X-Ray Vision." *Smithsonian Magazine*, 14 Mar. 2012, https://www.smithsonianmag.com/history/clarence-dally-the-man-who-gave-thomas-edison-x-ray-vision-123713565/. Accessed 28 June 2023.

[569] *Marie Curie and the Science of Radioactivity.* https://history.aip.org/exhibits/curie/radinst3.htm. Accessed 28 June 2023.

[570] Lamb, Evelyn. "That Time Marie Curie Was Just Trying to Get the Electricity Running." *Scientific American Blog Network*, 30 May 2017, https://blogs.scientificamerican.com/roots-of-unity/that-time-marie-curie-was-just-trying-to-get-the-electricity-running/. Accessed 28 June 2023.

[571] Vaughan, Don. "Radium Girls: The Women Who Fought for Their Lives in a Killer Workplace." *Encyclopedia Britannica*, https://www.britannica.com/story/radium-girls-the-women-who-fought-for-their-lives-in-a-killer-workplace. Accessed 28 June 2023.

[572] "Radium Girls: The Women Who Fought for Their Lives in a Killer Workplace." *Encyclopedia Britannica*, https://www.britannica.com/story/radium-girls-the-women-who-fought-for-their-lives-in-a-killer-workplace. Accessed 28 June 2023.

[573] WHO, World Health Organization: "Ionizing Radiation, Health Effects and Protective Measures." *World Health Organization: WHO*, 29 Apr. 2016, https://www.who.int/news-room/fact-sheets/detail/ionizing-radiation-health-effects-and-protective-measures. Accessed 28 June 2023.

[574] "Radiation Health Effects." *US EPA*, 12 Nov. 2014, https://www.epa.gov/radiation/radiation-health-effects. Accessed 29 June 2023.

[575] Institute of Medicine (US) Committee on Battlefield Radiation Exposure Criteria, et al. "Radiation Unit Conversion Chart." *NCBI Bookshelf*, 1 Jan. 1999, https://www.ncbi.nlm.nih.gov/books/NBK224062/. Accessed 29 June 2023.

[576] "Definition of Gray." *Radiation Emergency Medical Management*, https://remm.hhs.gov/gray_definition.htm. Accessed 29 June 2023.

[577] *Advisory Committee On Human Radiation Experiments Final Report.* https://ehss.energy.gov/ohre/roadmap/achre/intro_9_6.html. Accessed 23 July 2023.

[578] Munroe, Randall. *Radiation Dose Chart.* http://nrl.mit.edu/sites/default/files/documents/xkcd%20Radiation%20Dose

%20Chart.pdf. Accessed 29 June 2023.

[579] Health Physics Society (HPS). "Regulatory Dose Limits." *Health Physics Society*, 2016, https://hps.org/publicinformation/ate/faqs/regdoselimits.html. Accessed 29 June 2023.

[580] *Banana Equivalent Dose.* http://health.phys.iit.edu/extended_archive/9503/msg00074.html. Accessed 29 June 2023.

[581] Edwards, Gordon and Canadian Coalition for Nuclear Responsibility. *About Radioactive Bananas.* Accessed 29 June 2023.

[582] Center for Devices and Radiological Health. "What Are the Radiation Risks from CT?" *U.S. Food and Drug Administration*, https://www.fda.gov/radiation-emitting-products/medical-x-ray-imaging/what-are-radiation-risks-ct. Accessed 29 June 2023.

[583] Reuters Staff. "How Much Radiation Is Dangerous?" *Reuters*, 15 Mar. 2011, https://www.reuters.com/article/us-how-much-radiation-dangerous/how-much-radiation-is-dangerous-idUSTRE72E79Z20110315. Accessed 29 June 2023.

[584] "Radiation Risk from Medical Imaging." *Harvard Health*, 22 Sept. 2010, https://www.health.harvard.edu/cancer/radiation-risk-from-medical-imaging. Accessed 29 June 2023.

[585] "Radiation Hazard Scale." *CDC*, 13 Apr. 2023, https://www.cdc.gov/nceh/radiation/emergencies/radiationhazardscale.htm. Accessed 29 June 2023.

[586] Rogers, Simon. "Radiation Exposure: A Quick Guide to What Each Level Means." *The Guardian*, 15 Mar. 2011, https://www.theguardian.com/news/datablog/2011/mar/15/radiation-exposure-levels-guide. Accessed 29 June 2023.

[587] *"Radiation Health Effects."* US EPA, 12 Nov. 2014, https://www.epa.gov/radiation/radiation-health-effects. Accessed 29 June 2023.

[588] CDC. "Health Effects of Radiation: Health Effects Depend on the Dose." *Centers for Disease Control and Prevention*, 7 Dec. 2015, https://www.cdc.gov/nceh/radiation/dose.html#how. Accessed 29 June 2023.

[589] Medical Bag. "The Radium Girls." *Medical Bag*, 1 Jan. 2014, https://www.medicalbag.com/home/features/profile-in-rare-diseases/the-radium-girls/. Accessed 29 June 2023.

[590] "Acute Radiation Syndrome." *CDC*, 30 Dec. 2022, https://www.cdc.gov/nceh/radiation/emergencies/ars.htm. Accessed 29 June 2023.

[591] *Acute Radiation Syndrome: A Fact Sheet for Physicians*, 8 Apr. 2022, https://www.cdc.gov/nceh/radiation/emergencies/arsphysicianfactsheet.htm. Accessed 1 July 2023.

[592] *Id.*

[593] Monmouth County Health Department. *RADIATION HEALTH BASICS.* Accessed 30 June 2023.

[594] Libretexts. "11.6: Penetrating Power of Radiation." *Libretexts*, 26 May 2019,
https://chem.libretexts.org/Bookshelves/Introductory_Chemistry/Chemistry_f
or_Changing_Times_(Hill_and_McCreary)/11%3A_Nuclear_Chemistry/11.0
6%3A_Penetrating_Power_of_Radiation. Accessed 30 June 2023.

[595] "Residual Radiation and Fallout." *Encyclopedia Britannica*, 26 July 1999,
https://www.britannica.com/technology/nuclear-weapon/Residual-radiation-
and-fallout. Accessed 30 June 2023.

[596] "Survivors of the Atomic Bomb Share Their Stories." *TIME.Com*,
https://time.com/after-the-bomb/. Accessed 29 June 2023.

[597] Hiroshima Peace Memorial Museum. "Hiroshima Peace Memorial
Museum." *5-2-3 Heat*,
https://hpmmuseum.jp/modules/exhibition/index.php?action=ItemView&ite
m_id=59&lang=eng. Accessed 29 June 2023.

[598] "Hiroshima Peace Memorial Museum." *2-2-5 Human Shadow Etched in
Stone*,
https://hpmmuseum.jp/modules/exhibition/index.php?action=ItemView&ite
m_id=112&lang=eng. Accessed 29 June 2023.

[599] "Black Rain." *Museum of Radiation and Radioactivity*,
https://www.orau.org/health-physics-museum/collection/nuclear-
weapons/hiroshima/black-rain.html. Accessed 29 June 2023.

[600] Palmer, Ewan. "Hiroshima Radioactive 'Black Rain' Victims Recognized
as Atomic Bomb Survivors." *Newsweek*, 14 July 2021,
https://www.newsweek.com/hiroshima-atomic-bomb-black-rain-victims-
court-1609465. Accessed 29 June 2023.

[601] "Radiation Injuries." *Atomicarchive.Com*,
https://www.atomicarchive.com/resources/documents/med/med_chp22.html.
Accessed 29 June 2023.

[602] *Atomic Bomb Casualty Commission, 1945-1982*.
http://www.nasonline.org/about-nas/history/archives/collections/abcc-1945-
1982.html. Accessed 29 June 2023.

[603] *Past Results and Future Studies – Radiation Effects Research Foundation
(RERF)*. https://www.rerf.or.jp/en/programs/roadmap_e/. Accessed 29 June
2023.

[604] "Hiroshima and Nagasaki: The Long Term Health Effects." *K=1 Project*,
https://k1project.columbia.edu/news/hiroshima-and-nagasaki. Accessed 29
June 2023.

[605] "Harry Daghlian." *Nuclear Museum*,
https://ahf.nuclearmuseum.org/ahf/profile/harry-daghlian/. Accessed 29 June
2023.

[606] "Harry Daghlian - Linda Hall Library." *The Linda Hall Library*,
https://www.lindahall.org/about/news/scientist-of-the-day/harry-daghlian.
Accessed 29 June 2023.

[607] "Harry Daghlian." *The Linda Hall Library*,
https://www.lindahall.org/about/news/scientist-of-the-day/harry-daghlian.

Accessed 29 June 2023.

[608] Grabianowski, Ed. "How Radiation Sickness Works." *HowStuffWorks*, 26 Apr. 2011, https://science.howstuffworks.com/radiation-sickness3.htm. Accessed 29 June 2023.

[609] Wellerstein, Alex. "Demon Core: The Strange Death of Louis Slotin." *The New Yorker*, 21 May 2016, https://www.newyorker.com/tech/annals-of-technology/demon-core-the-strange-death-of-louis-slotin. Accessed 29 June 2023.

[610] "Demon Core: The Strange Death of Louis Slotin." *The New Yorker*, 21 May 2016, https://www.newyorker.com/tech/annals-of-technology/demon-core-the-strange-death-of-louis-slotin. Accessed 30 June 2023.

[611] RP Alba Ltd. *Radiation Dose Rate Converter*. RP Alba Ltd., https://rp-alba.com/index.php?filename=radiationDoseRateConverter.php. Accessed 30 June 2023.

[612] "The Death of Louis Slotin." *Nuclear Museum*, https://ahf.nuclearmuseum.org/death-louis-slotin/. Accessed 30 June 2023.

[613] West, Julian G. "The Atomic-Bomb Core That Escaped World War II." *The Atlantic*, 2 Apr. 2018, https://www.theatlantic.com/technology/archive/2018/04/tickling-the-dragons-tail-plutonium-time-bomb/557006/. Accessed 30 June 2023.

[614] United States General Accounting Office. *Operation Crossroads: Personnel Radiation Exposure Estimates Should Be Improved*. Dec. 1985.

[615] "Operation Crossroads: A Deadly Illusion." *The National World War II Museum*, 4 July 2021, https://www.nationalww2museum.org/war/articles/operation-crossroads-atomic-bomb-aftermath. Accessed 30 June 2023.

[616] The MIT Press Reader. "The Devastating Effects of Nuclear Weapons." *The MIT Press Reader*, 2 Mar. 2022, https://thereader.mitpress.mit.edu/devastating-effects-of-nuclear-weapons-war/. Accessed 30 June 2023.

[617] "Fallout from a Nuclear Detonation: Description and Management." *Radiation Emergency Medical Management*, https://remm.hhs.gov/nuclearfallout.htm. Accessed 30 June 2023.

[618] "Half-Life." *Encyclopedia Britannica*, 20 July 1998, https://www.britannica.com/science/half-life-radioactivity. Accessed 30 June 2023.

[619] University of Alabama at Birmingham. "Accepted Half-Lives of Commonly Used Radioisotopes." *UAB Radiation Safety Procedures Manual, Ninth Edition*, 2007, https://www.uab.edu/ehs/images/docs/rad/Accepted_HalfLives_Commonly_Used_Radioisotopes.pdf. Accessed 30 June 2023.

[620] "DOE Explains...Deuterium-Tritium Fusion Reactor Fuel." *Energy.Gov*, https://www.energy.gov/science/doe-explainsdeuterium-tritium-fusion-reactor-fuel. Accessed 30 June 2023.

[621] Munroe, Randall. *Radiation Dose Chart.*

http://nrl.mit.edu/sites/default/files/documents/xkcd%20Radiation%20Dose
%20Chart.pdf. Accessed 29 June 2023.

[622] RP Alba Ltd. *Radiation Dose Rate Converter*. RP Alba Ltd., https://rp-alba.com/index.php?filename=radiationDoseRateConverter.php. Accessed 30
June 2023.

[623] "Xcel Energy Monticello Power Plant Tritium Leak." *MN Dept. of Health*,
https://www.health.state.mn.us/communities/environment/air/tritiumleak.htm
l. Accessed 30 June 2023.

[624] "Radionuclide Basics: Tritium." *US EPA*, 15 Apr. 2015,
https://www.epa.gov/radiation/radionuclide-basics-tritium. Accessed 30 June
2023.

[625] "CDC Radiation Emergencies." *Radioisotope Brief: Plutonium-239 (Pu-239)*, 17 Mar. 2023,
https://www.cdc.gov/nceh/radiation/emergencies/isotopes/plutonium.htm.
Accessed 30 June 2023.

[626] Guillemette, Mélissa. "Chalk River: The Forgotten Nuclear Accidents."
The Walrus, 13 July 2022, https://thewalrus.ca/nuclear-accidents/. Accessed
1 July 2023.

[627] "Chalk River: The Forgotten Nuclear Accidents." *The Walrus*, 13 July
2022, https://thewalrus.ca/nuclear-accidents/. Accessed 1 July 2023.

[628] Planas, Oriol. *Chalk River Nuclear Accident, (Ontario) Canada*. 26 Oct.
2010, https://nuclear-energy.net/nuclear-accidents/chalk-river. Accessed 1
July 2023.

[629] "International Nuclear and Radiological Event Scale (INES)." *IAEA*,
https://www.iaea.org/resources/databases/international-nuclear-and-radiological-event-scale. Accessed 1 July 2023.

[630] "Chernobyl." *Chernobyl Disaster - World Nuclear Association*,
https://world-nuclear.org/information-library/safety-and-security/safety-of-plants/chernobyl-accident.aspx. Accessed 1 July 2023.

[631] *Chernobyl Disaster - World Nuclear Association*, https://world-nuclear.org/information-library/safety-and-security/safety-of-plants/chernobyl-accident.aspx. Accessed 1 July 2023.

[632] *Id.*

[633] White, Robyn. "Chernobyl Aftermath: How Long Will Exclusion Zone Be
Uninhabitable?" *Newsweek*, 14 Oct. 2022,
https://www.newsweek.com/chernobyl-aftermath-how-long-will-exclusion-zone-uninhabitable-1751834. Accessed 1 July 2023.

[634] Blakemore, Erin. "The Chernobyl Disaster: What Happened, and the
Long-Term Impacts." *National Geographic*, 17 May 2019,
https://www.nationalgeographic.com/culture/article/chernobyl-disaster.
Accessed 1 July 2023.

[635] "Fukushima Daiichi Accident." *World Nuclear Association*, https://world-nuclear.org/information-library/safety-and-security/safety-of-plants/fukushima-daiichi-accident.aspx. Accessed 1 July 2023.

[636] Kemp, Ellie. "How Many People Died in the Fukushima Disaster?"

Manchester Evening News, 7 June 2023, https://www.manchestereveningnews.co.uk/news/tv/how-many-people-died-fukushima-27074927. Accessed 1 July 2023.

[637] "Japan Confirms First Fukushima Worker Death from Radiation." *BBC News*, 5 Sept. 2018, https://www.bbc.com/news/world-asia-45423575. Accessed 1 July 2023.

[638] "Fukushima Accident." *Encyclopedia Britannica*, 20 Apr. 2011, https://www.britannica.com/event/Fukushima-accident. Accessed 1 July 2023.

[639] Kiger, Patrick J. "Hisashi Ouchi Suffered an 83-Day Death By Radiation Poisoning." *HowStuffWorks*, 8 Aug. 2022, https://science.howstuffworks.com/hisashi-ouchi.htm. Accessed 1 July 2023.

[640] Fisman, D. N. "Hemophagocytic Syndromes and Infection." *Emerging Infectious Diseases*, vol. 6, no. 6, Dec. 2000, https://doi.org/10.3201/eid0606.000608.

[641] International Atomic Energy Agency. "LESSONS LEARNED FROM THE JCO NUCLEAR CRITICALITY ACCIDENT IN JAPAN IN 1999 ." *IAEA*, International Atomic Energy Agency, http://www-ns.iaea.org/downloads/iec/tokaimura-report.pdf. Accessed 1 July 2023.

[642] Eaves, Elisabeth. "How the Unlucky Lucky Dragon Birthed an Era of Nuclear Fear." *Bulletin of the Atomic Scientists*, 28 Feb. 2018, https://thebulletin.org/2018/02/how-the-unlucky-lucky-dragon-birthed-an-era-of-nuclear-fear/. Accessed 1 July 2023.

[643] "Lucky Dragon and Atomic Tuna Memorial." *Atlas Obscura*, 21 Jan. 2010, https://www.atlasobscura.com/places/lucky-dragon-and-atomic-tuna-memorial. Accessed 1 July 2023.

[644] Kiger, Patrick J. "7 Surprising Facts about the Nuclear Bomb Tests at Bikini Atoll." *HISTORY*, 12 May 2022, https://www.history.com/news/nuclear-bomb-tests-bikini-atoll-facts. Accessed 1 July 2023.

[645] "7 Surprising Facts about the Nuclear Bomb Tests at Bikini Atoll." *HISTORY*, 12 May 2022, https://www.history.com/news/nuclear-bomb-tests-bikini-atoll-facts. Accessed 1 July 2023.

[646] Rust, Susanne. "How the U.S. Betrayed the Marshall Islands, Kindling the Next Nuclear Disaster." *Los Angeles Times*, 10 Nov. 2019, https://www.latimes.com/projects/marshall-islands-nuclear-testing-sea-level-rise/. Accessed 1 July 2023.

[647] Rust, Susanne. "How the U.S. Betrayed the Marshall Islands, Kindling the Next Nuclear Disaster." *Los Angeles Times*, 10 Nov. 2019, https://www.latimes.com/projects/marshall-islands-nuclear-testing-sea-level-rise/. Accessed 1 July 2023.

[648] *Radioisotope Brief: Strontium-90 (Sr-90)*, 17 Mar. 2023, https://www.cdc.gov/nceh/radiation/emergencies/isotopes/strontium.htm. Accessed 1 July 2023.

[649] ABC News. "World Wakes Up to Horrific Scientific History." *ABC News*,

https://abcnews.go.com/International/story?id=80970&page=1. Accessed 2 July 2023.

[650] Campbell, Murray. "Project Sunshine's Dark Secret." *The Globe and Mail*, 6 June 2001, https://www.theglobeandmail.com/news/national/project-sunshines-dark-secret/article1031823/. Accessed 2 July 2023.

[651] Atomic Archive. *Total Nuclear Testing Yields.* https://www.atomicarchive.com/almanac/test-sites/testing-yields.html. Accessed 2 July 2023.

[652] "Worldwide Effects of Atomic Weapons: Project SUNSHINE." *RAND*, 1 Jan. 1953, https://www.rand.org/pubs/reports/R0251.html.

[653] U.S. Environmental Projection Agency. *EPA FACTS ABOUT STRONTIUM-90.* https://semspub.epa.gov/work/HQ/175430.pdf. Accessed 2 July 2023.

[654] International Commission on Nuclear Non-Proliferation and Disarmament. Eliminating Nuclear Threats. Nov. 2009, http://www.icnnd.org/reference/reports/ent/pdf/ICNND_Report-EliminatingNuclearThreats.pdf. Accessed 2 July 2023.

[655] United States Air Force. *Air Force Manual 10-206.* Department of the Air Force, 18 June 2018, https://static.e-publishing.af.mil/production/1/af_a3/publication/afman10-206/afman10-206.pdf. Accessed 6 July 2023.

[656] Gault, Matthew. *The Bizarre Mystery of the Only Armed Nuke America Ever Lost.* 29 Aug. 2022, https://www.vice.com/en/article/y3p3xw/the-bizarre-mystery-of-the-only-armed-nuke-america-ever-lost. Accessed 6 July 2023.

[657] Strategic Air Command. *Peace... Is Our Profession: Alert Operations and the Strategic Air Command, 1957-1991.* Office of the Historian, 7 Dec. 1991, https://www.afgsc.af.mil/Portals/51/Docs/SAC%20Alert%20Operations%20Lo-Res.pdf?ver=2016-09-27-114343-960. Accessed 23 July 2023.

[658] Leone, Dario. "Remembering Operation Chrome Dome, the 1960's Airborne Alert Missions Flown by B-52 Strategic Bombers Armed with Thermonuclear Weapons." *The Aviation Geek Club*, 15 Mar. 2020, https://theaviationgeekclub.com/remembering-operation-chrome-dome-the-1960s-airborne-alert-missions-flown-by-b-52-strategic-bombers-armed-with-thermonuclear-weapons/. Accessed 7 July 2023.

[659] "The Perils of Chrome Dome." *Air & Space Forces Magazine*, 25 July 2011, https://www.airandspaceforces.com/article/0811dome/. Accessed 7 July 2023.

[660] U.S. Department of Energy. "Unrecovered Nuclear Weapons and Classified Components." *Federation of American Scientists*, https://sgp.fas.org/othergov/doe/cg-hr-3/appb.pdf. Accessed 7 July 2023.

[661] "Brush with Catastrophe: The Day the U.S. Almost Nuked Itself." *Pieces of History*, 22 Jan. 2021, https://prologue.blogs.archives.gov/2021/01/22/brush-with-catastrophe-the-day-the-u-s-almost-nuked-itself/. Accessed 6 July 2023.

[662] Maggelet, Michael H. *Goldsboro- 19 Steps Away from Detonation.* https://nuclearweaponsaccidents.blogspot.com/2013/03/goldsboro-19-steps-away-from-detonation.html#comment-form. Accessed 6 July 2023.

[663] U.S. Department of Energy. "Unrecovered Nuclear Weapons and Classified Components." *Federation of American Scientists,* https://sgp.fas.org/othergov/doe/cg-hr-3/appb.pdf. Accessed 7 July 2023.

[664] Philipps, Dave. "Decades Later, Sickness Among Airmen After a Hydrogen Bomb Accident." *The New York Times,* 19 June 2016, https://www.nytimes.com/2016/06/20/us/decades-later-sickness-among-airmen-after-a-hydrogen-bomb-accident.html. Accessed 7 July 2023.

[665] U.S. Department of Energy. "Unrecovered Nuclear Weapons and Classified Components." *Federation of American Scientists,* https://sgp.fas.org/othergov/doe/cg-hr-3/appb.pdf. Accessed 7 July 2023.

[666] "The Perils of Chrome Dome." *Air & Space Forces Magazine,* 25 July 2011, https://www.airandspaceforces.com/article/0811dome/. Accessed 7 July 2023.

[667] "The Broken Arrow Project: February 13, 1950 - British Columbia, Canada." *The Broken Arrow Project: Visualizing the Dangers of Maintaining the U.S. Nuclear Arsenal,* https://scalar.usc.edu/works/brokenarrowproject/1950---british-colombia.54. Accessed 7 July 2023.

[668] Berinato, Christopher. "That's So Savannah: Bomb at the Beach Remains Hauntingly Mysterious on Tybee Island." *Savannah Morning News,* 14 Apr. 2021, https://www.savannahnow.com/story/lifestyle/2021/04/14/what-happened-bomb-dropped-over-tybee-island-wassaw-sound-b-47-nuclear-radiation-thats-so-savannah/7190932002/. Accessed 7 July 2023.

[669] *"Complete List of All U.S. Nuclear Weapons."* https://nuclearweaponarchive.org/Usa/Weapons/Allbombs.html. Accessed 7 July 2023.

[670] *Scorpion (SSN-589).* https://www.history.navy.mil/browse-by-topic/ships/submarines/scorpion-ssn-589.html. Accessed 7 July 2023.

[671] U.S. Department of Energy. "Unrecovered Nuclear Weapons and Classified Components." *Federation of American Scientists,* https://sgp.fas.org/othergov/doe/cg-hr-3/appb.pdf. Accessed 7 July 2023.

[672] *Id.*

[673] "Titan II." *Missile Threat,* 22 Mar. 2017, https://missilethreat.csis.org/missile/titan-ii/. Accessed 7 July 2023.

[674] Gorvett, Zaria. "The Nuclear Mistakes That Nearly Caused World War Three." *BBC,* 10 Aug. 2020, https://www.bbc.com/future/article/20200807-the-nuclear-mistakes-that-could-have-ended-civilisation. Accessed 7 July 2023.

[675] "False Alarm: 1979 NORAD Incident Was One of Several Cold War Nuclear Scares - UPI.Com." *UPI,* 8 Nov. 2019, https://www.upi.com/Top_News/US/2019/11/08/False-alarm-1979-NORAD-scare-was-one-of-several-nuclear-close-calls/7491573181627/. Accessed 7

July 2023.

676 "False Warnings of Soviet Missile Attacks Put U.S. Forces on Alert in 1979-1980." *National Security Archive*, https://nsarchive.gwu.edu/briefing-book/nuclear-vault/2020-03-16/false-warnings-soviet-missile-attacks-during-1979-80-led-alert-actions-us-strategic-forces. Accessed 7 July 2023.

677 U. S. Government Accountability Office. "The World Wide Military Command and Control System--Major Changes Needed in Its Automated Data Processing Management and Direction." *U.S. GAO*, https://www.gao.gov/products/lcd-80-22. Accessed 7 July 2023.

678 Aksenov, Pavel. "Stanislav Petrov: The Man Who May Have Saved the World." *BBC News*, 26 Sept. 2013, https://www.bbc.com/news/world-europe-24280831. Accessed 7 July 2023.

679 *Stanislav Petrov (U.S. National Park Service)*. https://www.nps.gov/people/stanislav_petrov.htm. Accessed 7 July 2023.

680 Myre, Greg. "Stanislav Petrov, 'The Man Who Saved The World,' Dies At 77." *NPR*, 18 Sept. 2017, https://www.npr.org/sections/thetwo-way/2017/09/18/551792129/stanislav-petrov-the-man-who-saved-the-world-dies-at-77. Accessed 7 July 2023.

681 Lewis, Robert. "Kyshtym Disaster." *Encyclopedia Britannica*, 2 Aug. 2017, https://www.britannica.com/event/Kyshtym-disaster. Accessed 7 July 2023.

682 Reuters. "Putin Signs Law Suspending INF Disarmament Treaty." *The Moscow Times*, 3 July 2019, https://www.themoscowtimes.com/2019/07/03/putin-signs-law-suspending-inf-disarmament-treaty-a66270. Accessed 7 July 2023.

683 Allison, Graham. *What Happened to the Soviet Superpower's Nuclear Arsenal? Clues for the Nuclear Security Summit.* Belfer Center for Science and International Affairs, Mar. 2012, https://www.belfercenter.org/sites/default/files/legacy/files/3%2014%2012%20Final%20What%20Happened%20to%20Soviet%20Arsenals.pdf. Accessed 7 July 2023.

684 *Id.*

685 "U-2." *Encyclopedia Britannica*, 20 July 1998, https://www.britannica.com/technology/U-2. Accessed 11 July 2023.

686 "Cuban Missile Crisis." *JFK Library*, https://www.jfklibrary.org/learn/about-jfk/jfk-in-history/cuban-missile-crisis. Accessed 11 July 2023.

687 "To the Brink: JFK and the Cuban Missile Crisis." *JFK Library*, https://www.jfklibrary.org/visit-museum/exhibits/past-exhibits/to-the-brink-jfk-and-the-cuban-missile-crisis. Accessed 11 July 2023.

688 "October 17, 1962 - Cuban Missile Crisis." *John F. Kennedy Presidential Library & Museum*, https://microsites.jfklibrary.org/cmc/oct17/. Accessed 11 July 2023.

689 "Actions of the USSR - The Cuban Missile Crisis - CCEA - GCSE History Revision - CCEA." *BBC Bitesize*,

https://www.bbc.co.uk/bitesize/guides/zwk7rwx/revision/4. Accessed 11 July 2023.

[690] "SM-78 Jupiter." *Missile Threat*, 20 Dec. 2016, https://missilethreat.csis.org/missile/jupiter/. Accessed 11 July 2023.

[691] "Kennedy's Dilemma and the Response of the USA - The Cuban Missile Crisis - CCEA - GCSE History Revision - CCEA." *BBC Bitesize*, https://www.bbc.co.uk/bitesize/guides/zwk7rwx/revision/5. Accessed 11 July 2023.

[692] "Radio and Television Report to the American People on the Soviet Arms Buildup in Cuba - Cuban Missile Crisis." *John F. Kennedy Presidential Library & Museum*, https://microsites.jfklibrary.org/cmc/oct22/doc5.html. Accessed 11 July 2023.

[693] "Chairman Khrushchev's Letter to President Kennedy, October 23, 1962 - Cuban Missile Crisis." *John F. Kennedy Presidential Library & Museum*, https://microsites.jfklibrary.org/cmc/oct23/doc6.html. Accessed 11 July 2023.

[694] "October 23, 1962 - Cuban Missile Crisis." *John F. Kennedy Presidential Library & Museum*, https://microsites.jfklibrary.org/cmc/oct23/. Accessed 11 July 2023.

[695] *Id.*

[696] "October 24, 1962 - Cuban Missile Crisis." *John F. Kennedy Presidential Library & Museum*, https://microsites.jfklibrary.org/cmc/oct24/. Accessed 11 July 2023.

[697] "October 25, 1962 - Cuban Missile Crisis." *John F. Kennedy Presidential Library & Museum*, https://microsites.jfklibrary.org/cmc/oct25/. Accessed 11 July 2023.

[698] "Fidel Castro's Letter - Cuban Missile Crisis." *John F. Kennedy Presidential Library & Museum*, https://microsites.jfklibrary.org/cmc/oct26/doc2.html. Accessed 11 July 2023.

[699] "Department of State Telegram Transmitting Letter From Chairman Khrushchev to President Kennedy, October 26, 1962 - Cuban Missile Crisis." *John F. Kennedy Presidential Library & Museum*, https://microsites.jfklibrary.org/cmc/oct26/doc4.html. Accessed 11 July 2023.

[700] "How the Death of a U.S. Air Force Pilot Prevented a Nuclear War." *HISTORY*, 26 Oct. 2012, https://www.history.com/news/the-cuban-missile-crisis-pilot-whose-death-may-have-saved-millions. Accessed 11 July 2023.

[701] "October 27, 1962 - Cuban Missile Crisis." *John F. Kennedy Presidential Library & Museum*, https://microsites.jfklibrary.org/cmc/oct27/. Accessed 11 July 2023.

[702] "How the Death of a U.S. Air Force Pilot Prevented a Nuclear War." *HISTORY*, 26 Oct. 2012, https://www.history.com/news/the-cuban-missile-crisis-pilot-whose-death-may-have-saved-millions. Accessed 11 July 2023.

[703] "October 27, 1962 - Cuban Missile Crisis." *John F. Kennedy Presidential*

Library & Museum, https://microsites.jfklibrary.org/cmc/oct27/. Accessed 11 July 2023.

[704] Davis, Nicola. "Soviet Submarine Officer Who Averted Nuclear War Honoured with Prize." *The Guardian*, 27 Oct. 2017, https://www.theguardian.com/science/2017/oct/27/vasili-arkhipov-soviet-submarine-captain-who-averted-nuclear-war-awarded-future-of-life-prize. Accessed 11 July 2023.

[705] Walsh, Bryan. "How Vasili Arkhipov Helped Prevent Nuclear War 60 Years Ago." *Vox*, 27 Oct. 2022, https://www.vox.com/future-perfect/2022/10/27/23426482/cuban-missile-crisis-basilica-arkhipov-nuclear-war. Accessed 11 July 2023.

[706] "Soviet Submarine Officer Who Averted Nuclear War Honoured with Prize." *The Guardian*, 27 Oct. 2017, https://www.theguardian.com/science/2017/oct/27/vasili-arkhipov-soviet-submarine-captain-who-averted-nuclear-war-awarded-future-of-life-prize. Accessed 11 July 2023.

[707] Walsh, Bryan. "How Vasili Arkhipov Helped Prevent Nuclear War 60 Years Ago." *Vox*, 27 Oct. 2022, https://www.vox.com/future-perfect/2022/10/27/23426482/cuban-missile-crisis-basilica-arkhipov-nuclear-war. Accessed 11 July 2023.

[708] *Id.*

[709] "Hotline Agreements." *Arms Control Association*, https://www.armscontrol.org/factsheets/Hotlines. Accessed 14 July 2023.

[710] "Duck and Cover." *The Library of Congress*, https://www.loc.gov/item/2022604365/. Accessed 12 July 2023.

[711] Wallerstein, Alex. "NUKEMAP by Alex Wellerstein." *NUKEMAP*, https://nuclearsecrecy.com/nukemap/. Accessed 12 July 2023.

[712] *Id.*

[713] "Rad Resilient City Initiative." *Johns Hopkins Center for Health Security*, https://centerforhealthsecurity.org/our-work/research-projects/completed-projects/rad-resilient-city-initiative#checklist. Accessed 13 July 2023.

[714] Federal Emergency Management Agency (FEMA). *Be Prepared for a Nuclear Explosion*. Federal Emergency Management Agency (FEMA), Mar. 2018, https://www.ready.gov/sites/default/files/2020-11/ready_nuclear-explosion_fact-sheet_0.pdf. Accessed 12 July 2023.

[715] "Wireless Emergency Alerts (WEA)." *Federal Communications Commission*, https://www.fcc.gov/consumers/guides/wireless-emergency-alerts-wea. Accessed 13 July 2023.

[716] U.S. Department of Homeland Security. *Nuclear Attack: Communicating in a Crisis*. https://www.dhs.gov/xlibrary/assets/prep_nuclear_fact_sheet.pdf. Accessed 14 July 2023.

[717] Federal Emergency Management Agency (FEMA). *Be Prepared for a Nuclear Explosion*. Federal Emergency Management Agency (FEMA), Mar. 2018, https://www.ready.gov/sites/default/files/2020-11/ready_nuclear-explosion_fact-sheet_0.pdf. Accessed 12 July 2023.

[718] "Nuclear Explosion and Radiation Emergencies." *American Red Cross*, https://www.redcross.org/get-help/how-to-prepare-for-emergencies/types-of-emergencies/nuclear-explosion-radiation-emergencies.html. Accessed 13 July 2023.

[719] "Rad Resilient City Initiative." *Johns Hopkins Center for Health Security*, https://centerforhealthsecurity.org/our-work/research-projects/completed-projects/rad-resilient-city-initiative#checklist. Accessed 13 July 2023.

[720] *Id.*

[721] "Nuclear Explosion and Radiation Emergencies." *American Red Cross*, https://www.redcross.org/get-help/how-to-prepare-for-emergencies/types-of-emergencies/nuclear-explosion-radiation-emergencies.html. Accessed 13 July 2023.

[722] U.S. Department of Homeland Security. *Nuclear Attack: Communicating in a Crisis*. https://www.dhs.gov/xlibrary/assets/prep_nuclear_fact_sheet.pdf. Accessed 14 July 2023.

[723] Malchow. "Radioactive Plume Avoidance and Aircraft Decontamination – Best Practices." *OSTI.GOV*, 28 Apr. 2020, https://www.osti.gov/servlets/purl/1616322. Accessed 23 July 2023.

[724] "Rad Resilient City Initiative." *Johns Hopkins Center for Health Security*, https://centerforhealthsecurity.org/our-work/research-projects/completed-projects/rad-resilient-city-initiative#checklist. Accessed 13 July 2023.

[725] "Build A Kit." *Ready.Gov*, https://www.ready.gov/kit. Accessed 14 July 2023.

[726] Bojana Stevich-Heemer, MS, PharmD, MEd, BCOP Associate Professor Lake Erie College of Osteopathic Medicine School of Pharmacy Erie, Pennsylvania. *Management of Radiation-Induced Nausea and Vomiting*. 17 Dec. 2019, https://www.uspharmacist.com/article/management-of-radiationinduced-nausea-and-vomiting. Accessed 14 July 2023.

[727] CDC. "How Potassium Iodide (KI) Works." *Centers for Disease Control and Prevention. https://www.cdc.gov/nceh/radiation/emergencies/pdf/infographic_ki.pdf* Accessed 24 July 2023.

[728] CDC. "Facts about Potassium Iodide." *Centers for Disease Control and Prevention*, 10 Nov. 2022, https://www.cdc.gov/nceh/radiation/emergencies/ki.htm. Accessed 14 July 2023.

[729] "I-131 Radiation Exposure from Fallout." *National Cancer Institute*, https://www.cancer.gov/about-cancer/causes-prevention/risk/radiation/i-131. Accessed 14 July 2023.

[730] "Facts About Prussian Blue." *CDC*, 7 Apr. 2022, https://www.cdc.gov/nceh/radiation/emergencies/prussianblue.htm. Accessed 14 July 2023.

[731] "How Prussian Blue Works." *NCEH*, 6 Feb. 2020, https://www.cdc.gov/nceh/multimedia/infographics/how_prussian_blue_works.html. Accessed 14 July 2023.

[732] *Radioisotope Brief: Cesium-137 (Cs-137)*, 15 Mar. 2023, https://www.cdc.gov/nceh/radiation/emergencies/isotopes/cesium.htm. Accessed 14 July 2023.

[733] "Facts About DTPA(Diethylenetriamine Pentaacetate)." *CDC*, 10 Mar. 2023, https://www.cdc.gov/nceh/radiation/emergencies/dtpa.htm. Accessed 14 July 2023.

[734] "Facts About Neupogen® (Filgrastim)." *CDC*, 3 Apr. 2023, https://www.cdc.gov/nceh/radiation/emergencies/neupogenfacts.htm. Accessed 14 July 2023.

[735] Elbein, Saul, and Sharon Udasin. "The Hill." *The Hill*, 17 Jan. 2023, https://thehill.com/policy/equilibrium-sustainability/3817111-the-best-place-to-hide-during-a-nuclear-blast/. Accessed 14 July 2023.

[736] Brahambhatt, Rupendra. "Sheltering Miles from a Nuclear Blast May Not Be Enough to Survive Unless You Know Where to Hide, New Calculations Show." *Insider*, 25 Feb. 2023, https://www.businessinsider.com/where-hide-during-nuclear-blast-room-corners-avoid-hallway-doors-2023-2. Accessed 14 July 2023.

[737] Beera, Suraj. "The Best Places to Live in the Event of Nuclear War: Where Will You Go?" *Medium*, 27 Aug. 2021, https://medium.com/@surajbeera/the-best-places-to-live-in-the-event-of-nuclear-war-where-will-you-go-86a80ece02ce. Accessed 14 July 2023.

[738] Bartleby. "Nikita Sergeyevich Khrushchev (1894–1971)." *Collection at Bartleby.Com*, 25 June 2022, https://www.bartleby.com/lit-hub/respectfully-quoted/nikita-sergeyevich-khrushchev-18941971-3/. Accessed 14 July 2023.

[739] "Address before the General Assembly of the United Nations, September 25, 1961." *JFK Library*, https://www.jfklibrary.org/archives/other-resources/john-f-kennedy-speeches/united-nations-19610925. Accessed 14 July 2023.

[740] *The Franck Report: A Report to the Secretary of War, June 1945*. https://sgp.fas.org/eprint/franck.html. Accessed 14 July 2023.

[741] Krzyzaniak, John. "Prominent Nuclear Scientists Did Not Recommend the Atomic Bombings of Japan." *Bulletin of the Atomic Scientists*, 6 Aug. 2020, https://thebulletin.org/2020/08/prominent-nuclear-scientists-did-not-recommend-the-atomic-bombings-of-japan/. Accessed 14 July 2023.

[742] "Albert Einstein." *Nuclear Museum*, https://ahf.nuclearmuseum.org/ahf/profile/albert-einstein/. Accessed 14 July 2023.

[743] "Einstein and the Nuclear Arms Race." *American Museum of Natural History*, https://www.amnh.org/exhibitions/einstein/peace-and-war/nuclear-arms-race. Accessed 14 July 2023.

[744] Issues. "The Lessons of J. Robert Oppenheimer." *Issues in Science and Technology*, 1 Oct. 2013, https://issues.org/br_tenner-5/. Accessed 14 July 2023.

[745] Anderson, Tim. *Oppenheimer's Dilemma*. Stanford, 19 Apr. 2016, http://large.stanford.edu/courses/2016/ph241/anderson1/. Accessed 14 July

2023.

[746] Rouzé, Michel. "J. Robert Oppenheimer." *Encyclopedia Britannica*, 20 July 1998, https://www.britannica.com/biography/J-Robert-Oppenheimer. Accessed 14 July 2023.

[747] "Antarctic Treaty (1959)." *Nuclear Arms Control Treaties*, https://www.atomicarchive.com/resources/treaties/antarctic.html. Accessed 14 July 2023.

[748] *United Nations Disarmament Commission – UNODA*. https://disarmament.unoda.org/institutions/disarmament-commission/. Accessed 14 July 2023.

[749] "Test Ban Treaty (1963)." *National Archives*, 30 Sept. 2021, https://www.archives.gov/milestone-documents/test-ban-treaty. Accessed 14 July 2023.

[750] United Nations Office for Outer Space Affairs. *The Outer Space Treaty*. https://www.unoosa.org/oosa/en/ourwork/spacelaw/treaties/introouterspacetreaty.html. Accessed 14 July 2023.

[751] "Tlatelolco Treaty." *The Nuclear Threat Initiative*, 15 Feb. 2023, https://www.nti.org/education-center/treaties-and-regimes/treaty-prohibition-nuclear-weapons-latin-america-and-caribbean-lanwfz-tlatelolco-treaty/. Accessed 14 July 2023.

[752] Abe, Nobuyasu. "The NPT at Fifty: Successes and Failures." *Journal for Peace and Nuclear Disarmament*, vol. 3, no. 2, July 2020, pp. 224–33, https://doi.org/10.1080/25751654.2020.1824500.

[753] "Milestones: 1961–1968." *Office of the Historian*, https://history.state.gov/milestones/1961-1968/npt. Accessed 14 July 2023.

[754] "Seabed Treaty (1971)." *Nuclear Arms Control Treaties*, https://www.atomicarchive.com/resources/treaties/seabed.html. Accessed 14 July 2023.

[755] "Strategic Arms Limitation Treaty I (1972)." *Nuclear Arms Control Treaties*, https://www.atomicarchive.com/resources/treaties/salt-I.html. Accessed 14 July 2023.

[756] "Milestones: 1969–1976." *Office of the Historian*, https://history.state.gov/milestones/1969-1976/salt. Accessed 14 July 2023.

[757] "Threshold Test Ban Treaty (TTBT)." *Arms Control Association*, https://www.armscontrol.org/treaties/threshold-test-ban-treaty. Accessed 14 July 2023.

[758] "Threshold Test Ban Treaty (1974)." *Nuclear Arms Control Treaties*, https://www.atomicarchive.com/resources/treaties/ttbt.html. Accessed 14 July 2023.

[759] *The Vladivostok Summit Meeting on Arms Control, November 23-24, 1974*. https://www.fordlibrarymuseum.gov/library/exhibits/vladivostok/vdaytwo.asp. Accessed 14 July 2023.

[760] *WashingtonPost.Com: Superpower Summits Archive*. https://www.washingtonpost.com/wp-

srv/inatl/longterm/summit/archive/nov74.htm. Accessed 14 July 2023.

[761] U.S. Department of State. *Treaty Between The United States of America and The Union of Soviet Socialist Republics on Underground Nuclear Explosions For Peaceful Purposes (and Protocol Thereto) (PNE Treaty)*. https://2009-2017.state.gov/t/isn/5182.htm. Accessed 14 July 2023.

[762] U.S. Department of State. *Convention on the Prohibition of Military or Any Other Hostile Use of Environmental Modification Techniques*. https://2009-2017.state.gov/t/isn/4783.htm. Accessed 14 July 2023.

[763] "Strategic Arms Limitations Talks/Treaty (SALT) I and II." *Office of the Historian*, https://history.state.gov/milestones/1969-1976/salt. Accessed 14 July 2023.

[764] "Strategic Arms Limitation Talks (SALT II)." *The Nuclear Threat Initiative*, 25 Oct. 2011, https://www.nti.org/education-center/treaties-and-regimes/strategic-arms-limitation-talks-salt-ii/. Accessed 14 July 2023.

[765] "Treaty of Rarotonga." *The Nuclear Threat Initiative*, 29 Nov. 2020, https://www.nti.org/education-center/treaties-and-regimes/south-pacific-nuclear-free-zone-spnfz-treaty-rarotonga/. Accessed 14 July 2023.

[766] "Treaty of Rarotonga." *United Nations Platform for Nuclear-Weapon-Free Zones*, https://www.un.org/nwfz/content/treaty-rarotonga. Accessed 14 July 2023.

[767] "NATO and the INF Treaty." *NATO*, https://www.nato.int/cps/en/natohq/topics_166100.htm. Accessed 14 July 2023.

[768] Center for Strategic & International Studies. "9M729 (SSC-8)." *Missile Threat*, 23 Oct. 2018, https://missilethreat.csis.org/missile/ssc-8-novator-9m729/. Accessed 14 July 2023.

[769] "The Intermediate-Range Nuclear Forces (INF) Treaty at a Glance." *Arms Control Association*, https://www.armscontrol.org/factsheets/INFtreaty. Accessed 14 July 2023.

[770] *Agreement on Notifications of ICBM and SLBM Launches*. https://nuke.fas.org/control/launch/intro.htm. Accessed 14 July 2023.

[771] "START I at a Glance." *Arms Control Association*, https://www.armscontrol.org/factsheets/start1. Accessed 14 July 2023.

[772] "Strategic Arms Reduction Treaty (START I)." *Center for Arms Control and Non-Proliferation*, 16 Nov. 2022, https://armscontrolcenter.org/strategic-arms-reduction-treaty-start-i/. Accessed 14 July 2023.

[773] "The Lisbon Protocol At a Glance." *Arms Control Association*, https://www.armscontrol.org/node/3289. Accessed 14 July 2023.

[774] "START II." *The Nuclear Threat Initiative*, 28 Jan. 2021, https://www.nti.org/education-center/treaties-and-regimes/treaty-between-united-states-america-and-union-soviet-socialist-republics-strategic-offensive-reductions-start-ii/. Accessed 14 July 2023.

[775] "Strategic Arms Reduction Treaty II." *Center for Arms Control and Non-Proliferation*, 16 Nov. 2022, https://armscontrolcenter.org/strategic-arms-reduction-treaty-ii/. Accessed 14 July 2023.

[776] "Treaty of Pelindaba." *United Nations Platform for Nuclear-Weapon-Free Zones*, https://www.un.org/nwfz/content/treaty-pelindaba. Accessed 14 July 2023.

[777] "Comprehensive Test Ban Treaty at a Glance." *Arms Control Association*, https://www.armscontrol.org/factsheets/test-ban-treaty-at-a-glance. Accessed 14 July 2023.

[778] "SORT." *The Nuclear Threat Initiative*, 25 Oct. 2011, https://www.nti.org/education-center/treaties-and-regimes/strategic-offensive-reductions-treaty-sort/. Accessed 15 July 2023.

[779] "The Strategic Offensive Reductions Treaty (SORT) At a Glance." *Arms Control Association*, https://www.armscontrol.org/factsheets/sort-glance. Accessed 15 July 2023.

[780] "Treaty on the Prohibition of Nuclear Weapons (TPNW)." *The Nuclear Threat Initiative*, 11 Oct. 2022, https://www.nti.org/education-center/treaties-and-regimes/treaty-on-the-prohibition-of-nuclear-weapons/. Accessed 15 July 2023.

[781] Roser, Max, et al. "Nuclear Weapons." *Our World in Data*, 2013, https://ourworldindata.org/nuclear-weapons. Accessed 15 July 2023.

[782] *U.S. Nuclear Weapon Enduring Stockpile.* https://nuclearweaponarchive.org/Usa/Weapons/Wpngall.html. Accessed 15 July 2023.

[783] "Nuclear Notebook: United States Nuclear Weapons, 2023." *Bulletin of the Atomic Scientists*, 16 Jan. 2023, https://thebulletin.org/premium/2023-01/nuclear-notebook-united-states-nuclear-weapons-2023/. Accessed 15 July 2023.

[784] Harris, Bryant. "Republicans Lay Battle Lines over Biden's Plan to Retire B83 Megaton Bomb." *Defense News*, 19 May 2022, https://www.defensenews.com/congress/budget/2022/05/19/republicans-lay-battle-lines-over-bidens-plan-to-retire-b83-megaton-bomb/. Accessed 15 July 2023.

[785] *"The B83 (Mk-83) Bomb."* Nuclear Weapon Archive. https://nuclearweaponarchive.org/Usa/Weapons/B83.html. Accessed 15 July 2023.

[786] "Defence Ministry Board Meeting." *President of Russia*, http://en.kremlin.ru/events/president/news/62401. Accessed 15 July 2023.

[787] "U.S. Withdraws From ABM Treaty; Global Response Muted." *Arms Control Association*, https://www.armscontrol.org/act/2002-07/news/us-withdraws-abm-treaty-global-response-muted. Accessed 15 July 2023.

[788] "GMD: Frequently Asked Questions." *Center for Arms Control and Non-Proliferation*, 16 Jan. 2019, https://armscontrolcenter.org/issues/missile-defense/gmd-frequently-asked-questions/. Accessed 15 July 2023.

[789] *Boeing: Ground-Based Midcourse Defense.* https://www.boeing.com/defense/missile-defense/ground-based-midcourse/index.page. Accessed 15 July 2023.

[790] Center for Strategic & International Studies. Missile Defense. "Ground-

Based Midcourse Defense (GMD) System." *Missile Threat*, 14 June 2018, https://missilethreat.csis.org/system/gmd/. Accessed 15 July 2023.

[791] "A-135 Anti-Ballistic Missile System." *MilitaryToday.Com*, https://www.militarytoday.com/missiles/a135.htm. Accessed 15 July 2023.

[792] "System A-235 / RTC-181M Complex 14TS033 / ROC 'Nudol.'" *Global Security*, https://www.globalsecurity.org/wmd/world/russia/a-235.htm. Accessed 15 July 2023.

[793] *Japan Looks to Partner with U.S. on Railgun Project.* https://www.nationaldefensemagazine.org/articles/2023/4/17/japan-looks-to-partner-with-us-on-railgun-project. Accessed 15 July 2023.

[794] *Israeli-Made High-Energy Laser Makes Debut.* https://www.nationaldefensemagazine.org/articles/2023/3/21/israeli-made-high-energy-laser-makes-debut. Accessed 15 July 2023.

[795] Thompson, Loren. "Lockheed Martin Laser Breakthroughs Could Signal A Turning Point For Missile Defense." *Forbes*, 13 Oct. 2022, https://www.forbes.com/sites/lorenthompson/2022/10/13/lockheed-martin-laser-breakthroughs-could-signal-a-turning-point-for-missile-defense/?sh=6b92ac7b2cf0. Accessed 15 July 2023.

[796] *JUST IN: Lockheed Martin Delivers High-Powered Laser Tech to Pentagon.* https://www.nationaldefensemagazine.org/articles/2022/9/15/lockheed-martin-delivers-high-powered-laser-tech-to-dod. Accessed 15 July 2023.

[797] "Inside the Lockheed Martin Laser Technology." *Lockheed Martin*, https://www.lockheedmartin.com/en-us/news/features/2022/inside-the-lockheed-martin-laser-technology-that-defeated-a-surrogate-cruise-missile.html. Accessed 15 July 2023.

[798] ABC News. "What's the Big Deal about Hypersonic Weapons and Why Are Major Powers Vying for This Capability?" *ABC News*, 28 Oct. 2021, https://www.abc.net.au/news/2021-10-28/hypersonic-missiles-explainer-china-us-tests/100577470. Accessed 15 July 2023.

[799] "LGM-30G Minuteman III." *Air Force*, https://www.af.mil/About-Us/Fact-Sheets/Display/Article/104466/lgm-30g-minuteman-iii/. Accessed 15 July 2023.

[800] "U.S. Hypersonic Weapons and Alternatives." *Congressional Budget Office*, https://www.cbo.gov/publication/58924. Accessed 15 July 2023.

[801] Talmazan, Yuliya, and Mosheh Gains. "Russian Hypersonic Missile Scientists Arrested on Treason Charges." *NBC News*, 18 May 2023, https://www.nbcnews.com/news/world/russia-hypersonic-missile-scientists-arrested-treason-patriots-ukraine-rcna84857. Accessed 15 July 2023.

[802] Boyd, Iain. "How Hypersonic Missiles Work and the Unique Threats They Pose – an Aerospace Engineer Explains." *The Conversation*, 15 Apr. 2022, https://theconversation.com/how-hypersonic-missiles-work-and-the-unique-threats-they-pose-an-aerospace-engineer-explains-180836. Accessed 15 July 2023.

[803] Sevastopulo, Demetri. "Chinese Hypersonic Weapon Fired a Missile over

South China Sea." *Financial Times*, 21 Nov. 2021,
https://www.ft.com/content/a127f6de-f7b1-459e-b7ae-c14ed6a9198c.
Accessed 15 July 2023.

[804] Dangwal, Ashish. "'Nuclear Ashes': Russia's 'Satan Missile' Is Ready For
Deployment; Moscow Says Can Wipe Out Entire US East Coast." *Latest
Asian, Middle-East, EurAsian, Indian News*, 23 June 2022,
https://www.eurasiantimes.com/nuclear-ashes-russias-satan-missile-is-ready-
for-deployment-moscow-says-can-wipe-out-entire-us-east-coast/. Accessed
23 July 2023.

[805] Roga, Sukanta. "CFD Analysis of Scramjet Engine Combustion Chamber
with Diamond-Shaped Strut Injector at Flight Mach 4.5." *Journal of Physics:
Conference Series*, vol. 1276, no. 1, Aug. 2019, p. 012041,
https://doi.org/10.1088/1742-6596/1276/1/012041.

[806] Defense Advanced Research Projects Agency. *DARPA'S Hypersonic Air-
Breathing Weapon Concept (HAWC) Achieves Successful Flight.* 27 Sept.
2021, https://www.darpa.mil/news-events/2021-09-27. Accessed 15 July
2023.

[807] Liebermann, Oren. "US Tested Hypersonic Missile in Mid-March but
Kept It Quiet to Avoid Escalating Tensions with Russia." *CNN*, 5 Apr. 2022,
https://www.cnn.com/2022/04/04/politics/us-hypersonic-missile-
test/index.html. Accessed 15 July 2023.

[808] Congressional Research Service, et al. *Defense Primer: LGM-35A
Sentinel Intercontinental Ballistic Missile (IF11681).* 10 Jan. 2023,
https://crsreports.congress.gov/product/details?prodcode=IF11681. Accessed
15 July 2023.

[809] Tirpak, John. "New B-21 Raider Being Accelerated With Overlapping
Development and Production." *Air & Space Forces Magazine*, 28 Apr. 2022,
https://www.airandspaceforces.com/new-b-21-raider-being-accelerated-with-
overlapping-development-and-production/. Accessed 16 July 2023.

[810] "B-2 Spirit." *Air Force*, https://www.af.mil/About-Us/Fact-
Sheets/Display/Article/104482/b-2-spirit/. Accessed 16 July 2023.

[811] Prisco, Jacopo. "B-2 Spirit: The $2 Billion Flying Wing." *CNN*, 23 Jan.
2020, https://www.cnn.com/style/article/b-2-spirit-stealth-
bomber/index.html. Accessed 16 July 2023.

[812] Roza, David. "USAF Analysis: China Hopes The US Can't Afford
Enough B-21s to Make A Difference." *Air & Space Forces Magazine*, 28
Apr. 2023, https://www.airandspaceforces.com/air-force-b-21-raider-china/.
Accessed 16 July 2023.

[813] "10 Facts About Northrop Grumman's B-21 Raider." *Northrop Grumman*,
3 Dec. 2022, https://www.northropgrumman.com/what-we-do/air/b-21-
raider/10-facts-about-northrop-grummans-b-21-raider/. Accessed 16 July
2023.

[814] "Planet of the Apes (1968)." *IMDb*, Video,
https://www.imdb.com/title/tt0063442/characters/nm0000032. Accessed 16
July 2023.

[815] *Planet of the Apes*. Directed by Franklin J. Schaffner, APJAC Productions / Twentieth Century Fox, 1968.

[816] Clark, W. H. "Chemical and Thermonuclear Explosives." *Bulletin of the Atomic Scientists*, Nov. 1961, https://books.google.com/books?id=gAkAAAAAMBAJ&pg=PA356&dq=chemical+and+thermonuclear+explosives&hl=en&sa=X&ved=2ahUKEwj92LbH55OAAxURlIkEHRTZCqAQ6AF6BAgNEAI#v=onepage&q=chemical%20and%20thermonuclear%20explosives&f=false. Google Books. Accessed 16 July 2023.

[817] Arnold, James R. "The Hydrogen-Cobalt Bomb." *Bulletin of the Atomic Scientists*, Oct. 1950, https://books.google.com/books?id=_Q0AAAAAMBAJ&printsec=frontcover&source=gbs_ge_summary_r&cad=0#v=onepage&q&f=false. Google Books. Accessed 16 July 2023.

[818] "Section 1.0 Types of Nuclear Weapons." *Nuclear Weapon Archive*, 1 May 1998, https://nuclearweaponarchive.org/Nwfaq/Nfaq1.html#nfaq1.6. Accessed 16 July 2023.

[819] Academic Accelerator. *Cobalt Bomb*. https://academic-accelerator.com/encyclopedia/cobalt-bomb. Accessed 16 July 2023.

[820] Tegmark, Max. "Experiment in Annihilation." *Future of Life Institute*, 2 Apr. 2016, https://futureoflife.org/nuclear/experiment-in-annihilation/. Accessed 16 July 2023.

[821] *Id.*

[822] "Understanding the Effects of ERWs and Salted Devices." *HDIAC*, 12 Sept. 2016, https://hdiac.org/articles/understanding-the-effects-of-erws-and-salted-devices/. Accessed 16 July 2023.

[823] "Status-6 / Kanyon - Ocean Multipurpose System." *Russian and Soviet Nuclear Forces*, https://www.globalsecurity.org/wmd/world/russia/status-6.htm. Accessed 16 July 2023.

[824] *Section 1.0 Types of Nuclear Weapons*. Nuclear Weapon Archive. 1 May 1998, https://nuclearweaponarchive.org/Nwfaq/Nfaq1.html#nfaq1.6. Accessed 16 July 2023.

[825] Clark, W. H. "Chemical and Thermonuclear Explosives." *Bulletin of the Atomic Scientists*, Nov. 1961, https://books.google.com/books?id=gAkAAAAAMBAJ&pg=PA356&dq=chemical+and+thermonuclear+explosives&hl=en&sa=X&ved=2ahUKEwj92LbH55OAAxURlIkEHRTZCqAQ6AF6BAgNEAI#v=onepage&q=chemical%20and%20thermonuclear%20explosives&f=false. Google Books. Accessed 16 July 2023.

[826] "Backgrounder on Dirty Bombs." *NRC Web*, https://www.nrc.gov/reading-rm/doc-collections/fact-sheets/fs-dirty-bombs.html. Accessed 16 July 2023.

[827] Bunn, Matthew. "The National Interest." *The National Interest*, 11 July 2014, https://nationalinterest.org/feature/isis-seizes-nuclear-material%E2%80%94-that%E2%80%99s-not-the-reason-worry-10849.

Accessed 16 July 2023.

[828] Kallenborn, Zachary. "The Biden Administration Overestimates Radiological Terrorism Risks and Underplays Biothreats." *Bulletin of the Atomic Scientists*, 17 Mar. 2023, https://thebulletin.org/2023/03/the-biden-administration-overestimates-radiological-terrorism-risks-and-underplays-biothreats/. Accessed 16 July 2023.

[829] National Consortium for the Study of Terrorism and Responses to Terrorism. "GTD Search Results." *Global Terrorism Database*, University of Maryland, https://www.start.umd.edu/gtd/search/Results.aspx?start_yearonly=&end_yearonly=&start_year=&start_month=&start_day=&end_year=&end_month=&end_day=&asmSelect0=&asmSelect1=&weapon=3&dtp2=all&success=yes&casualties_type=b&casualties_max=. Accessed 16 July 2023.

[830] "International Convention on the Suppression of Acts of Nuclear Terrorism." *The Nuclear Threat Initiative*, 15 Mar. 2023, https://www.nti.org/education-center/treaties-and-regimes/international-convention-suppression-acts-nuclear-terrorism/. Accessed 16 July 2023.

[831] "Neutron Bomb." *Encyclopedia Britannica*, 20 July 1998, https://www.britannica.com/technology/neutron-bomb. Accessed 16 July 2023.

[832] "Nuclear Weapons Primer." *Wisconsin Project on Nuclear Arms Control*, https://www.wisconsinproject.org/nuclear-weapons/. Accessed 16 July 2023.

[833] "The Neutron Bomb." *Air & Space Forces Magazine*, 30 Oct. 2017, https://www.airandspaceforces.com/article/the-neutron-bomb/. Accessed 16 July 2023.

[834] Pike, John. *AFAP (Artillery Fired Atomic Projectile)*. https://www.globalsecurity.org/wmd/systems/afap.htm. Accessed 16 July 2023.

[835] "*W70*." https://www.globalsecurity.org/wmd/systems/w70.htm. Accessed 16 July 2023.

[836] Lawrence Livermore National Laboratory. *1975 W79 Development*. https://www.llnl.gov/sites/www/files/1975.pdf. Accessed 16 July 2023.

[837] Franco, Samantha. "Sprint Missile: The ABM That Traveled So Fast a Plasma Ring Formed Around It." *Warhistoryonline*, 10 Mar. 2023, https://www.warhistoryonline.com/weapons/sprint-missile.html. Accessed 16 July 2023.

[838] "Red China's 'Capitalist Bomb': Inside the Chinese Neutron Bomb Program." *Institute for National Strategic Studies*, 1 Jan. 2015, https://inss.ndu.edu/Media/News/Article/652871/red-chinas-capitalist-bomb-inside-the-chinese-neutron-bomb-program/. Accessed 23 July 2023.

[839] Yuan, Xianbao, et al. "Development and Application of an Aerosol Model Under a Severe Nuclear Accident." *Frontiers in Energy Research*, vol. 10, https://doi.org/10.3389/fenrg.2022.852501. Accessed 23 July 2023.

[840] "Nuclear Winter." *Encyclopedia Britannica*, 20 July 1998, https://www.britannica.com/science/nuclear-winter. Accessed 17 July 2023.

[841] "The Atomic Age." *American Experience*, 5 Apr. 2019, https://www.pbs.org/wgbh/americanexperience/features/lasvegas-atomic-age/. Accessed 17 July 2023.

[842] Baroudos, Constance. "Nuclear Weapons Enable Peace." *Lexington Institute*, 6 May 2015, https://www.lexingtoninstitute.org/nuclear-weapons-enable-peace/. Accessed 16 July 2023.

[843] Rauchhaus, Robert. "Evaluating the Nuclear Peace Hypothesis: A Quantitative Approach." *The Journal of Conflict Resolution*, vol. 53, no. 2, 2009, pp. 258–77, https://doi.org/10.2307/20684584.

[844] Goddard, Taegan. "Pax Americana." *Taegan*, 3 July 2023, https://politicaldictionary.com/words/pax-americana/. Accessed 16 July 2023.

[845] Rauchhaus, Robert. "Evaluating the Nuclear Peace Hypothesis: A Quantitative Approach." *The Journal of Conflict Resolution*, vol. 53, no. 2, 2009, pp. 258–77, https://doi.org/10.2307/20684584.

[846] Sagan, Scott Douglas, and Kenneth Neal Waltz. *The Spread of Nuclear Weapons: An Enduring Debate*. W. W. Norton, 2013.

[847] "Nuclear Emulation: Pakistan's Nuclear Trajectory." *Carnegie Endowment for International Peace*, https://carnegieendowment.org/2019/01/22/nuclear-emulation-pakistan-s-nuclear-trajectory-pub-78215. Accessed 17 July 2023.

[848] Wivel, Anders. "Security Dilemma." *Encyclopedia Britannica*, 22 Mar. 2017, https://www.britannica.com/topic/security-dilemma. Accessed 23 July 2023.

www.ingramcontent.com/pod-product-compliance
Lightning Source LLC
Chambersburg PA
CBHW032048020426
42335CB00011B/234

* 9 7 9 8 9 8 8 7 6 0 3 1 3 *